COASTAL BOTTOM BOUNDARY LAYERS AND SEDIMENT TRANSPORT

ADVANCED SERIES ON OCEAN ENGINEERING

Series Editor-in-Chief
Philip L-F Liu
Cornell University, USA

Vol. 1 The Applied Dynamics of Ocean Surface Waves
by Chiang C Mei (MIT, USA)

Vol. 2 Water Wave Mechanics for Engineers and Scientists
by Robert G Dean (Univ. Florida, USA) *and Robert A Dalrymple* (Univ. Delaware, USA)

Vol. 3 Mechanics of Coastal Sediment Transport
by Jørgen Fredsøe and Rolf Deigaard (Tech. Univ. Denmark, Denmark)

Vol. 4 Coastal Bottom Boundary Layers and Sediment Transport
by Peter Nielsen (Univ. Queensland, Australia)

Forthcoming titles:

Water Waves Propagation Over Uneven Bottoms
by Maarten W Dingemans (Delft Hydraulics, The Netherlands)

Ocean Outfall Design
by Ian R Wood (Univ. Canterbury, New Zealand)

Tsunami Run-up
by Philip L-F Liu (Cornell Univ.), *Costas Synolakis* (Univ. Southern California), *Harry Yeh* (Univ. Washington) *and Nobu Shuto* (Tohoku Univ.)

Physical Modules and Laboratory Techniques in Coastal Engineering
by Steven A Hughes (Coastal Engineering Research Center, USA)

Advanced Series on Ocean Engineering — Volume 4

COASTAL BOTTOM BOUNDARY LAYERS AND SEDIMENT TRANSPORT

Peter Nielsen
The University of Queensland

Published by

World Scientific Publishing Co. Pte. Ltd.
P O Box 128, Farrer Road, Singapore 9128
USA office: Suite 1B, 1060 Main Street, River Edge, NJ 07661
UK office: 73 Lynton Mead, Totteridge, London N20 8DH

First published 1992
First reprint 1994

COASTAL BOTTOM BOUNDARY LAYERS AND SEDIMENT TRANSPORT

ISBN 981-02-0472-8
981-02-0473-6 (pbk)

Printed in Singapore.

To Felicity and the boys

PREFACE

Coastal process modelling as an art and a science has progressed significantly over the last four decades. The main areas of progress have been the development of more realistic models of natural waves and the discovery of the wave thrust or radiation stress which enables the quantitative description of phenomena like longshore currents and wave setup.

Progress has been slower in the area of quantitative morphodynamic modelling. The reason is that the quantitative link has been missing between the main flow and the rate of morphological change. The required link consists of efficient models of the boundary layer flow and the resulting sediment transport. These are the topics of the present book.

The book has two main objectives. The first is to provide a review of coastal bottom boundary layer flow and sediment transport by summarising the presently available experimental data. The second objective is to provide the basic sediment transport models which can be used as building blocks for comprehensive models of coastal sediment transport.

In order to address both of these objectives, the treatment of major topics like oscillatory boundary layer flow and sediment suspension are given in two parts each. The first parts each attempt to review the experimental evidence without reference to any particular theory. The second parts subsequently deal with state of the art of modelling.

It is hoped that the book will be useful for Earth Scientists and Engineers who work in the general area of coastal process modelling and to graduate students in the areas of Marine Geology, Coastal Morphodynamics and Coastal Engineering. For such students, the book should help bridge the gap in the literature between general texts on Coastal Hydrodynamics and descriptive texts on coastal processes. It is also hoped that the student's supervisors will find something new and useful in the book.

Indexing and many cross references are provided in order to make the book more efficient as a handbook for practising professionals.

The structure of the book can be summarised as follows.

Section 1.1 gives a general introduction to coastal bottom boundary layers and the flow above them. This includes a discussion of the equations of motion for incompressible boundary layer flow. The major new contribution in this area is the introduction of a set of equations which are similar to the classical Reynolds equations for turbulent, steady flow. The new equations, however, consider explicitly the periodic velocity component as well as the time-average and the random component. These equations, like the Reynolds equations for steady turbulent flow, are very helpful in analysing the mechanisms of momentum transfer in combined wave-current flows.

Section 1.2 presents a review of wave boundary layers. It is followed by a discussion of oscillatory boundary layer models in Section 1.3.

Section 1.4 discusses the wave-generated boundary layer currents including boundary layer streaming and the surf zone undertow.

Wave-current boundary layer interaction is the subject of Section 1.5. Experimental data on wave energy dissipation and on wave boundary layer structure in the presence of currents indicate that currents of typical, relative strengths have very little effect on the wave boundary layer structure. On the other hand, the current boundary layer structure is usually strongly influenced by the presence of waves. A flexible, traditional modelling framework is suggested for currents in the presence of waves. It is realised however, that the traditional eddy viscosity based models are theoretically unsatisfactory. Therefore, a new type of model is suggested in Section 1.5.9.

Chapter 2 considers the initiation of sediment motion and the quasi-steady processes of sediment transport over flat sand beds under waves. Emphasis is placed on discussing similarity and differences between these processes and steady bed-load and sheet-flow sediment transport.

Small scale coastal bedforms are discussed in Chapter 3 together with the hydraulic roughness of beds of loose sand under waves.

Chapter 4 deals with the motion of suspended sediment particles in coastal flows. Emphasis is placed on the important mechanism of sediment trapping by vortices. At the same time, it is shown that a pure wave motion has but negligible effect on sediment settling.

Suspended sediment distributions is the topic of Chapter 5. Section 5.2 presents the nature of suspended sediment concentrations in coastal flows through a review of the existing experimental data. The concept of pickup functions for suspended sediment in unsteady flows is discussed in Section 5.3. Practical pickup function formulae are derived, partly in analogy with pickup functions from steady flow and, partly based on time-averaged suspended sediment concentrations in unsteady flows. Section 5.4 discusses the mechanisms which distribute the suspended sediment through the water column. It is shown that pure gradient diffusion is an unsatisfactory model of suspended sediment distributions in most natural flows. The reason is that the sediment distribution process often includes mechanisms which have mixing lengths of the same order of magnitude as the overall scale of the concentration distribution. These mechanisms are referred to as "convective". Because of the inadequacy of pure gradient diffusion as a model of such large scale mixing, a new, quantitative framework is introduced for convective processes. Recognizing that most natural sediment suspension processes include elements of both convection and gradient diffusion, a new combined convection-diffusion model is developed in Sections 5.4.5 through 5.4.8. The new combined convection-diffusion model needs some calibration before all combinations of flow type and bed sediment can be modelled quantitatively. Qualitatively however, the new model can predict the observed differences between sediment concentration profiles in different flows and between the concentration profiles of different sand sizes in the same flow.

The topic of Chapter 6 is sediment transport model building. Models of coastal sediment transport are seen as being essentially of two varieties. The classical concentration-times-velocity integral models (*cu-integral models*) and the *particle trajectory models*. It is found that the particle trajectory type of models are generally simpler to construct and they provide more consistently accurate predictions of coastal sediment transport rates. In particular, the surprisingly weak grain size dependence of observed shore normal sediment transport rates over rippled beds is predicted by the simplest particle integral models, the *grab and dump models*.

Coastal sediment transport modelling is developing very rapidly. It is therefore quite possible that todays state of the art models could be superseded tomorrow. It is less likely however, that all of today's unanswered questions will be answered as quickly. Hence, an account of what is presently unknown may be of more lasting value than a statement of what is known. Chapter 7 is therefore an account of the most important unanswered questions in the subject area of this book. It is hoped that this may provide some inspiration to other researchers in the field.

Acknowledgements. I would like to thank my former and present employers, The New South Wales Public Works Department, Dr Ian S F Jones of The University of Sydney, and the Department of Civil Engineering, University of Queensland for their support during the preparation of this book. Several colleagues and friends, and most of all my wife Felicity have helped improve the book by offering numerous valuable suggestions.

Brisbane
March 1992

Peter Nielsen.

CONTENTS

NOTATION

A	[L]	Orbital amplitude of fluid just above the boundary layer.
A_E	[-]	Eulerian, turbulent scale ratio, p 197.
C_o	[-]	Reference sediment concentration, usually $\overline{c}(o)$.
C_1	[-]	Kajiura's wave friction coefficient, p 23.
C_D	[-]	Drag coefficient, p 100 and Fig 4.2.1, p 164.
C_D'	[-]	Instantaneous drag coefficient, p 167.
C_M	[-]	Added mass coefficient, p 7.
c	[LT^{-1}]	Wave celerity.
$c(z,t)$	[-]	Volumetric suspended sediment concentration.
$\overline{c}(z)$	[-]	Time-average of $c(z,t)$.
c_b	[-]	Bed-load sediment concentration.
c_g	[LT^{-1}]	Group velocity.
c_{ga}	[LT^{-1}]	Group velocity seen by a fixed observer.
c_{gr}	[LT^{-1}]	Group velocity relative to mean current.
c_{max}	[-]	Sediment concentration in the stationary bed.
c_s	[-]	Suspended sediment concentration, only in Section 5.2.1.
D	[L]	Water depth.
D_E	[MT^{-3}]	Wave energy dissipation rate, p 27.
$D(z,t)$	[-]	Velocity defect function, see p 19.
$D_n(z)$	[-]	Velocity defect function, p 20.
$D_1(z)$	[-]	Velocity defect function, first harmonic, p 21.
d	[L]	Sediment grain diameter.
d_{50}	[L]	Median grain diameter.
d_{90}	[L]	Grain diameter exceeded by 10% by weight of sample.
$E\{\}$		Expected value of stochastic variable.
E_f	[MLT^{-3}]	Wave energy flux per metre of wave crest.
$F(z)$	[-]	Convective distribution function, p 237.
$F(\omega)$	[-]	Frequency response function for $\tau(o,t)$, p 23 and p 125.

F_D	$[MLT^{-2}]$	Drag force, p 100.
F_i	$[MLT^{-2}]$	Inertial force, p 7.
F_L	$[MLT^{-2}]$	Lift force, p 101.
F_p	$[MLT^{-2}]$	Pressure force, p 100.
f	[-]	Friction factor in steady flow, p 149.
$\bar{\bar{f}}$	[-]	Dissipation factor used by Carstens et al, p 27.
$\bar{f_1}$	[-]	Energy dissipation factor used by Lofquist, p 27.
f_e	[-]	Wave energy dissipation factor, p 27.
f_w	[-]	Wave friction factor, p 23.
$f_{2.5}$	[-]	Grain roughness friction factor, p 105.
g	$[LT^{-2}]$	Acceleration due to gravity, $\mathbf{g} = (o,-g)$.
H	[L]	Wave height.
I	$[MLT^{-3}]$	Immersed weight longshore sediment transport rate.
i	[-]	Imaginary unit, $\sqrt{-1}$.
$Im\{\}$	[-]	Imaginary part of complex number.
K	$[LT^{-1}]$	Permeability.
k	$[L^{-1}]$	Wave number $2\pi/L$.
k	[-]	Dissipation factor used by Bagnold, p 27.
L	[L]	Wave length in Chapter 1, vertical convection scale in Chapters 5 and 6.
L_B	[L]	Bed-load thickness, p 111.
L_E	[L]	Eulerian turbulent length scale.
L_s	[L]	Scale of suspended sediment distribution, p 217.
l	[L]	Thickness of wave-dominated layer, p 86.
l_m	[L]	Mixing length, p 201.
l_x, l_y	[L]	Horizontal distance travelled by sediment particle.
n	[-]	Solid fraction of a volume of bed sediment.
P_n	$[LT^{-1}]$	Amplitude of the *n-th* harmonic of $p(t)$.
p	[-]	Velocity distribution parameter, p 44.
$p(t)$	$[LT^{-1}]$	Pickup function.
$\bar{p}$	$[LT^{-1}]$	Time-averaged pickup function.
$p(x,y,z,t)$	$[ML^{-2}T^{-2}]$	Pressure.

$\overline{p}$	$[ML^{-2}T^{-2}]$	Time-averaged pressure.
$\tilde{p}$	$[ML^{-2}T^{-2}]$	Phase-averaged pressure, see Eq (1.1.17).
p'	$[ML^{-2}T^{-2}]$	Random pressure component.
$Q(t)$	$[L^2T^{-1}]$	Depth-integrated sediment transport rate, p 201.
$\overline{Q}$	$[L^2T^{-1}]$	Time-average of $Q(t)$.
Q_B	$[L^2T^{-1}]$	Depth-integrated bed-load transport rate.
q	$[LT^{-1}]$	Local sediment flux, $\boldsymbol{q} = c\,\boldsymbol{u} = (q_x, q_y, q_z)$.
q_C	$[LT^{-1}]$	Convective sediment flux.
q_D	$[LT^{-1}]$	Diffusive sediment flux.
R	[-]	Reynolds number.
R	[L]	Orbit radius.
$Re\{\}$	[-]	Real part of complex number.
r	[L]	Hydraulic roughness, Eq (1.5.9).
S_{xx}	$[MT^{-2}]$	Shore normal wave radiation stress.
s	[-]	Relative density of sediment.
T	[T]	Wave period.
T_a	[T]	Wave period seen by fixed observer.
T_E	[T]	Eulerian turbulent time scale.
T_L	[T]	Lagrangian integral time scale.
T_p	[T]	Spectral peak wave period.
T_r	[T]	Period seen by observer who follows the mean current.
t, t_n	[T]	Time.
t^i, t^u, t^d	[T]	Time of entrainment event.
U	$[LT^{-1}]$	Depth-averaged current velocity in Section 1.5.6.
$U(z), U_\infty$	$[LT^{-1}]$	Amplitude of $u(z,t)$, respectively $u_\infty(t)$.
U_1	$[LT^{-1}]$	Amplitude of u_1.
U_B	$[LT^{-1}]$	Average bed-load velocity, p 114.
$\boldsymbol{U_r}$	$[LT^{-1}]$	Dimensionless, relative velocity vector, p 173.
$\boldsymbol{u}$	$[LT^{-1}]$	Water velocity vector, $\boldsymbol{u} = (u, v, w)$.
$\boldsymbol{u_r}$	$[LT^{-1}]$	Relative sediment velocity vector, $\boldsymbol{u_r} = \boldsymbol{u_s} - \boldsymbol{u}$, p 163 .
$\boldsymbol{u_s}$	$[LT^{-1}]$	Sediment velocity vector $\boldsymbol{u_s} = (u_s, v_s, w_s)$.
u	$[LT^{-1}]$	Horizontal, velocity in direction of wave propagation.

u_*	[LT^{-1}]	Friction velocity $\sqrt{\tau/\rho}$.
u_1	[LT^{-1}]	First harmonic of u.
u_d	[LT^{-1}]	Defect velocity, Eq (1.1.3).
$\overline{u}_E$	[LT^{-1}]	Eulerian time-averaged velocity.
$\overline{u}_L$	[LT^{-1}]	Lagrangian time-averaged velocity.
$\overline{u}_r$	[LT^{-1}]	Reference current velocity.
u_s	[LT^{-1}]	Horizontal velocity of sediment particle.
$\overline{u}_{stokes}$	[LT^{-1}]	Stokes drift, p 53.
$\overline{u}$	[LT^{-1}]	Time-averaged horizontal velocity.
$\tilde{u}$	[LT^{-1}]	Periodic component of u, p 11.
u'	[LT^{-1}]	Random component of u.
$\overline{u_*}$	[LT^{-1}]	Average friction velocity, $\overline{\lvert u_*\rvert\, u_*} = \overline{\tau}(o)/\rho$.
$\hat{u}_*$	[LT^{-1}]	Peak wave friction velocity, $= \sqrt{½ f_w}\ A\omega$.
$u_\infty(t)$	[LT^{-1}]	Velocity just above the wave boundary layer.
V	[L^3]	Particle volume.
V^d, V^i, V^u	[L]	Volume of sand picked up per unit area.
$Var\{\ \}$		Variance of stochastic variable.
v	[LT^{-1}]	Horizontal velocity, perpendicular to wave direction.
$\overline{v_*}$	[LT^{-1}]	Time averaged friction velocity in the y-direction.
w	[LT^{-1}]	Vertical water velocity.
$\overline{w}$	[LT^{-1}]	Time averaged, vertical velocity.
$\tilde{w}$	[LT^{-1}]	Periodic component of w, p 11.
w'	[LT^{-1}]	Random component of w.
w'_{rms}	[LT^{-1}]	Root mean square value of w'.
w^*	[LT^{-1}]	Profile shape ($\overline{c}(z)$) parameter $w_o L/\varepsilon_s$, see p 254.
w_c	[LT^{-1}]	Convection velocity of sediment, p 236.
w_o	[LT^{-1}]	Still water settling velocity, $\mathbf{w}_o = (0, -w_o)$.
w_s	[LT^{-1}]	Vertical sediment particle velocity.
w_t	[LT^{-1}]	Convection velocity of turbulence, p 19.
x	[L]	Horizontal coordinate in the wave direction.
y	[L]	Horizontal coordinate perpendicular to the x-direction.
z	[L]	Vertical coordinate from flat bed or ripple crest level.

z_o	[L]	Zero-intercept level of log velocity profile, p 65.
z_1	[L]	See Fig 1.3.2, p 44.
z_a	[L]	Zero-intercept level of log velocity profile, p 80.
z_b	[L]	Bed level.
z_e	[L]	Entrainment level of suspended particle, p 237.
z_r	[L]	Level of reference current.
α	[-]	Constant, Eq (1.5.28), or Eq (4.3.8).
β_n	[-]	Constant, Eq (5.4.23).
β_x, β_z	[-]	Constants, Eqs (4.5.37) and (4.5.22).
γ	[-]	Constant, Fig 4.2.1, pp 164 and 167.
γ_b	[-]	Wave height to water depth ratio at the break point.
δ	[L]	Boundary layer thickness, δ_d, δ_j, $\delta_{.05}$, δ_*, $\delta_{.01}$, p 30.
δ_{lam}	[L]	Laminar sublayer thickness, p 66.
δ_s	[L]	Stokes Length $\sqrt{2\,\nu/\omega}$.
δ_T	[T^{-1}]	Periodic delta function, p 230.
ε	[-]	Perturbation parameter, p 163.
ε_s	[L^2T^{-1}]	Sediment diffusivity, p 203.
λ	[L]	Ripple length.
λ	[-]	Linear sediment concentration, p 96.
ζ	[L]	Same as z.
η	[L]	Ripple height.
$\bar{\eta}$	[L]	Time-averaged water surface elevation.
θ	[-]	Shields parameter, Section 2.2.3.
θ'	[-]	Skin friction- or effective Shields parameter.
θ_c	[-]	Critical Shields parameter, p 107.
θ_r	[-]	Effective Shields parameter over ripples p 228.
$\theta_{2.5}$	[-]	Grain roughness Shields parameter, p 105.
θ_Φ	[-]	See p 149.
κ	[-]	von Karman's constant ($\approx$ 0.4).
ν	[L^2T^{-1}]	Kinematic (laminar) viscosity.
ν_1	[L^2T^{-1}]	Eddy viscosity derived from $u_1(z,t)$, p 34.
ν_c	[L^2T^{-1}]	Eddy viscosity felt by $\bar{u}$ in combined flow, p 69.

ν_{c1}	$[L^2T^{-1}]$	Eddy viscosity felt by $\bar{u}$ in combined flow, p 92.
ν_t	$[L^2T^{-1}]$	Eddy viscosity.
ν_{TR}	$[L^2T^{-1}]$	Turbulent wave eddy viscosity, p 33.
ν_w	$[L^2T^{-1}]$	Eddy viscosity felt by $\tilde{u}$ in combined flow, p 69. or in purely oscillatory flow, p 31.
ξ	[-]	z/z_1
ρ, ρ_{i-1}	[-]	Correlation coefficients.
ρ	$[ML^{-3}]$	Fluid density.
ρ_s	$[ML^{-3}]$	Sediment density.
σ	$[ML^{-2}T^{-2}]$	Normal stress.
$\overline{\sigma}$	$[ML^{-2}T^{-2}]$	Time-averaged normal stress.
σ_e	$[ML^{-2}T^{-2}]$	Effective normal stress, dispersive stress.
σ, σ_w	$[LT^{-1}]$	Turbulence intensities.
$\tau(z,t)$	$[ML^{-2}T^{-2}]$	Horizontal shear stress in the x-direction ($=\tau_x$).
$\tau(o,t)$	$[ML^{-2}T^{-2}]$	Bed shear stress.
$\overline{\tau}(z)$	$[ML^{-2}T^{-2}]$	Time-averaged shear stress.
$\tilde{\tau}(z,t)$	$[ML^{-2}T^{-2}]$	Periodic shear stress component, see Eq (1.1.17), p 11.
$\hat{\tau}$	$[ML^{-2}T^{-2}]$	Peak bed shear stress under waves, p 23.
τ_L	[T]	Lagrangian microscale.
τ_y	$[ML^{-2}T^{-2}]$	Shear stress in the y-direction.
τ_R	$[ML^{-2}T^{-2}]$	Turbulent shear stress, p 17.
τ_w	$[ML^{-2}T^{-2}]$	Wind shear stress.
$\Phi(t)$	[-]	Dimensionless sediment transport rate, $\dfrac{Q(t)}{d\sqrt{(s-1)\,g\,d}}$.
$\overline{\Phi}$	[-]	Time average of $\Phi(t)$.
Φ_B	[-]	Dimensionless bed-load transport rate.
$\Phi_{T/2}$	[-]	Average of $\Phi(t)$ through half a sine wave.
φ	[-]	Phase angle, or angle between current and waves.
φ_B	[-]	Phase shift between bedload transport and $u_\infty(t)$.
φ_d	[-]	Dynamic angle of repose.
φ_s	[-]	Static angle of repose.

φ_τ	[-]	Phase shift between bed shear stress and $u_\infty(t)$.
φ_ν	[-]	Argument of complex eddy viscosity, p 35.
ψ	[-]	Mobility number, p 103.
ψ	[-]	Phase angle.
ω	$[T^{-1}]$	Radian frequency, $2\pi/T$.
ω_a	$[T^{-1}]$	Radian frequency seen by fixed observer.
ω_p	$[T^{-1}]$	Spectral peak angular frequency.

General operators

$\bar{}$	Time-average, steady component.
$\tilde{}$	Periodic component, Eq (1.1.17).
'	Random component, first derivative or skin friction.
''	Second derivative or form drag.
^	Peak value.

Subscripts

b	Break point value.
c	Pertaining to current in combined flow.
o	Deep water quantity.
rms	Root mean square value.
s	Pertaining to sediment.
w	Pertaining to waves in combined flow.

COASTAL BOTTOM BOUNDARY LAYERS AND SEDIMENT TRANSPORT

CHAPTER 1

BOTTOM BOUNDARY LAYER FLOW

1.1 THE BOUNDARY LAYER AND THE FLOW ABOVE IT

1.1.1 Introduction

With respect to sediment transport modelling, the most important part of the flow is the bottom boundary layer through which the main flow influences the bed.

The present chapter therefore describes the types of boundary layer flows which occur in the coastal environment.

For simplicity, the bottom topography is taken for granted and considered to be stationary. The effects of the boundary layer flow on mobile sand beds, and the resulting topographical changes will be considered in the following chapters.

The bottom boundary layer is intuitively defined as the layer inside which the flow is significantly influenced by the bed. There are various ways of defining the thickness δ of this layer in quantitative terms. However, in general terms the boundary layer thickness obeys the formula

$$\delta \propto \sqrt{\nu_t T} \qquad (1.1.1)$$

where ν_t is the eddy viscosity and T is the flow period. Thus, for fixed eddy viscosity, the boundary layer for tidal flow with a period of 12 hours will be approximately sixty six times thicker than that of a ten second wave motion, and while the tidal boundary layer thickness is often equal to the water depth the wave boundary layer is generally only a small fraction of the depth. See Figure 1.1.1.

The ability of a certain flow component to transport sediment is mainly a function of the shear stresses it induces at the bed. Therefore, since thinner boundary layers mean larger shear stresses for a certain free stream velocity, the waves will tend to dominate over the tide with respect to sediment entrainment and bedform formation. However, because of the "one step forwards and one step backwards" nature of the wave motion, the currents may still be very important transporters of wave entrained sediment unless the sediment concentrations vary periodically in step with the wave motion.

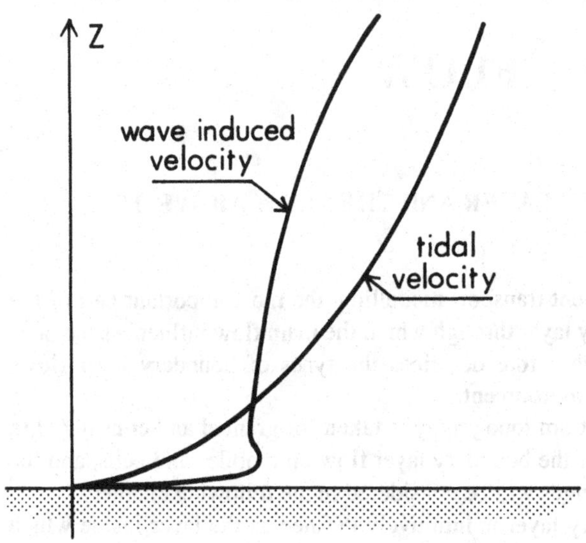

Figure 1.1.1: The tidal boundary layer is much thicker than the wave boundary layer. So, even if the tidal velocity u_{tide} is much larger than the wave-induced velocity amplitude $U(z)$ near the surface, the waves will dominate the situation at the bed.

1.1.2 Natural Flows

In natural situations the flow outside the boundary layer is a complicated function of time which we often describe in terms of its harmonic components, for example, a long term average, tidal components and a spectrum of wind-generated waves. However, for sediment transport calculations at the present state of the art, it is reasonable to simplify this picture considerably. Thus, in the following, we shall generally consider the tides and even rip currents and longshore currents as quasi-steady flows, and ignore effects of the Earth's rotation. The waves will generally be thought of as monocromatic with height H and radian frequency $\omega\,(=2\,\pi/T)$.

The treatment of the wave motion itself is outside the scope of this book, so the wave-induced velocity $u_\infty(t)$ just above the bottom boundary layer will be taken for granted and generally expressed as a simple harmonic function of time.

$$u_\infty(t) \;=\; U_\infty \cos \omega t \;=\; A\omega \cos \omega t \qquad (1.1.2)$$

where A is the water particle semi-excursion. Wave-induced velocities vary only slowly with the elevation z near the bed, see Figure 1.1.2, and for most sediment transport purposes variation in the horizontal x and y directions can be neglected.

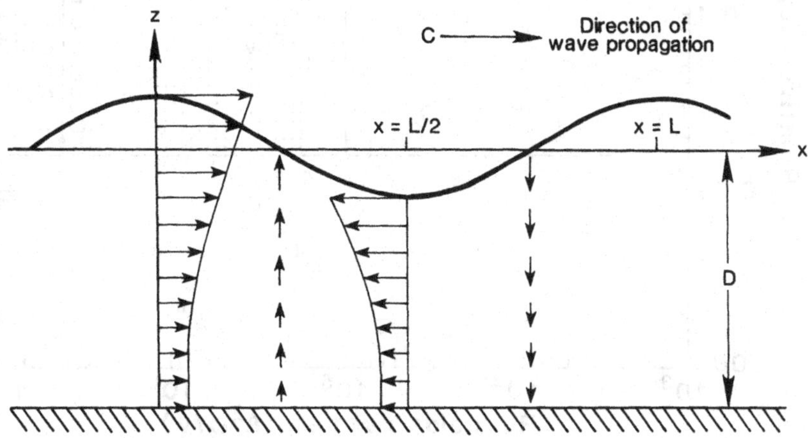

Figure 1.1.2: Water motion under a progressive wave in an inviscid fluid. In the context of wave boundary layer flow the quasi-constant ($\partial/\partial z \approx 0$) velocity near the bed is often referred to as 'the free stream velocity' $u_\infty(t)$.

Jonsson (1966) concluded from dimensional analysis that the structure of oscillatory boundary layers depends mainly on the Reynolds number $A^2\,\omega/\nu$, and on the relative bed roughness r/A. It is therefore natural to ask the following question at the start of our study: What are the likely ranges of the parameters $A^2\,\omega/\nu$ *and* r/A under natural conditions? and to keep the answers in mind when choosing theoretical models and designing laboratory experiments.

Figure 1.1.3 represents an attempt towards answering this question in terms of the presently available experimental data.

The indicated limit $r/A \approx 0.5$ for the use of horizontally uniform descriptions: $u = u(z,t)$ as opposed to more detailed descriptions of the form

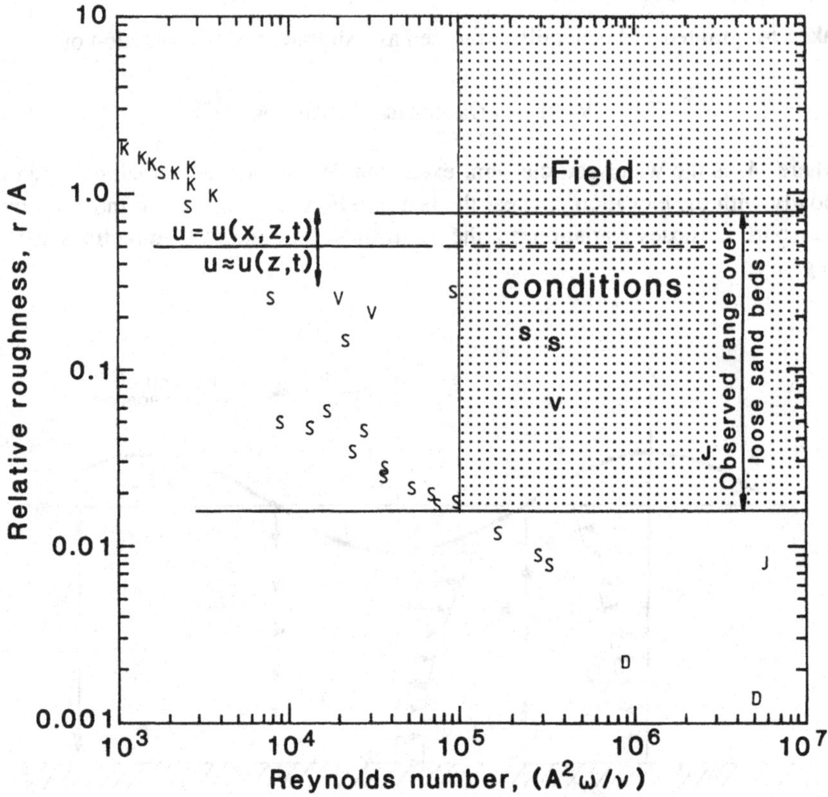

Figure 1.1.3: An overview of the ranges of the Reynolds number $A^2\,\omega/\nu$ and the relative roughness r/A under likely field conditions and in some previous laboratory experiments. The range shown for observed relative roughness values over loose sand beds corresponds to all of the laboratory data of Carstens et al (1969) and Lofquist (1986). If data from artificially flat beds at very low flow velocities are excluded, the lower limit of r/A is 0.08. Legend, K: Kemp & Simons, flume (1982, 1983); S: Sleath, flume and tunnel (1982,1987); V: van Doorn, flume and tunnel (1981,1982); J: Jonsson & Carlsen, tunnel (1976); D: Jensen, tunnel (1989).

$u = u(x, z, t)$ is based on considerations of the data of Kemp and Simons (1982, 1983) and those of van Doorn (1981, 1982). The former show pronounced differences between the velocity structure over the bedform crests and over the troughs, while the latter show reasonably uniform behaviour through most of the boundary layer.

The value of *0.5* is only meant as an indication. The choice of an upper limit of r/A for the application of horizontally averaged models does of course, in the end, depend on the amount of detail one needs to consider.

The observed range from laboratory experiments of the relative roughness of natural sand beds, is derived from all of the dissipation data of Carstens et al (1969) and the bed shear stress data of Lofquist (1986). Values of r/A are obtained by applying Equation (1.2.22) "in reverse" to the authors' friction factors.

Rippled sand beds generally gave $r/A > 0.2$, while values of r/A below *0.08* were found only for flat beds with very little sand movement, see Section 3.6.

The lower limit of observed r/A-values of *0.08* for natural (flat) sand beds is surprisingly high. It corresponds to roughness values of the order a hundred grain diameters, and there is some doubt regarding its interpretation in relation to effective sediment transporting stresses, as discussed in Section 2.4. The roughness of natural sand beds is discussed in detail in Section 3.6.

Very large relative roughness values may apply in nature over rocky areas and over reef platforms.

The message from the presently available data for oscillatory flow over loose sand is, in essence, that the range of Reynolds numbers and relative roughness values likely to be found under field conditions are

$$A^2\,\omega/\nu \;>\; 10^5$$

corresponding to a limiting condition of, for example, $(A, T) \;=\; (0.3m, 6s)$, and

$$r/A > 0.08$$

From Figure 1.1.3 it can be seen that there is a serious shortage of experimental data on boundary layer structure in this range.

1.1.3 Flow in laboratory models

Due to the great practical difficulties involved with good field measurements a lot of experimental work on coastal hydrodynamics and sediment transport is carried out in the laboratory with three different kinds of facilities.

The first and most common kind includes wave flumes and wave basins where most aspects of prototype wave motion outside the boundary layer can be modelled in accordance with Froude's model law. However, because the size of most wave flumes and basins is fairly limited it can be difficult to obtain adequately large values of the Reynolds number $A^2\omega/\nu$ for modelling boundary layer phenomena.

Therefore, a different type of apparatus was suggested by Lundgren and Soerensen (1956), and later applied in many studies, namely the oscillating water tunnel. This is essentially a large U-tube where the flow is driven by a piston in one of the vertical legs, see Figure 1.1.4.

The horizontal test section of such tunnels can be several metres long so that Reynolds numbers in excess of 10^6 can be obtained. The orbital motion in the test section of the U-tube differs from real wave-induced flow by being totally uniform in the x-direction ($\partial u_\infty/\partial x = 0$) and by having no vertical orbital motion, but these dissimilarities are often of no concern.

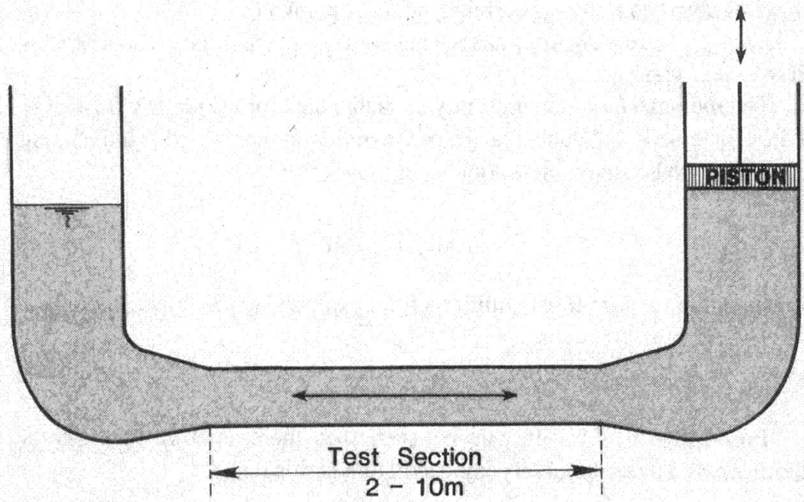

Figure 1.1.4: An oscillating water tunnel can be used to model many characteristics of the wave boundary layer, but the vertical velocities of a wave motion and the horizontal variation of the free stream are not modelled.

Due to the considerable cost of large wave flumes and tunnels a somewhat simpler and cheaper type of facility was introduced by Bagnold (1946) and later applied with modifications by many others. Instead of oscillating the bulk of the water, Bagnold oscillated the bed in it's own plane through otherwise still water, see Figure 1.1.5.

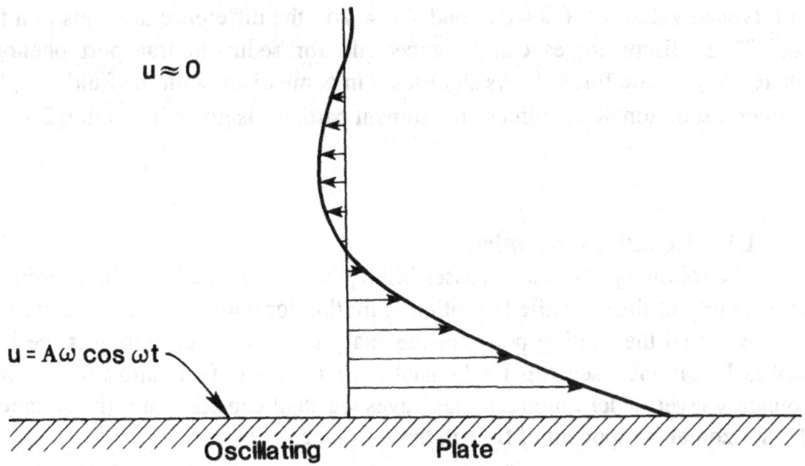

Figure 1.1.5: Oscillating plates or trays can be used for modelling oscillatory boundary layers and sediment transport. However, distortion of the inertia/pressure forces on sediment particles is of some concern.

The flow over such an oscillating bed is, for all practical purposes, similar to the velocity defect

$$u_d(z,t) = u_\infty(t) - u(z,t) \tag{1.1.3}$$

in a tunnel, but the two types are in some respects dissimilar as far as sediment transport experiments are concerned.

The reason is that the forces resulting from fluid pressure gradients on a resting sand particle are exaggerated on the oscillating bed. A grain at rest in the reference frame which follows the tray with velocity $-u_\infty(t)$ will experience an inertial force of magnitude

$$F_i = \rho\,(s + C_M)\,V\frac{du_\infty}{dt} \tag{1.1.4}$$

where ρ is the fluid density, s is the specific sediment density, V is the particle volume and C_M is the added mass coefficient. However, the corresponding force on a particle on a fixed bed in moving fluid is only

$$F_P = \rho\,(1 + C_M)\,V\frac{du_\infty}{dt} \tag{1.1.5}$$

with typical values of $C_M = 0.5$ and $s = 26$ the difference amounts to a factor two. Thus, discrepancies can be expected for sediment transport phenomena where the pressure force F_P is significant in comparison with the fluid drag force. Further discussion of the forces on sediment particles is given in Section 2.1.

1.1.4 Equations of motion

The following section discusses briefly the background and the conditions of applicability of the simplified equation of motion for bottom boundary layer flow.

As usual the starting point for the analysis of the fluid motion is the Navier Stokes Equations, see e g Le Mehaute (1976), p 61. In relation to the bottom boundary layer under unidirectional waves we shall consider only the equation for the horizontal component of flow in the x-z plane

$$\frac{\partial u}{\partial t} + u\frac{\partial u}{\partial x} + w\frac{\partial u}{\partial z} = -\frac{1}{\rho}\frac{\partial p}{\partial x} + \nu\left(\frac{\partial^2 u}{\partial x^2} + \frac{\partial^2 u}{\partial z^2}\right) \tag{1.1.6}$$

where u and w are the velocities in the x and z directions respectively, ρ is the fluid density, p is the pressure, and ν is the kinematic viscosity of the fluid.

The flow inside the boundary layer can often be considered to be essentially horizontal ($w \approx 0$). The equation of motion can then be further simplified to

$$\rho\left(\frac{\partial u}{\partial t} + u\frac{\partial u}{\partial x}\right) = -\frac{\partial p}{\partial x} + \frac{\partial \tau}{\partial z} \tag{1.1.7}$$

where τ is the viscous shear stress ($\tau = \rho\,\nu\,du/dz$).

This equation is however, still difficult to solve because of the non-linear, convective acceleration term. It is therefore worthwhile discussing when that term can be omitted. In broad terms this is possible when the velocity is horizontally uniform i e, when $u = u(z, t)$.

The first requirement for obtaining horizontal uniformity is that the free stream velocity u_∞ is uniform. This condition is fulfilled exactly over oscillating plates and in oscillating water tunnels. However, under real waves the variation from wave crest to wave trough generates a convective acceleration of magnitude

$$\left|u_\infty\frac{\partial u_\infty}{\partial x}\right| = A\omega\frac{2\pi A\omega}{L} = \frac{2\pi A}{L}\left|\frac{\partial u_\infty}{\partial t}\right| \tag{1.1.8}$$

where L is the wave length. Hence the relative importance of the convective term originating from u_∞ is given by the factor $2\pi A/L$.

The second criterion for horizontal uniformity in the boundary layer is that non-uniformities introduced by individual roughness elements should be restricted to a layer which is considerably thinner than the boundary layer itself , see Figure 1.1.6.

Since the scale of the disturbances introduced by the individual roughness elements is the bed roughness r, this may be expressed by $\delta/r >> 1$ which corresponds to $A/r >> 1$, since δ/r is an increasing function of A/r.

More precise information about the boundary layer thickness relative to the roughness height can be gained from Figure 28 of Sleath (1987). The Figure shows, that over a bed of three dimensional roughness elements, the ratio of boundary layer thickness to roughness size is given approximately by

$$\delta_{.05}/r \;=\; 0.26(A/r)^{0.70} \qquad (1.1.9)$$

corresponding to $\delta_{.05} = r$ for $A/r = 6.9$ and $\delta_{.05} = 10r$ for $A/r \;=\; 184$. By Sleath's (1987) definition the top of the boundary layer ($z = \delta_{.05}$) is where the velocity defect amplitude becomes less than 5% of $A\omega$.

When the criteria for horizontal uniformity are met, the non-linear, convective acceleration term can be omitted, and the equation of motion becomes

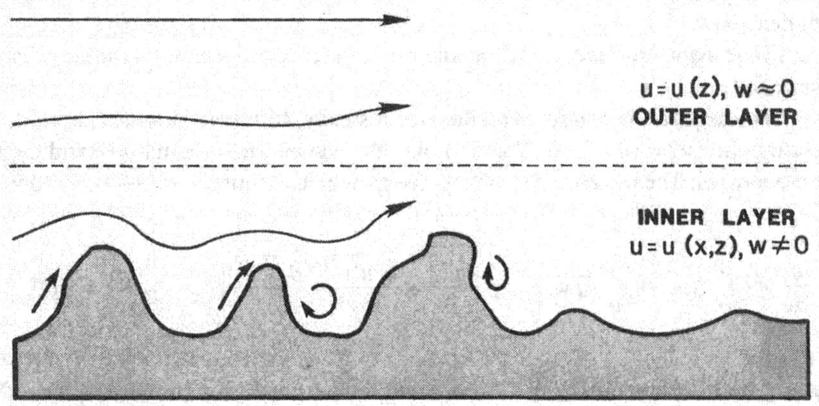

Figure 1.1.6: Even when the free stream velocity is horizontally uniform, the flow near the bed may be non uniform throughout an inner layer of thickness similar to the boundary layer thickness.

$$\rho \frac{\partial u}{\partial t} = -\frac{\partial p}{\partial x} + \frac{\partial \tau}{\partial z} \tag{1.1.10}$$

This equation of motion for horizontally uniform flow can be simplified further under the assumption of hydrostatic pressure distribution in the boundary layer. That is when the vertical accelerations are negligible compared to the acceleration of gravity. Then we can utilize the fact that the shear stresses vanish outside the boundary layer so that

$$\rho \frac{\partial u_\infty}{\partial t} = -\frac{\partial p}{\partial x} \tag{1.1.11}$$

and rewrite Equation (1.1.10) in the form

$$\rho \frac{\partial}{\partial t}(u - u_\infty) = \frac{\partial \tau}{\partial z} \tag{1.1.12}$$

1.1.5 Reynolds equations for combined wave current flows

In the following section we shall derive equivalent equations to the classical Reynolds equations for combined wave current flows. That is, for flows which contain a periodic component $\tilde{u}$ as well as the familiar $\bar{u} + u'$ for steady, turbulent flows.

These equations are useful in the analysis of wave current boundary layer interaction.

The classical Reynolds equations for a steady, turbulent flow are derived by inserting $(u,v,w) = (\bar{u}+u', \bar{v}+v', \bar{w}+w')$ into the Navier Stokes equations and taking time-averages. The equation describing the flow in the x-direction can be written

$$\rho\left(\frac{\partial}{\partial x}\bar{u}^2 + \frac{\partial}{\partial y}(\bar{u}\,\bar{v}) + \frac{\partial}{\partial z}(\bar{u}\,\bar{w})\right) + \rho\left(\frac{\partial \overline{u'^2}}{\partial x} + \frac{\partial \overline{u'v'}}{\partial y} + \frac{\partial \overline{u'w'}}{\partial z}\right) = -\frac{\partial \bar{p}}{\partial x} + \rho\nu\nabla^2 \bar{u} \tag{1.1.13}$$

see, e g, Le Mehaute (1976) p 77. For a two dimensional flow in the xz-plane, this can be written in the form

$$\frac{\partial \bar{\sigma}}{\partial x} + \frac{\partial \bar{\tau}}{\partial z} = 0 \tag{1.1.14}$$

where the total, time-averaged stresses $(\overline{\sigma}, \overline{\tau})$ are given by

$$(\overline{\sigma}, \overline{\tau}) = (-\overline{p} - \rho\overline{u}^2 - \rho\overline{u'^2},\ \rho\nu\frac{d\overline{u}}{dz} - \rho\overline{u}\,\overline{w} - \rho\overline{u'w'}) \qquad (1.1.15)$$

Thus, one purpose of the Reynolds equations is to identify stress (or momentum transfer) contributions from the different flow components. The shear stress component $-\rho\overline{u'w'}$ from the random velocity components is generally referred to as the Reynolds stress.

We shall now derive the corresponding equations for a combined wave-current motion where the velocity in each direction may include a periodic component as well as the steady and the random components

$$(u, v, w) = (\overline{u} + \widetilde{u} + u', \overline{v} + \widetilde{v} + v', \overline{w} + \widetilde{w} + w') \qquad (1.1.16)$$

The tilde denotes the periodic component, which is the phase average over several (N) wave periods minus the time average

$$\widetilde{u}(z,t) = \frac{1}{N}\sum_{j=1}^{N} u(z, t + jT) - \overline{u}(z) \qquad (1.1.17)$$

We note that with these definitions we have $\overline{\widetilde{x}} = \overline{x'} = \widetilde{x'} = 0$, and $\overline{\overline{x}\widetilde{y}} = \overline{\overline{x}y'} = \overline{\widetilde{x}y'} = \overline{x}\widetilde{y'} = \widetilde{\widetilde{x}y'} = 0$, while $\widetilde{\widetilde{x}\widetilde{y}} = \widetilde{x}\widetilde{y} - \overline{\widetilde{x}\widetilde{y}}$. Inserting the expressions (1.1.16) into the "horizontal" Navier Stokes equation (1.1.6) gives

$$\frac{\partial(\overline{u}+\widetilde{u}+u')}{\partial t} + (\overline{u}+\widetilde{u}+u')\frac{\partial(\overline{u}+\widetilde{u}+u')}{\partial x} + (\overline{v}+\widetilde{v}+v')\frac{\partial(\overline{u}+\widetilde{u}+u')}{\partial y} + (\overline{w}+\widetilde{w}+w')\frac{\partial(\overline{u}+\widetilde{u}+u')}{\partial z}$$

$$= -\frac{1}{\rho}\frac{\partial(\overline{p}+\widetilde{p}+p')}{\partial x} + \nu\nabla^2(\overline{u}+\widetilde{u}+u') \qquad (1.1.18)$$

In order to get the governing equation for the time-averaged velocity $\overline{u}$, we extract all non-trivial steady contributions from this equation and find

$$\overline{u}\frac{\partial\overline{u}}{\partial x} + \overline{v}\frac{\partial\overline{u}}{\partial y} + \overline{w}\frac{\partial\overline{u}}{\partial z} + \overline{\widetilde{u}\frac{\partial\widetilde{u}}{\partial x}} + \overline{\widetilde{v}\frac{\partial\widetilde{u}}{\partial y}} + \overline{\widetilde{w}\frac{\partial\widetilde{u}}{\partial z}} + \overline{u'\frac{\partial u'}{\partial x}} + \overline{v'\frac{\partial u'}{\partial y}} + \overline{w'\frac{\partial u'}{\partial z}}$$

$$= \frac{-1}{\rho}\frac{\partial\overline{p}}{\partial x} + \nu\nabla^2\overline{u} \qquad (1.1.19)$$

This equation is now modified by using the continuity equation

$$\frac{\partial}{\partial x}(\overline{u}+\widetilde{u}+u') + \frac{\partial}{\partial y}(\overline{v}+\widetilde{v}+v') + \frac{\partial}{\partial z}(\overline{w}+\widetilde{w}+w') = 0 \qquad (1.1.20)$$

Thus, by adding $\overline{u}\left(\frac{\partial \overline{u}}{\partial x}+\frac{\partial \overline{v}}{\partial y}+\frac{\partial \overline{w}}{\partial z}\right) = 0$ to the first three terms in Equation (1.1.19) and reorganising, these terms can be rewritten as $\frac{\partial \overline{u}^2}{\partial x} + \frac{\partial \overline{u}\,\overline{v}}{\partial y} + \frac{\partial \overline{u}\,\overline{w}}{\partial z}$. By performing the analogous operations on the middle and the last three terms on the right hand side of Equation (1.1.19) the complete, time-averaged equation becomes

$$\frac{\partial}{\partial x}\overline{u}^2 + \frac{\partial}{\partial y}(\overline{u}\,\overline{v}) + \frac{\partial}{\partial z}(\overline{u}\,\overline{w}) + \frac{\partial}{\partial x}\overline{\widetilde{u}^2} + \frac{\partial}{\partial y}(\overline{\widetilde{u}\widetilde{v}}) + \frac{\partial}{\partial z}(\overline{\widetilde{u}\widetilde{w}}) + \frac{\partial}{\partial x}\overline{u'^2} + \frac{\partial}{\partial y}\overline{(u'v')} + \frac{\partial}{\partial z}\overline{(u'w')}$$

$$= \frac{-1}{\rho}\frac{\partial \overline{p}}{\partial x} + \nu\left(\frac{\partial^2 \overline{u}}{\partial x^2}+\frac{\partial^2 \overline{u}}{\partial y^2}+\frac{\partial^2 \overline{u}}{\partial z^2}\right) \qquad (1.1.21)$$

By collecting all terms which represent momentum flux in the vertical direction under the label $\overline{\tau}$ this gives, by analogy with Equation (1.1.15),

$$\overline{\tau} = \rho\nu\frac{\partial \overline{u}}{\partial z} - \rho\overline{u}\,\overline{w} - \rho\overline{\widetilde{u}\widetilde{w}} - \rho\overline{u'w'} \qquad (1.1.22)$$

This expression is of central interest with respect to the modelling of current profiles in the presence of waves. The third term on the right hand side will generally be dominant, except very close to the bed, and this has important implications for the shape of $\overline{u}(z)$-profiles with colinear waves superimposed, see Section 1.5.9.

The corresponding expression related to a current perpendicular to the wave motion is simply

$$\overline{\tau_{yz}} = \rho\nu\frac{\partial \overline{v}}{\partial z} - \rho\overline{v}\,\overline{w} - \rho\overline{u'w'} \qquad (1.1.23)$$

because $\widetilde{v} \equiv 0$.

If the flow is horizontally uniform we have $\overline{w} = 0$ by continuity, so that the second term on the right hand side of Equations (1.1.22) and (1.1.23) disappear. In order to derive the equivalent equation to Equation (1.1.21) for the periodic flow component, we take phase averages (with zero mean in accordance with Equation (1.1.17)) on both sides of Equation (1.1.18) and make use of the continuity equation as above. Then the following equation results

$$
\begin{aligned}
&\frac{\partial \tilde{u}}{\partial t} + 2\frac{\partial}{\partial x}(\bar{u}\,\tilde{u}) + \frac{\partial}{\partial y}(\bar{u}\,\tilde{v} + \tilde{u}\,\bar{v}) + \frac{\partial}{\partial z}(\bar{u}\,\tilde{w} + \tilde{u}\,\bar{w}) \\
&\quad + \frac{\partial}{\partial x}\widetilde{(\tilde{u})^2} + \frac{\partial}{\partial y}\widetilde{(\tilde{u}\tilde{v})} + \frac{\partial}{\partial z}\widetilde{(\tilde{u}\tilde{w})} \\
&\quad + \frac{\partial}{\partial x}\widetilde{u'^2} + \frac{\partial}{\partial y}\widetilde{(u'v')} + \frac{\partial}{\partial z}\widetilde{(u'w')} \\
&= \frac{-1}{\rho}\frac{\partial \tilde{p}}{\partial x} + \nu\left(\frac{\partial^2 \tilde{u}}{\partial x^2} + \frac{\partial^2 \tilde{u}}{\partial y^2} + \frac{\partial^2 \tilde{u}}{\partial z^2}\right)
\end{aligned}
\tag{1.1.24}
$$

Again we are particularly interested in the terms which represent vertical flux of horizontal momentum, so we extract all such terms and put them under the collective label $\tilde{\tau}$

$$
\tilde{\tau} = \rho\nu\frac{\partial \tilde{u}}{\partial z} - \rho\,\bar{u}\tilde{w} - \rho\,\tilde{u}\,\bar{w} - \rho\,\widetilde{(\tilde{u}\tilde{w})} - \rho\,\widetilde{(u'w')} \tag{1.1.25}
$$

The velocity product terms in Equations (1.1.22) and (1.1.25) play the same role for wave-current motion as the Reynolds stresses $-\rho\,\overline{u'w'}$ do for turbulent steady flows. These velocity product terms will be used to discuss the total shear stresses and the eddy viscosity concept for oscillatory flows and for combined wave-current motions in Sections 1.2.8-1.2.9, and Section 1.5.3.

1.2 THE NATURE OF OSCILLATORY BOUNDARY LAYERS

1.2.1 Introduction

The wave boundary layer is intuitively defined as the layer close to the bottom, where the wave-induced water motion is noticeably affected by the boundary. This layer is normally very thin, generally a few millimetres over a smooth, solid bed and a few centimetres over a flat bed of loose sand. Bedforms, like ripples, may change the structure of the boundary layer by introducing strong rhythmic vortices. Hence, the boundary layer over sharp crested ripples will extend to a height of four or five ripple heights, or a total of about *50* centimetres under field conditions.

Although the water motion induced by natural waves is not simple harmonic, it is instructive and useful to study the simple harmonic, oscillatory boundary layer, which corresponds to $u_\infty = A\omega \cos \omega t$, and use it as an approximation to natural wave boundary layers.

The total picture of the velocity variation in such a layer is at first sight very complicated because both amplitude and relative phase change with the distances from the bed, see Figure 1.2.1. However, in the following section we shall see that a much clearer picture can be obtained by applying simple transformations to the data.

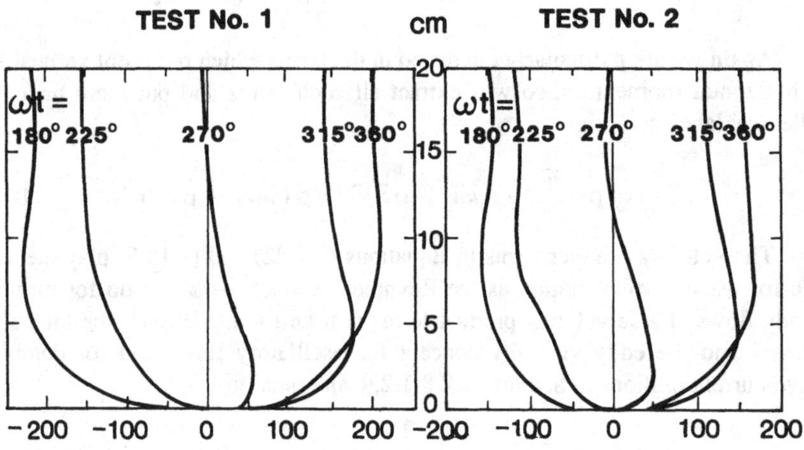

Figure 1.2.1: Instantaneous velocities $u(z,t)$ $[cm/s]$ plotted against elevation from top of the roughness elements. Numbers on the curves refer to the phase of the free stream velocity $u_\infty(z,t) = A\,\omega \cos\omega t$. The measurements were made in an oscillating water tunnel by Jonsson and Carlsen (1976). Note that the velocity near the bed turns before the free stream velocity and that the velocity amplitude is largest in the range $5cm < z < 10cm$ not at $z \rightarrow \infty$.

It is a typical feature of oscillatory boundary layers that the velocity close to the boundary and the bed shear stress $\tau(o, t)$ are ahead of the free stream velocity $u_\infty(t)$ as shown by Figure 1.2.1.

The bed shear stress in a simple harmonic, laminar flow is simple harmonic and leads u_∞ by 45°, but in a turbulent flow the variation with time is much more complicated. The deviation from the simple harmonic behaviour increases with the ratio between the bed roughness, r and the semi-excursion A. Thus, for small r/A the variation of $\tau(t)$ is still quite smooth and rather like a simple harmonic. This is the case for the measurements of Jonsson and Carlsen (1976) Test 1 where r/A was only *0.008*, see Figure 1.2.2. In this case the phase shift between u_∞ and the bed shear stress is still fairly well defined and we see that it is somewhat smaller than the 45° of smooth laminar flow.

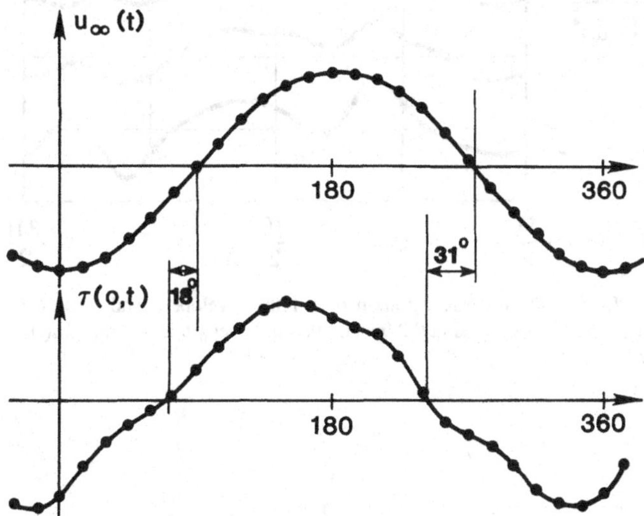

Figure 1.2.2: Time variations of the bed shear stress $\tau(o, t)$ for rough turbulent flow over relatively small roughness elements $r/A = 0.008$. After Jonsson and Carlsen (1976).

For flow over fully developed sand ripples, the ratio r/A is of the order of magnitude one, and the flow near the bed is dominated by the rhythmic formation and release of strong vortices. Lofquist (1986) measured instantaneous values of $\tau(o, t)$ under such conditions. Figure 1.2.3 shows some of his results and we see that the behaviour is completely different from that of a simple harmonic and also from that of $\sin(\omega t - \varphi)|\sin(\omega t - \varphi)|$ which has been assumed in several "theoretical" studies.

Another typical feature of oscillatory boundary layers is the "overshoot" near the bed. That is, there are elevations where the velocity amplitude U exceeds $A\omega$, see Figure 1.2.4.

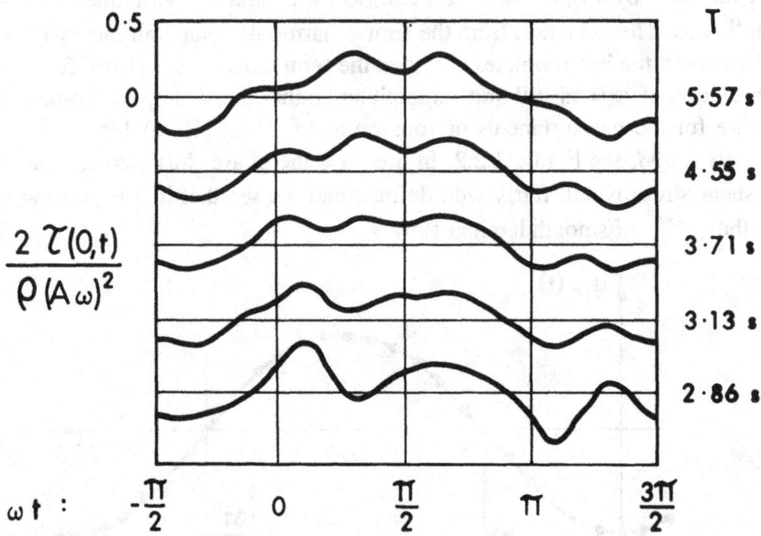

Figure 1.2.3: Shear stress variation over fully developed sand ripples. For these experiments A was fixed at *0.24m* and the ripple length at *0.32m*. $u_\infty(t)$ varied as sin ωt.

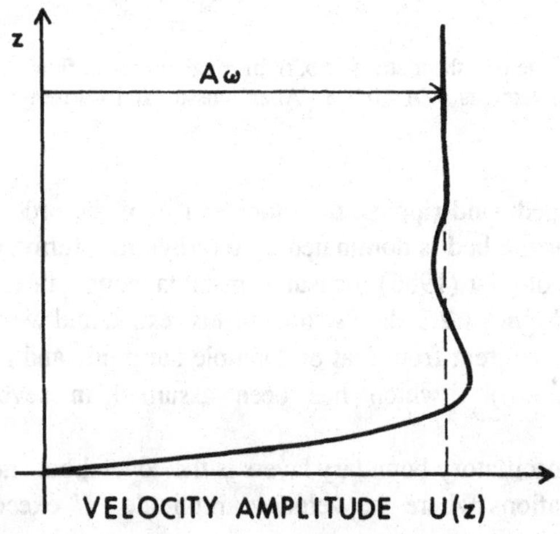

Figure 1.2.4: The local velocity amplitude $U(z)$ oscillates around the free stream value $A\omega$ reflecting the fact that the velocity defect has the nature of a dampened wave, which propagates away from the bed.

The velocity overshoot occurs because the velocity defect $u_\infty(t) - u(z, t)$ has the nature of a damped wave which alternately adds to and subtracts from the free stream velocity $u_\infty(t)$, see Section 1.2.4.

1.2.2 The shear stress distribution

The simplicity of the equation of motion (1.1.12) is really somewhat deceptive because the humble τ, which represents the shear stress, contains a whole set of complexities of its own, which is indicated by Equation (1.1.25).

However complex, the total shear stress $\tau(z, t)$ in a horizontally uniform flow is nevertheless a measurable quantity in as far as Equation (1.1.12) is valid. By integrating this equation and using the fact that $\tau(\infty, t) = 0$ we find

$$\tau(z,t) = \rho \int_z^\infty \frac{\partial}{\partial t}(u_\infty - u)dz \qquad (1.2.1)$$

which says that the shear stress at the level z is equal to the fluid density times the total acceleration defect above z.

Sleath (1987) discussed shear stresses and related quantities in great detail for turbulent oscillatory flows. One of his most striking findings was that the total shear stress calculated from Equation (1.2.1) was, for his experiments, about a factor ten larger than the periodic, turbulent Reynolds stress defined by

$$\tau_R(z, t) = -\rho\, \widetilde{(u'w')}$$

where u' and w' are horizontal and vertical, turbulent velocity fluctuations respectively, see Figure 1.2.13.

Hence, as Sleath pointed out, the turbulent fluctuations u' and w' are "mere spectators" to the oscillatory boundary layer processes. Their contribution to the momentum transfer is totally overshadowed by the analogous contribution $-\rho\, \tilde{u}\, \tilde{w}$ from the periodic velocity components. This latter contribution in turn was found to agree rather closely with the values of τ calculated from the defect integral (1.2.1).

For the shear stress amplitude $|\tau(z, t)|$ the data of Sleath (1987), and of several previous authors, show that it tends to decrease roughly exponentially with increasing distance from the bed. The vertical decay scale is approximately the boundary layer thickness δ, i e,

$$|\tau(z,t)| \approx |\tau(0,t)|\, e^{-z/\delta}$$

The magnitude of the bed shear stress is proportional to $\rho(A\omega)^2$ and to a complicated function of the Reynolds number $A^2\omega/\nu$ and the relative roughness r/A which will be considered in Section 1.2.5.

1.2.3 Turbulence structure in oscillatory flow

When discussing the turbulence intensity in oscillatory flows it is important to distinguish between the time averages $u'_{rms} = (\overline{u'^2})^{0.5}$ and $w'_{rms} = (\overline{w'^2})^{0.5}$ and the phase-averaged (time dependent) $(\widetilde{u'^2})^{0.5}$ and $(\widetilde{w'^2})^{0.5}$.

With respect to u'_{rms} and w'_{rms}, Sleath (1987) found, for the relative roughness range $(0.03 < r/A < 0.25)$, that both decrease with increasing z. This is in contrast to steady flows where the turbulence intensity practically constant throughout the boundary layer, see Figure 1.5.7. For smaller relative roughness, a layer of constant turbulence intensity becomes noticeable close to the bed. For almost smooth beds $(r/A < 0.0005)$ this layer extends almost throughout the boundary layer indicating a similarity with steady flow, as in some of the experiments of Jensen (1989).

The two components u'_{rms} and w'_{rms} are of similar magnitude, and their maximum value which occurs at the bed is approximately equal to the friction velocity $\hat{u}_* = \sqrt{\tfrac{1}{2} f_w}\, A\omega$, f_w is the friction factor defined in Equation 1.2.18.

Sleath (1991) pointed out that the turbulence decay with distance from the bed in fairly rough oscillatory boundary layers is analogous to the decay of grid turbulence. He recommended the formula

$$\frac{1}{w'_{rms}} = 6.29\, A^{-3/2}\, r^{-1/2}\, T\, z \qquad (1.2.2)$$

which is in good agreement with the data except within the above mentioned layer of constant turbulence intensity immediately above the bed. The thickness and importance of this constant-intensity-layer is generally small, but may be judged from Figure 4 of Sleath (1987).

An example of the distribution of turbulence intensities in a turbulent, oscillatory boundary layer is shown in Figure 1.5.7.

With respect to the time-dependent turbulence intensities $(\widetilde{u'^2})^{0.5}$ and $(\widetilde{w'^2})^{0.5}$, both Sleath (1987) and Jensen (1989) found that the amplitudes were greatest near the bed and that the peak values occured later at higher elevations.

The general picture is one of parcels of turbulence propagating upwards from the bed. On the basis of his data, Sleath derived the expression

$$w_t \approx \frac{\omega\,\delta_{.05}}{2.27} \tag{1.2.3}$$

for the speed of vertical convection of turbulence, where the boundary layer thickness, $\delta_{.05}$ may be estimated from Equation (1.1.9).

1.2.4 Laminar oscillatory flow over a smooth bed

Although natural sand beds are never perfectly flat and natural flows tend to be turbulent, it is worthwhile studying the case of smooth, laminar flow in detail because many of its features are present in natural flows and because its structure gives clues towards efficient methods of analysing natural flows.

We base the analysis on the linear equation of motion (1.1.12) which means that the flow is assumed essentially horizontal and uniform in the x-direction. In order to simplify the mathematical treatment we represent the free stream velocity $u_\infty(t)$ by the complex exponential

$$u_\infty(t) = A\omega\, e^{i\omega t} \tag{1.2.4}$$

The real part $A\omega \cos \omega t$ represents the physical velocity, see Figure 1.2.5.

For laminar flow the shear stress is proportional to the local velocity gradient and the fluid viscosity

$$\tau(z, t) = \rho\nu \frac{\partial u}{\partial z} \tag{1.2.5}$$

so the equation of motion (1.1.12) can be written

$$\frac{\partial}{\partial t}(u - u_\infty) = \nu \frac{\partial^2 u}{\partial z^2} \tag{1.2.6}$$

Here we may introduce the non-dimensional velocity defect function $D(z, t)$ defined by

$$u_d(z, t) = u_\infty(t) - u(z, t) = A\omega D(z, t) \tag{1.2.7}$$

in terms of which the equation of motion takes the form of the diffusion equation

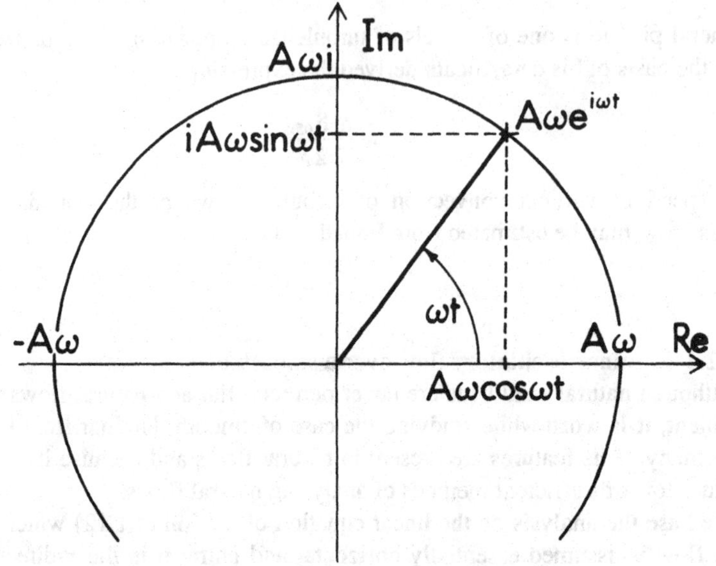

Figure 1.2.5: The complex velocity $A\omega\, e^{i\omega t}$ has the constant modulus $A\omega$ and moves around a circle with angular velocity ω. The real part $A\omega \cos \omega t$ which represents the physical velocity oscillates between $A\omega$ and $-A\omega$.

$$\frac{\partial D}{\partial t} = \nu \frac{\partial^2 D}{\partial z^2} \tag{1.2.8}$$

This is easily solved by separation of variables and assuming that the velocity defect function has the form

$$D(z,t) = \sum_{1}^{\infty} D_n(z)\, e^{in\omega t} \tag{1.2.9}$$

where $D_n(z)$ must then satisfy

$$in\omega D_n = \nu \frac{\partial^2 D_n}{\partial z^2} \tag{1.2.10}$$

This has solutions of the form

$$D_n(z) = A_n\, e^{z\sqrt{in\omega/\nu}} + B_n\, e^{-z\sqrt{in\omega/\nu}} \tag{1.2.11}$$

Since the velocity defect must vanish for $z \to \infty$, A_n must be zero for all n, and the boundary condition at the bed where the velocity itself vanishes

$$u(o,t) = u_\infty(t) - A\omega\, D(o,t) = A\omega\, e^{i\omega t} - A\omega \sum B_n\, e^{in\omega t} = 0 \qquad (1.2.12)$$

gives $B_1 = 1$ and $B_n = 0$ for $n \neq 1$. Hence, the complete solution is

$$u(z,t) = A\omega[1 - D_1(z)]\, e^{i\omega t} \qquad (1.2.13)$$

$$= A\omega[1 - \exp(-[1+i]\frac{z}{\sqrt{2\nu/\omega}})]\, e^{i\omega t} \qquad (1.2.14)$$

The complex velocity defect function $D_1(z)$ gives the velocity different phases as well as different magnitudes at different elevations, see Figure 1.2.6.

The shear stress distribution is found by inserting the expression (1.2.14) for the velocity field into Newton's formula (1.2.5) which gives

$$\tau(z,t) = \rho\nu A\omega\,(1+i)\sqrt{\omega/2\nu}\,\exp[-(1+i)\frac{z}{\sqrt{2\nu/\omega}}]\, e^{i\omega t}$$

This shows that the shear stress magnitude decreases exponentially away from the bed with a decay length scale of $\sqrt{2\nu/\omega}$. The bed shear stress is

$$\tau(o,t) = \rho\nu A\omega(1+i)\sqrt{\omega/2\nu}\, e^{i\omega t} = \rho\sqrt{\omega\nu}\, A\omega\, e^{i(\omega t + \frac{\pi}{4})} \qquad (1.2.15)$$

so we see that the bed shear stress in smooth, laminar oscillatory flow leads the free stream velocity by $\pi/4$ radians or 45°.

The value $\hat{\tau} = \rho\sqrt{\omega\nu}\, A\omega$ for the maximum bed shear stress shows that, with the definition $\hat{\tau} = \frac{1}{2}\rho f_w (A\omega)^2$ from Jonsson (1966), the wave friction factor for smooth, laminar flow is

$$f_{w,\mathrm{lam}} = \frac{2}{\sqrt{A^2\,\omega/\nu}} \qquad (1.2.16)$$

Equation (1.2.15) for the bed shear stress corresponding to $u_\infty(t) = A\omega\, e^{i\omega t}$ is also valid for the individual harmonic components of an arbitrary free stream

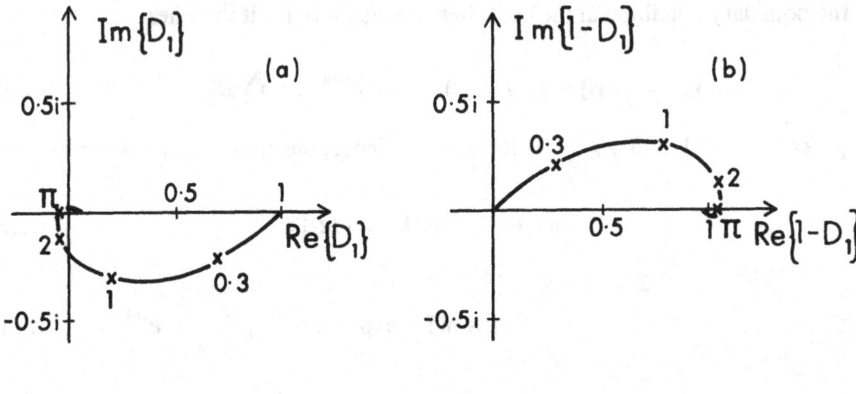

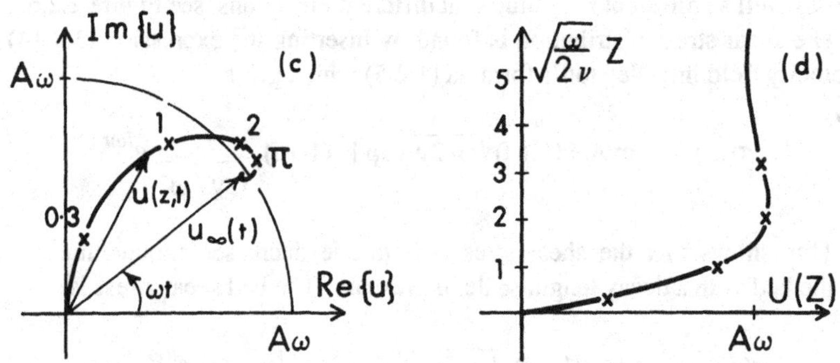

Figure 1.2.6: Velocity variations with elevation in simple harmonic, oscillatory, laminar flow over a smooth bed.

a: The defect function $D_1(z)$ moves along a logarithmic spiral starting at 1 and approaching 0 as z increases. Numbers on the curves refer to the non-dimensional elevation $z\sqrt{\omega/2\nu} = z/\delta_s$.

b: Corresponding variation of $1-D_1(z)$ which is the complex ratio between $u(z,t)$ and $u_\infty(t)$. See Equation (1.2.13).

c: In the simple case of laminar flow over a smooth bed where $u(z,t)$ is simple harmonic $u(z,t) = A\omega[1-D_1(z)]\,e^{i\omega t}$ we can construct $u(z,t)$ geometrically by using the circle from Figure 1.2.5 and the spiral above (b). The velocity at the bed leads u_∞ by 45°.

d: The variation of the velocity amplitude $U(z) = A\omega\,|1-D_1(z)|$ with elevation. The maximum value (about $1.07A\omega$) occurs for $z\sqrt{\omega/2\nu}$. ≈ 2.28.

velocity $u_\infty(t)$, so the frequency response function for the bed shear stress $\tau(o,t)$ from input $u_\infty(t)$ is

$$F(\omega) = \rho \sqrt{\omega \nu}\, e^{i\pi/4} \tag{1.2.17}$$

1.2.5 The wave friction factor

The water in streams and under waves interacts with the bed sediment mainly through the bed shear stress $\tau(o,t)$. Hence the determination of $\tau(o,t)$ is a crucial step in all sediment transport calculations. Therefore, considerable effort has been put into the study of $\tau(o,t)$ under waves. Both Jonsson (1966) and Kajiura (1968) developed semi-empirical and theoretical formulae based on their early flow models.

Jonsson defined the wave friction factor f_w in relation to the maximum of $\tau(o,t)$ by

$$\hat{\tau} = \frac{1}{2}\rho f_w (A\omega)^2 \tag{1.2.18}$$

This definition may be applied to natural flows where $\tau(o,t)$ is not necessarily simple harmonic. Kajiura, on the other hand, considered only the fundamental (sinusoidal) mode of the flow and wrote the bed shear stress in the form

$$\tau(o,t) = \rho C_1 (A\omega)^2 e^{i\omega t} \tag{1.2.19}$$

He accounted for the phase shift φ between bed shear stress τ and the free stream velocity u_∞ by allowing C_1 to be complex $C_1 = |C_1|\, e^{i\varphi}$.

Kamphuis (1975) presented very comprehensive measurements of $\hat{\tau}$ on flat beds of glued-down sand which are summarized in Figure 1.2.7.

Jonsson (1966) showed from dimensional analysis that the wave friction factor can be expected to depend on the Reynolds number $A^2\omega/\nu$ and on the relative bed roughness r/A

$$f_w = f_w\left(\frac{A^2\omega}{\nu}, \frac{r}{A}\right) \tag{1.2.20}$$

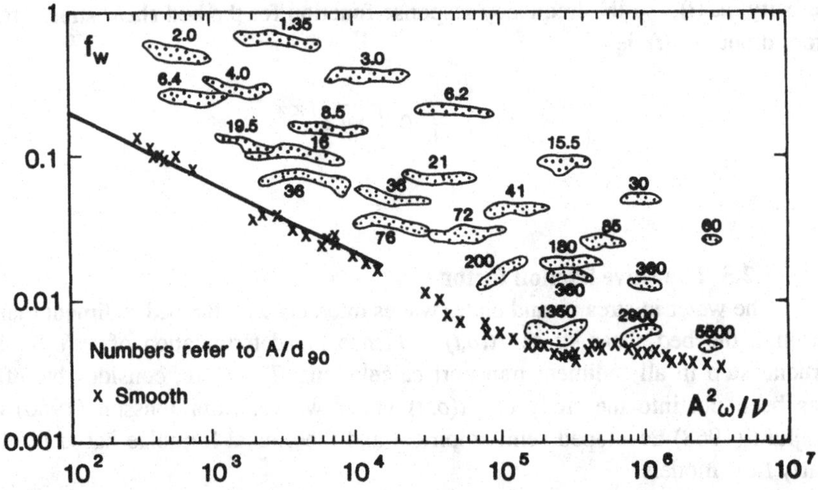

Figure 1.2.7: Measured values of f_w from oscillatory flow over flat beds of fixed sand grains, after Kamphuis, 1975. The Nikuradse roughness for these sand beds was taken to be $2d_{90}$.

Somewhat simpler formulae are adequate for the fully developed, rough turbulent regime where $A^2\omega/\nu \to \infty$ while r/A is finite, and for smooth conditions $(r/A \to 0)$.

In Figure 1.2.7 the crosses correspond to experiments with a smooth bed. It can be seen that, for Reynolds numbers up to about $3\cdot 10^5$ the smooth bed wave friction factor is well described by

$$f_w = \frac{2}{\sqrt{A^2\omega/\nu}} \tag{1.2.16}$$

which is the theoretical result for smooth laminar flow. The range $3\cdot 10^5 < A^2\omega/\nu < 6\cdot 10^5$ is a transition zone where f_w increases. It is followed by the fully developed, smooth turbulent regime where f_w decreases again, although at a slower rate than in the laminar regime. Justesen (1988) suggests

$$f_w = 0.024\,(A^2\omega/\nu)^{-0.123} \quad \text{for} \quad 10^6 < A^2\omega/\nu < 10^8 \tag{1.2.21}$$

for the wave friction factor in the fully developed, smooth turbulent regime.

Kamphuis' rough bed data are arranged in groups with fixed relative roughness in Figure 1.2.7, the numbers referring to the ratio d_{90}/A. . Note that data points towards the upper right hand corner tend to fall on horizontal lines corresponding to constant f_w for constant d_{90}/A.. We say that these data, for which $f_w = f_w(r/A)$, are in the fully developed, *Rough Turbulent Regime*. Several authors have proposed formulae for f_w in this regime, of which the most commonly used is

$$f_w = \exp\left[5.213 \left(\frac{r}{A}\right)^{0.194} - 5.977 \right] \tag{1.2.22}$$

This formula was suggested by Swart (1974) as an explicit approximation to an implicit, semi-empirical formula given by Jonsson (1966).

Kajiura (1968) and Jonsson (1980) have both suggested upper limits for the value of f_w. Kajiura suggested *0.25*, Jonsson *0.30*. These limits were, however, inspired by observations of the energy dissipation factor, f_e (defined in Section 1.2.6) by Bagnold (1946). Bagnold recommended a constant limiting value of f_e of *0.24* for large relative roughness. However, the recent measurements by Sleath (1985) yielded f_e-values in excess of *0.5*, and Simons et al (1988) measured f_e-values far in excess of *1.0*.

It is well known that f_e can be expected to be somewhat smaller than f_w for very rough beds ($r/A \approx 1$) as illustrated by Sleath (1984) p 200. The reason is, that for large r/A a considerable part of $\tau(o, t)$ is due to the pressure gradients acting on the individual roughness elements. These pressure gradients are in quadrature with u_∞ and therefore do not contribute to the energy dissipation rate which is given by $D_E = \overline{\tau(o,t)\, u_\infty(t)}$.

Figure 1.2.8 shows the presently available f_w and f_e data for rough, turbulent flow over beds with known Nikuradse roughness. The full line corresponds to Swart's formula (1.2.22) and we see that it tends to over predict f_w for small r/A. On the basis of this data set it might be justified to adjust the coefficients in Swart's formula to

$$f_w = \exp\left[5.5 \left(\frac{r}{A}\right)^{0.2} - 6.3\right] \tag{1.2.23}$$

which corresponds to the dotted curve.

The fully developed, rough turbulent regime in steady flow is bounded by the condition $u_* r/\nu > 70$, where u_* is the friction velocity (see Section 1.5.2). In

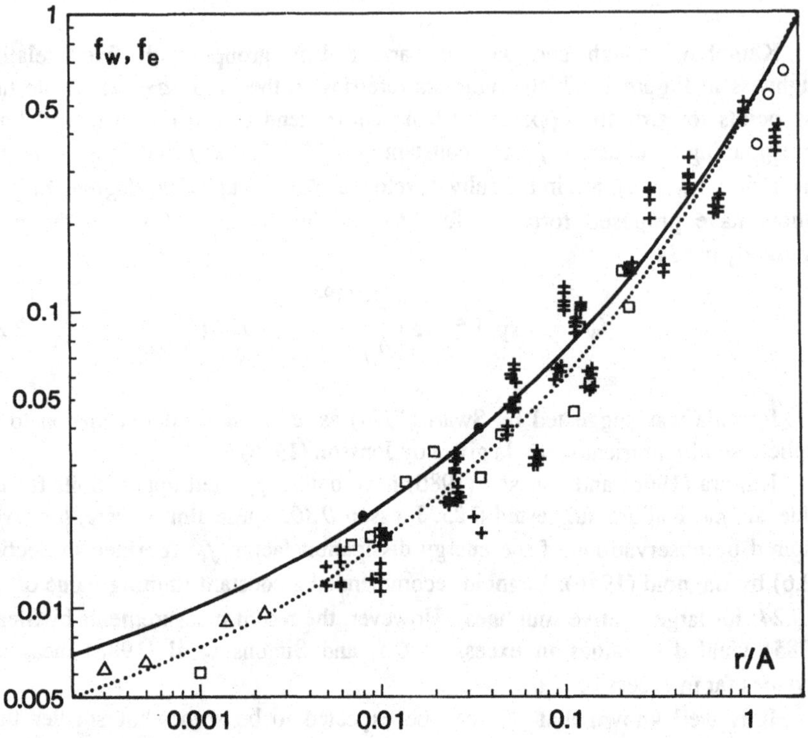

Figure 1.2.8: Observed wave friction factors for rough turbulent flow. + : Riedel (1972) with $r = 2d_{90}$, o: Kemp and Simons (1982), □ Sleath (1987), △ Jensen (1989), •: Jonsson and Carlsen (1976).

analogy with this, Kamphuis (1975) suggested a criterion of the form

$$\hat{u}_* \, r/\nu \;=\; \sqrt{f_w/2}\; A\omega \, r/\nu \;\geq\; constant \qquad (1.2.24)$$

for oscillatory flow. However, he found that a single constant would not apply throughout the whole range of relative roughness. Instead he suggested two values, namely *200* for fairly rough beds $(r/A \geq 0.01)$ while the value *70* from steady flow applies asymptotically for $r/A \to 0$.

While the peak value $\hat{\tau}$ is reasonably described by the formulae above, the time dependence of the bed shear stress is not so well understood in general.

Two examples of the variation of $\tau(o,t)$ with time were given in Figures 1.2.2 and 1.2.3 for two cases of turbulent flow with different relative roughness.

These show how $\tau(o,t)$ is fairly similar to the smooth laminar solution (1.2.16) for small r/A. That is, the variation is almost sinusoidal but the phase lead, relative to u_∞, is less than the "laminar" 45°. Jonsson and Carlsen's Test 1 shows a phase lead of about 25°.

Lofquist (1980, 1986) reported comprehensive measurements of $\tau(o,t)$ over natural sand beds. The sequence shown in Figure 1.2.3 corresponds to a constant value of the velocity amplitude $A\omega$ with the peak acceleration $A\omega^2$ growing from top to bottom. The curves show that larger $A\omega^2$ leads to a greater number of more pointed peaks in $\tau(o, t)$. Lofquist explained how the peaks are related to the growth and release of lee vortices.

1.2.6. Wave energy dissipation due to bed friction

The time-averaged rate of *energy dissipation* due to bed friction is given by

$$D_E = \overline{\tau(o,t)\, u_\infty(t)} \tag{1.2.25}$$

and Jonsson (1966) defined the energy dissipation factor f_e by

$$D_E = \frac{2}{3\pi}\rho f_e (A\omega)^3 \tag{1.2.26}$$

Thus Jonsson's f_e is related to Kajiura's C_1, defined in Equation (1.2.19), by

$$f_e = \frac{3\pi}{4}|C_1| \cos\varphi = \frac{3\pi}{4}\mathrm{Re}\{C_1\} \tag{1.2.27}$$

Bagnold (1946), Carstens et al (1969) and Lofquist (1986) all used different definitions and terminology. Their energy dissipation factors k, $\overline{\overline{f}}$ and $\overline{f_1}$ are related to f_e by

$$f_e = 3k = 4\overline{\overline{f}} = \frac{3\pi}{4}\overline{f_1} \tag{1.2.28}$$

The two friction related coefficients f_w and f_e are different according to their definitions. However, the experimental scatter of measurements of one or the other over natural sand beds is so large that for practical purposes f_w and f_e can be assumed equal. This is illustrated by the data in Figure 1.2.9 which shows f_w plotted against f_e.

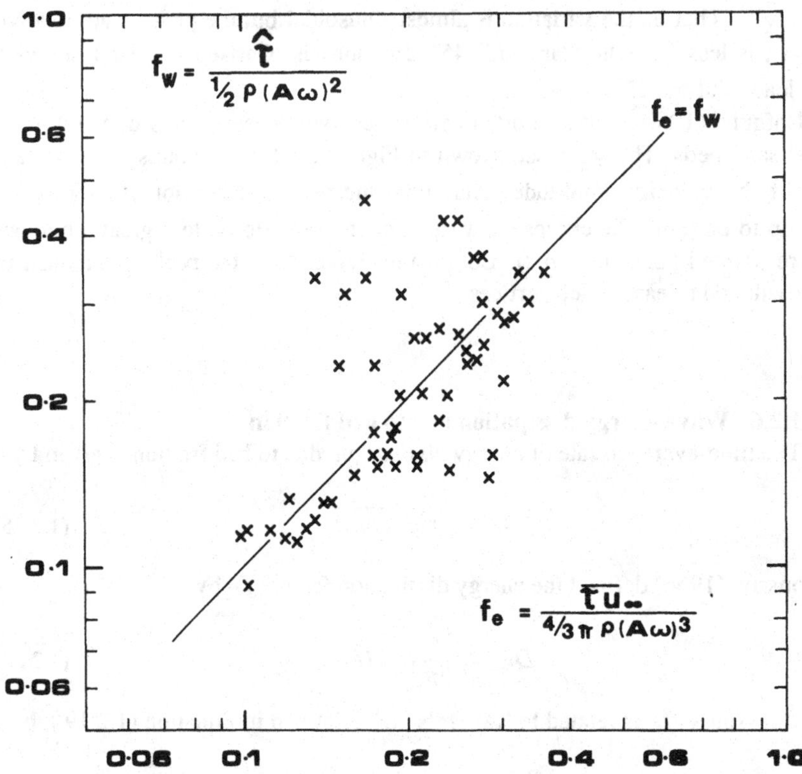

Figure 1.2.9: The friction factor f_w plotted against the energy dissipation factor f_e from measurements by Lofquist (1986) over rippled sand beds.

The range of measured f_e values is quite large. Laboratory measurements by Bagnold (1946), Carstens et al (1969), Kemp & Simons (1981), Sleath (1985), Lofquist (1986) and Simons et al (1988) range from 0.03 to 40 while field measurements by Bretschneider (1954), and Iwagaki & Kakinuma (1967) range from *0.02* to *2.46*.

It should be noted that other mechanisms of energy dissipation than friction may be present in natural flows. These include viscous dissipation in mud bottoms and losses due to percolation through a sand bed. For a review of these effects, see Dean & Dalrymple (1991) and Sleath (1984).

1.2.7 Boundary layer thickness for oscillatory flow

While the qualitative meaning of the term boundary layer is straightforward, the opinions about the most appropriate quantitative definition are varied.

Sleath (1987) intuitively defined the top of the boundary layer as the position where the amplitude of the velocity defect has dropped to a certain small fraction of $A\omega$. He chose $|D| = 0.05$ which, for a smooth laminar boundary layer flow corresponds to $|D_1(\delta_{.05})| = \exp(-\delta_{.05}\sqrt{\omega/2\nu}) = 0.05$ (see Equation 1.2.14) or $\delta_{.05} \approx 3\sqrt{2\nu/\omega} = 3\delta_s$, where δ_s is the Stokes length,

$$\delta_s = \sqrt{\frac{2\nu}{\omega}}$$

Jonsson (1966) used a different type of definition. He defined the top of the boundary layer as the minimum elevation where $u(z, t)$ equals $u_\infty(t)$ when the latter is maximum. Jonsson's boundary layer is quite thin i e, his definition corresponds to $\delta_j = \frac{\pi}{2}\sqrt{2\nu/\omega}$ for smooth laminar flow or approximately half of $\delta_{.05}$.

Kajiura (1968) worked with the displacement thickness defined as

$$\delta_d = \frac{1}{A\omega}\text{Max}\left\{\int_o^\infty (u_\infty - u)\, dz\right\} \tag{1.2.29}$$

This again is a fairly thin boundary layer since for smooth laminar flow it corresponds to $\delta_d = \sqrt{\nu/\omega}$ and hence $|D_1(\delta_d)| = \exp(-\sqrt{2}/2) \approx 0.49$.

However, the displacement thickness has the advantage that it is related in a simple way to the other important boundary layer parameter, namely, the wave friction factor. Their interrelation stems from the fact that the above definition for δ_d is very similar to the integrated momentum Equation (1.2.1) which in turn defines the friction factor through Equation (1.2.18). Combining these equations leads to

$$\delta_d = \frac{1}{2} f_w A \tag{1.2.30}$$

This formula is exact for simple harmonic flows with the form $u(z, t) = [1 - D_1(z)]\, u_\infty(t)$ (Equation 1.2.13). However, it also provides a useful estimate of δ_d in general. This is important because the two major data sets of oscillatory boundary layer flow over natural sand beds, Carstens et al (1969) and

Lofquist (1986) provide only f_w (or f_e) but no details of the velocity distribution. Therefore, δ_d (or any other vertical scale for the boundary layer) cannot be determined directly from the data, only via Equation (1.2.30).

The practical limit for measuring boundary layer structures with present day technology lies around the level where the velocity defect is one percent of the free stream velocity amplitude. This level, $\delta_{.01}$ relates to the other measures of boundary layer thickness as follows

$$\delta_{.01} \approx 1.5\,\delta_{.05} \approx 3\,\delta_j \approx 4.5\,\delta_s \approx 6.4\,\delta_d \tag{1.2.31}$$

The definitions and interrelations are also illustrated in Figure 1.2.10.

Another boundary layer length scale which has the advantage of being related to the friction factor is

$$\delta_* = \hat{u}_* / \omega = \sqrt{f_w/2}\,A \tag{1.2.32}$$

which, for flow of the form $u(z, t) = [1 - D_1(z)]\,u_\infty(t)$, is related to the displacement thickness by $\delta_* = \delta_d\,(f_w/2)^{-0.5}$, and since f_w is generally of the order 0.2 or less we see that δ_* is fairly large compared to the other δ-values.

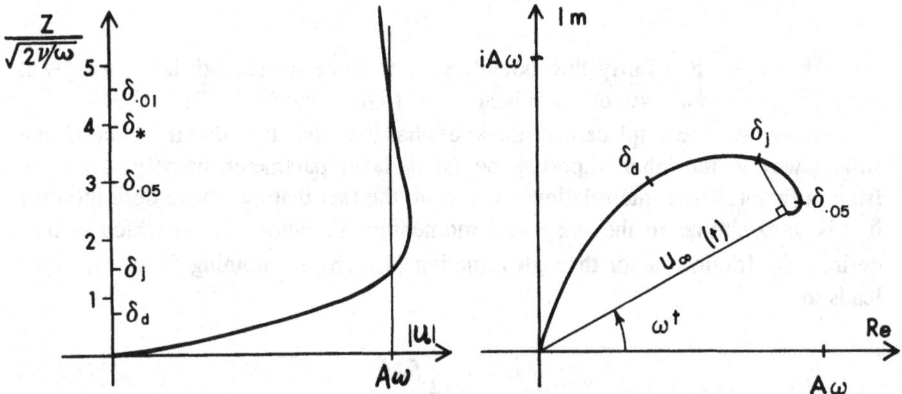

Figure 1.2.10: Comparison of the different definitions of boundary layer thickness for the case of smooth, laminar flow with $A\omega^2/\nu = 10^3$. The laminar length scale $\sqrt{2\nu/\omega}$ is called the "Stokes length".

1.2.8. Eddy viscosity in oscillatory flow

The eddy viscosity concept, which was introduced by Boussinesq more than a century ago, is a useful tool for obtaining simple flow models. However, as we shall see in the following section, some unusual eddy viscosities are required for the description of oscillatory boundary layer flows.

The simplified equation of motion for horizontally uniform flow

$$\rho \frac{\partial}{\partial t}(u - u_\infty) = \frac{\partial \tau}{\partial z} \tag{1.1.12}$$

contains two unknowns namely u and τ. It can therefore only be solved when the relation between these two is known or assumed to have a certain form. We know the relationship for laminar flows where it is given by Newton's formula (1.2.5) but for turbulent flows it is not well understood.

Many schemes (turbulence closure schemes) have been suggested for getting around this problem and the simplest of these involves the use of the eddy viscosity concept, which was first introduced by Boussinesq. It is defined by analogy with Newton's formula for laminar shear stress i e

$$\tau = \rho \, \nu_t \frac{\partial u}{\partial z} \tag{1.2.33}$$

which for a steady, uniform turbulent flow corresponds to

$$\nu_t = \frac{-\overline{u'w'}}{\frac{d\overline{u}}{dz}} + \nu \tag{1.2.34}$$

in terms of the Reynolds stress $-\rho \overline{u'w'}$ and it is a general feature of steady turbulent flows that the first term dominates over the molecular viscosity except inside the laminar sublayer.

For a uniform, oscillatory flow with zero net flow $(u = \tilde{u} + u')$, the analogous expression for the eddy viscosity becomes (in accordance with Equation 1.1.25)

$$\nu_t = \frac{\tilde{\tau}}{\rho \frac{\partial \tilde{u}}{\partial z}} = \frac{-(\widetilde{\tilde{u}\tilde{w}}) - (\widetilde{u'w'})}{\frac{\partial \tilde{u}}{\partial z}} + \nu \tag{1.2.35}$$

where the meaning of the " ~ " operator is as defined by Equation (1.1.17).

We note that by this definition ν_t is not a purely turbulent quantity. It contains a dominant, deterministic contribution $-(\tilde{\tilde{u}}\tilde{w})_{\sim}$ which was found by Sleath (1987) to be an order of magnitude larger than $-(u'w')$. This caused Sleath to express concern about its use. However, Equation (1.2.35) is the most useful definition of ν_t because it enables us to write the equation of motion (1.1.12) in the linear form

$$\frac{\partial \tilde{u}}{\partial t} = \frac{\partial \tilde{u}_\infty}{\partial t} + \frac{\partial}{\partial z}\left(\nu_t \frac{\partial \tilde{u}}{\partial z}\right) \tag{1.2.36}$$

and it is the definition which has been used in all existing eddy viscosity based models of oscillatory boundary layer flow.

Except for laminar flow, and for one particular class of turbulent flows, which will be discussed in Section 1.2.9, the eddy viscosity for oscillatory boundary layers is a function of the distance from the bed and it can generally not be considered constant (at least not a real valued constant) over time for fixed z. The reason is that the shear stress and the velocity gradient tend not to be zero at the same phase see Figure 1.2.11.

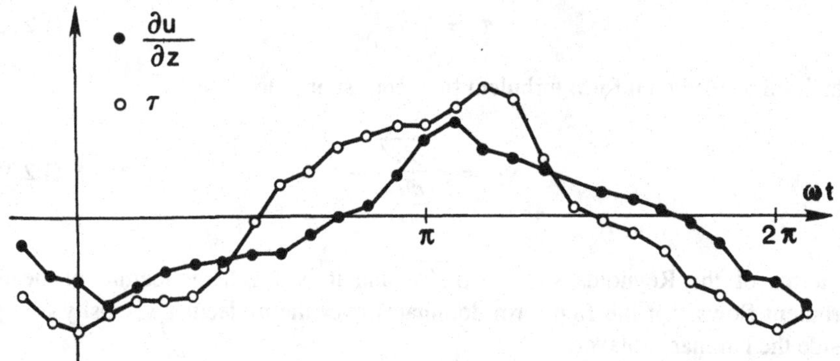

Figure 1.2.11: Time dependence of shear stress and velocity gradient, both phase averaged. From Jonsson & Carlsen (1976) Test 1, elevation above ripple crest: *4.5cm.*

This was realised previously by Horikawa & Watanabe (1968), Jonsson & Carlsen (1976) and Sleath (1987). An example of the corresponding variation of the eddy viscosity derived directly from the definition (1.2.33) is shown in Figure 1.2.12. This eddy viscosity takes on negative and even infinite values.

Sleath (1987) studied the turbulent stress component

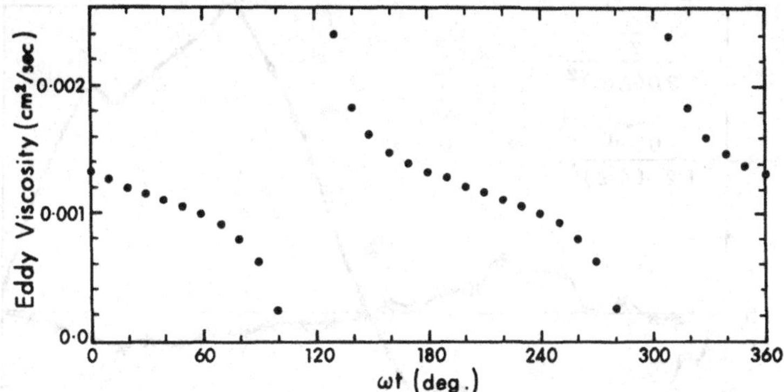

Figure 1.2.12: Eddy viscosity derived from velocity measurements using Equation (1.2.33), as function of wave phase. After Horikawa & Watanabe (1968).

$$\nu_{TR} = \frac{-\widetilde{(u'w')}}{\dfrac{\partial \tilde{u}}{\partial z}} \qquad (1.2.37)$$

as well as the total given by Equation (1.2.35). He found that the turbulent stress component contributed only about ten percent of the total and it showed a very different variation with phase, see Figure 1.2.13. In contrast, the contribution $-(\widetilde{\tilde{u}\,\tilde{w}})$ from the phase-averaged flow corresponded rather closely to the total shear stress derived from the momentum equation (1.2.1).

Sleath (1987) also studied the time average of ν_{TR} and of the total ν_t. He found that $\overline{\nu_{TR}}$ was "firmly negative" over several roughness heights near the bed, then went positive and continued to increase away from the bed.

For the time average $\overline{\nu_t}$ of the total eddy viscosity he found that it was generally positive and tended to increase with distance from the bed. However, the general magnitude was significantly smaller than the equivalent for steady flow. That is, in terms of the average friction velocity $\overline{u_*} = \overline{(|\tau|/\rho)^{0.5}}$ he found

$$\overline{\nu_t} \approx K\,\overline{u_*}\,z \;, \quad 0.10 < K < 0.13 \qquad (1.2.38)$$

where K is analogous to von Karman's constant κ ($\approx$ 0.4) for steady flow, but is seen to be smaller by a factor of 3 to 4. These results were obtained for relative roughness of the order $r/A \approx 0.01$.

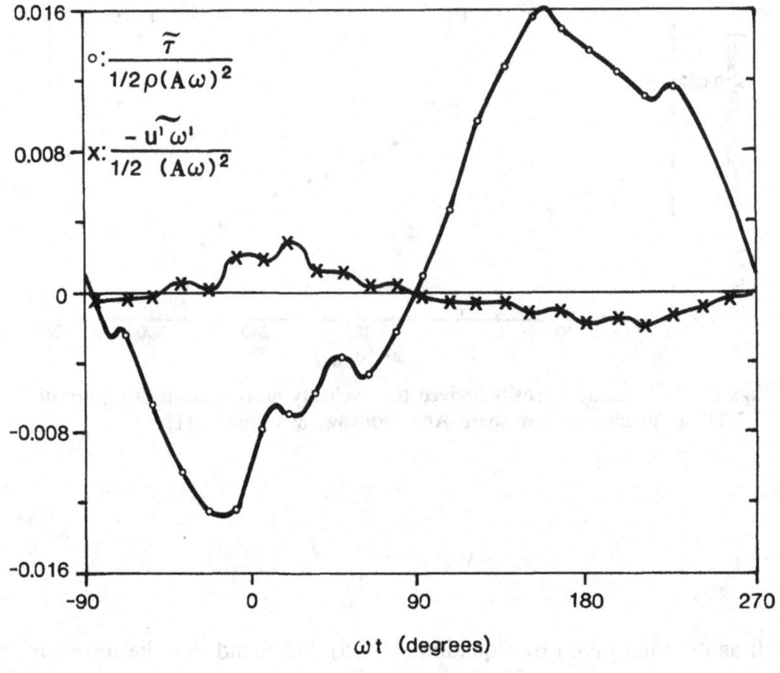

Figure 1.2.13: Reynolds stress $-\rho(\widetilde{u'w'})$ compared to the total shear stress $\tilde{\tau}$ calculated from Equation (1.2.1). After Sleath (1987), Test 4.

1.2.9. Eddy viscosity for sinusoidal flow

In order to obtain manageable, analytical descriptions for oscillatory boundary layer flow, it is generally deemed acceptable to consider only the fundamental harmonic component $u_1(z,t)$ of the horizontal velocity and the corresponding shear stress component $\tau_1(z,t)$. When this simplification is applied, the eddy viscosity also becomes somewhat simpler than for the general situation, described above.

In terms of the first harmonics $u_1(z,t)$ and $\tau_1(z,t)$ we may define an eddy viscosity ν_1 by

$$\nu_1 = \frac{\tau_1}{\rho \frac{\partial u_1}{\partial z}} \tag{1.2.39}$$

which, with the velocity gradient and the shear stress written in the complex forms

$$\frac{\partial u_1}{\partial z} = \left|\frac{\partial u_1}{\partial z}\right| e^{\,i(\omega t + \varphi)} \qquad (1.2.40)$$

and

$$\tau_1 = |\tau_1|\, e^{\,i(\omega t + \psi)} \qquad (1.2.41)$$

gives

$$\nu_1 = |\nu_1|\, e^{\,i\varphi_\nu} = \frac{|\tau_1|}{\left|\frac{\partial u_1}{\partial z}\right|} e^{\,i(\psi - \varphi)} \qquad (1.2.42)$$

Thus, treating the oscillatory boundary layer flow as simple harmonic with a correspondingly simple harmonic shear stress (or analysing only the fundamental harmonic modes) leads to an eddy viscosity $\nu_1(z)$ which is a function of z but not of t.

However, making provision for the shear stress to be out of phase with the velocity gradient requires ν_1 to be complex with an argument φ_ν equal to the phase shift between τ_1 and $\partial u_1/\partial z$.

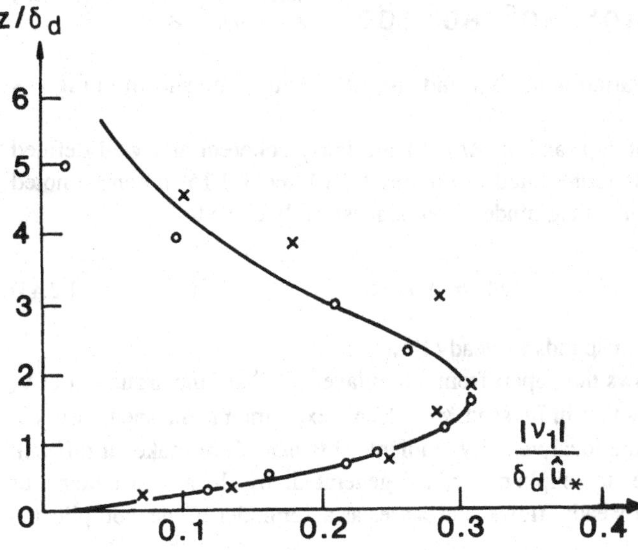

Figure 1.2.14: Eddy viscosity magnitude based on the fundamental modes $u_1(z,t)$ and $\tau_1(z,t)$ of Jonsson & Carlsen's data. Elevations are measured from the level of the theoretical bed, and the displacement thickness δ_d and the friction velocity $\hat{u}_*$ are defined by Equations (1.2.30) and (1.2.24) respectively. x: Test 1, o: Test 2.

We note that if one chooses to use real valued cosine functions instead of the complex exponentials in (1.2.40) and (1.2.41) then, the result of the phase shift between shear stress and velocity gradient is a real valued eddy viscosity which varies strongly with time and takes both negative and infinite values as indicated by Figure 1.2.12.

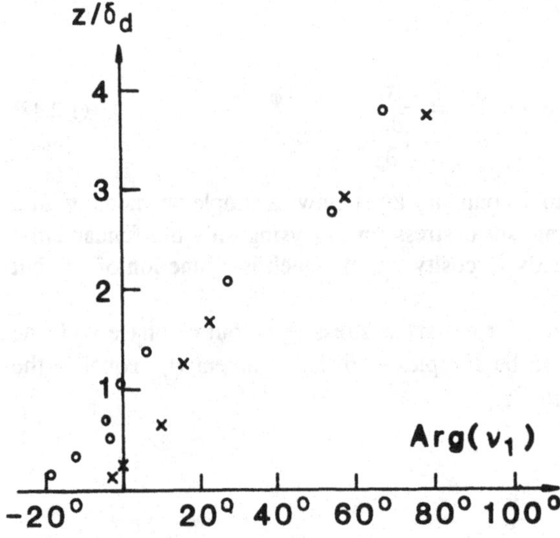

Figure 1.2.15: Argument variation for the complex eddy viscosity ν_1. Data from Jonsson & Carlsen (1976). Test I: x, *T=8.4s, A=2.85m, r/A* = 0.008, Test II: o, *T=7.2s, A=1.79m, r/A* = 0.035.

Examples of the variation of $|\nu_1|$ and $Arg(\nu_1)$ with z are shown in Figures 1.2.14 and 1.2.15.

The behaviours of $|\nu_1|$ and of $Arg(\nu_1)$ are fairly coherent and well defined for the two experiments represented in Figures 1.2.14 and 1.2.15. It can be noted that the initial trend for the magnitude of ν_1 is reasonably close to

$$|\nu_1| = \kappa \, \hat{u}_* \, z \tag{1.2.43}$$

with $\kappa \approx 0.4$, which corresponds to steady flow.

Figure 1.2.15 shows that, apart from a thin layer ($z < \delta_d$), the argument of ν_1 is positive which means that in Jonsson & Carlsen's experiments the shear stress τ_1 tended to be ahead of the local velocity gradient. This behaviour makes it difficult or indeed impossible to explain ν_1 as generated by local, instantaneous parameters, as done for steady flows, and serves as a reminder of the complicated

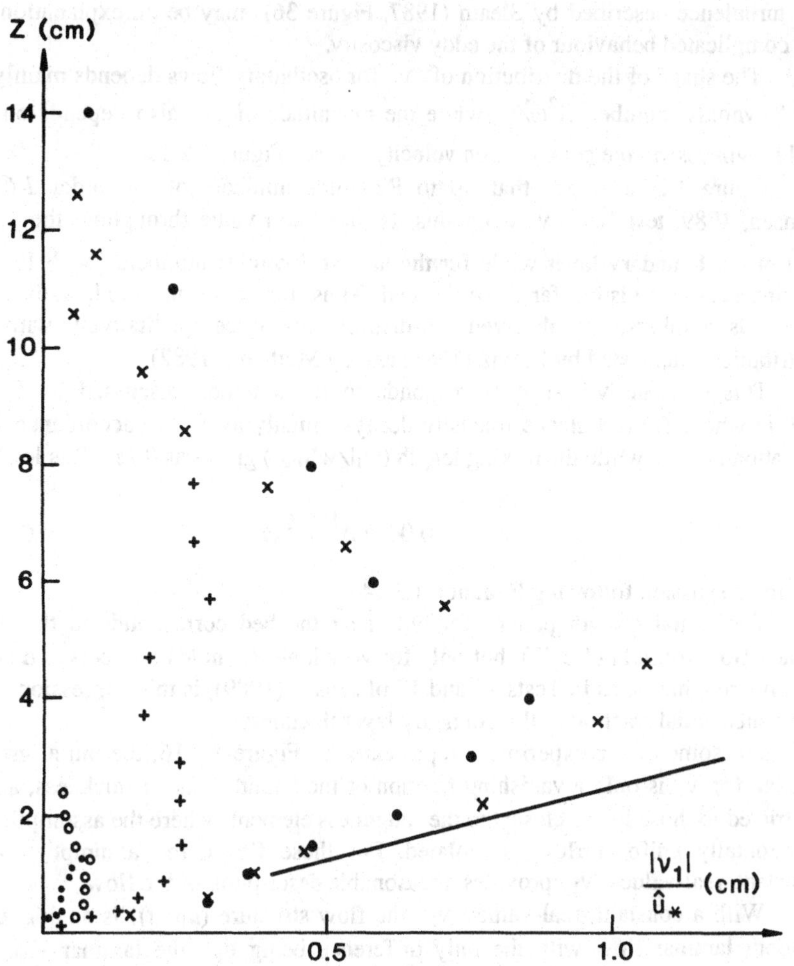

Figure 1.2.16: Variation of $|v_1|/\hat{u}_*$ for a range of flow conditions. The straight line corresponds to Equation (1.2.43). x : Jensen (1989) test 10, *(A, T, r, δ_{05}) = (3.1m, 9.72s, 0.0mm, 60mm);* + : Jensen (1989) test 12, *(A, T, r, δ_{05}) = (1.58m, 9.72s, 0.84mm, 50mm)*; ●: Jonsson & Carlsen test 2 *(A, T, r, δ_{05}) = (1.79m, 7.2s, 63mm, 150mm)* ; o : van Doorn (1982) M-series *(A, T, r, δ_{05}) = (0.33m, 2.0s, 21mm, 26mm)* ; •: van Doorn (1982) S-series *(A, T, r, δ_{05}) = (0.10m, 2.0s, 21mm, 14.5mm).*

nature of ν_t which is indicated by Equation (1.2.35). The convective behaviour of the turbulence described by Sleath (1987, Figure 36) may be an explanation for the complicated behaviour of the eddy viscosity.

The shape of the distribution of ν_1 for oscillatory flows depends mainly on the Reynolds number, $A^2\omega/\nu$ while the magnitude of ν_1 also depends on the bed roughness via the peak friction velocity $\hat{u}_*$ see Figure 1.2.16.

Figure 1.2.16 shows that up to Reynolds numbers of the order $1.6{\cdot}10^6$ (Jensen, 1989, test 10) ν_1 maintains its maximum value throughout the upper part of the boundary layer while for the largest Reynolds numbers ($\gtrsim 3\cdot 10^6$), a decline becomes visible far from the bed. Thus, for small and moderately large Reynolds numbers, the observed ν_1-distributions agree qualitatively with the distributions suggested by Brevik (1981) and by Myrhaug (1982).

This constant-$|\nu_1|$-range corresponds to the situation discussed by Sleath (1991) where the turbulence intensity decays roughly as z^{-1} in accordance with Equation (1.2.2), while the mixing length ($|\nu_1|/w'_{rms}$) grows as $0.1z$. This leads to

$$|\nu_1| \approx 0.0025\, A^{3/2}\, r^{1/2}\, \omega \tag{1.2.44}$$

see the discussion following Equation 1.3.14.

The initial growth pattern for $|\nu_1|$ near the bed corresponds to the usual steady flow formula (1.2.43), but only for very high Reynolds numbers and small relative roughness, as in Tests 10 and 13 of Jensen (1989), is this expression valid for a substantial fraction of the boundary layer thickness.

For some of the experimens represented in Figure 1.2.16, the initial growth region for ν_1 is only a vanishing fraction of the boundary layer thickness, and is restricted to those levels closest to the roughness elements where the assumption of horizontally uniform flow is violated. For these flows, the assumption of a constant, real-valued ν_1 provides a reasonable description of the flow.

With a constant, real-valued ν_1 the flow structure ($u_1(z,t)$) is analogous to smooth laminar flow with the only difference being that the laminar viscosity ν is replaced the much larger constant ν_1.

In order to interpret the structure of oscillatory boundary layers with constant, real-valued ν_1, we draw on the mathematical framework derived for smooth laminar flow in Section 1.2.4. In Section 1.2.4 it was found that the velocity defect function defined by

$$u_1(z,t) = [1 - D_1(z)]\, A\omega\, e^{i\omega t} \tag{1.2.13}$$

was given by

$$D_1(z) \;=\; \exp\left[-(1+i)\frac{z}{\sqrt{2\nu/\omega}}\right] \tag{1.2.45}$$

for smooth laminar flow. Thus, the real and imaginary parts of the complex logarithm $\ln(D_1) \;=\; \ln|D_1| \;+\; i\,\mathrm{Arg}\,D_1$, are identical:

$$\ln|D_1| \;\equiv\; \mathrm{Arg}\,D_1 \;=\; -\frac{z}{\sqrt{2\,\nu/\omega}} \tag{1.2.46}$$

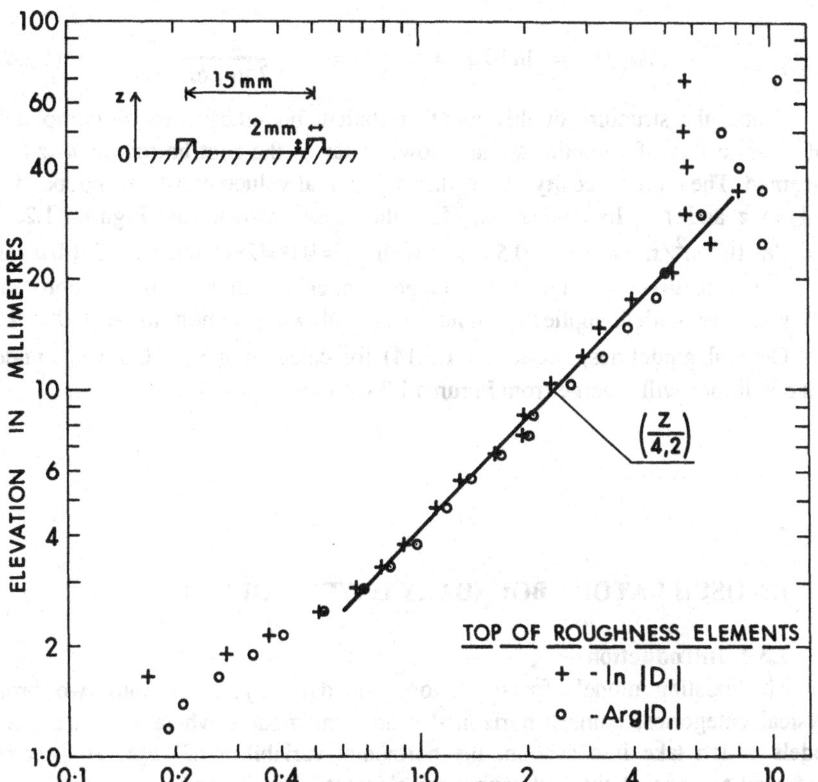

Figure 1.2.17: $ArgD_1$ and $\ln|D_1|$ for Test S0ORA from van Doorn (1982). $A = 10\,cm$, $r = 2.1\,cm$, $T = 2.0\,s$. The identity of $ArgD_1$ and $\ln|D_1|$ and their shared proportionality with z shows that the boundary layer structure is analogous to smooth laminar flow. A deviation from this "smooth laminar analogy" is visible below $z = 3mm$, but at these elevations, the assumption of horizontal uniformity ($u = u(z,t)$) is not valid anyway.

Hence, it is interesting to plot experimental values of ArgD_1 and $\ln|D_1|$ together against z for turbulent flows in order to get a comprehensive picture of the behaviour of $u(z, t)$ and especially of its similarities and dissimilarities with the smooth laminar solution.

This has been done in Figure 1.2.17 for a set of very detailed velocity measurements over a fairly rough bed.

The data show that, except for the first half millimetre above the roughness crest, the identity Arg$D_1 \equiv \ln|D_1|$ holds in analogy with smooth, laminar flow. Furthermore, both quantities are linearly proportional to z, so we may write

$$\mathrm{Arg}D_1 = \ln|D_1| = -\frac{z}{z_1} = -\frac{z}{\sqrt{2\nu_1/\omega}} \tag{1.2.47}$$

Hence, the structure of this rough, turbulent boundary layer is completely analogous to that of smooth, laminar flow, as far as the first harmonic $u_1(z,t)$ is concerned. The eddy viscosity ν_1 is therefore a real valued constant, independent of both z and t. Its value is, for the case shown in Figure 1.2.17, $\nu_1 = 28 \cdot 10^{-6} m^2/s$, or $\nu_1 = 0.5\,\omega\, z_1^2$ with $z_1 = 0.0042m$, and $\omega = 3.14 rad/s$

The relative roughness range over which the "constant-eddy-viscosity-model" applies is found, in the following section, to be $0.06 < r/A < 1$. General guidelines, (Equation 1.3.14) for calculating $\nu_1 = 0.5\,\omega\, z_1^2$ under such conditions will emerge from Figures 1.3.4 and 1.3.5.

1.3 OSCILLATORY BOUNDARY LAYER MODELS

1.3.1 Introduction

The existing models for oscillatory boundary layers fall into two broad physical categories, namely horizontally uniform models where $u = u(z,t)$ and models which take into account the horizontal variability of $u(x,z,t)$ between crests and troughs of the bed roughness elements. The latter group is by far the smallest although realistic modelling of flow over the commonly observed sand ripples obviously calls for models which can describe localised vortex formation.

Longuet-Higgins (1981) developed an essentially inviscid model based on conformal mapping of a sharp crested ripple profile and the discrete vortex method. The model yields a reasonable description of $\tau(o,t)$, but the results are numerical only. Also, because of the assumed sharp ripple crest the vortex shedding is continuous throughout each half cycle while observations of flow over natural ripples tend to show that separation only occurs during flow deceleration.

A different numerical model was developed by Sleath (1982). His ripple profiles had rounded crests and separation did not always occur. This model also reproduced $\tau(o,t)$ quite well, but the fact that it is laminar calls for some caution when extrapolating the results.

Recently, a different type of numerical solution was presented by Blondeaux and Vittori (1990) which gives a good reproduction of the vortex shedding and the resulting sediment clouds over vortex ripples.

Horizontally uniform models are much simpler. They can however, only be literally valid at elevations which are well clear of the top of the roughness elements, i e for $z >> r$. Hence, unless $\delta >> r$ they are at most relevant as descriptions of the horizontal average of the flow. Estimates of the ratio δ / r can be found from Equation (1.1.9).

1.3.2 Quasi-steady models of oscillatory boundary layers

Horizontally uniform models of rough turbulent oscillatory flow can be further subdivided into three main categories.

The first category consists of the quasi-steady models which assume that the velocity distribution is at all times logarithmic throughout a boundary layer thickness which may be constant or time-dependent. The classic model of Jonsson (1966) is quasi-steady in the above mentioned sense.

This model provided surprisingly good estimates (essentially Equation (1.2.22)) of the wave friction factor, see Figure 1.2.8, page 26.

The good performances of this formula for f_w at large relative roughness was unexpected and, in fact, the more recent, quasi-steady model of Fredsoe (1984) fails to predict f_w as accurately for $r/A > 0.03$. This limit for the use of "logarithmic velocity models" was predicted by Kajiura (1968) and has been experimentally verified by the work of Jensen (1989).

The viability of the assumption of a logarithmic velocity distribution at all phases of the flow depends on the relative roughness r/A. Some velocity distribution data which may give an impression are those of Jonsson & Carlsen (1976). They correspond to relative roughness of *0.008* and *0.035* respectively and are shown in Figure 1.2.1. These data show that the assumption of a constant shape of the velocity distribution is not very realistic. Better agreement is shown by the smooth bed data of Jensen (1989), see Figure 1.3.1.

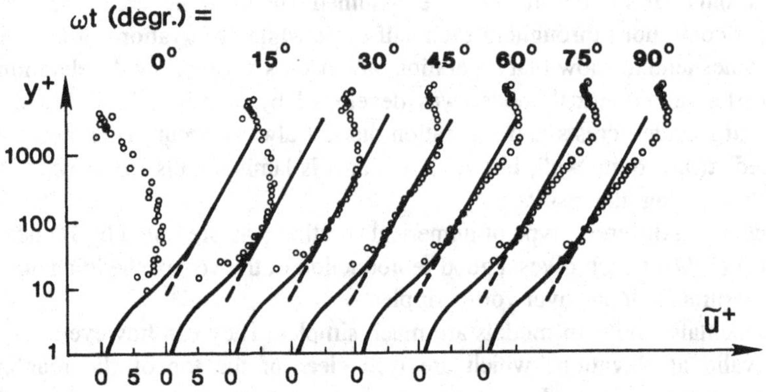

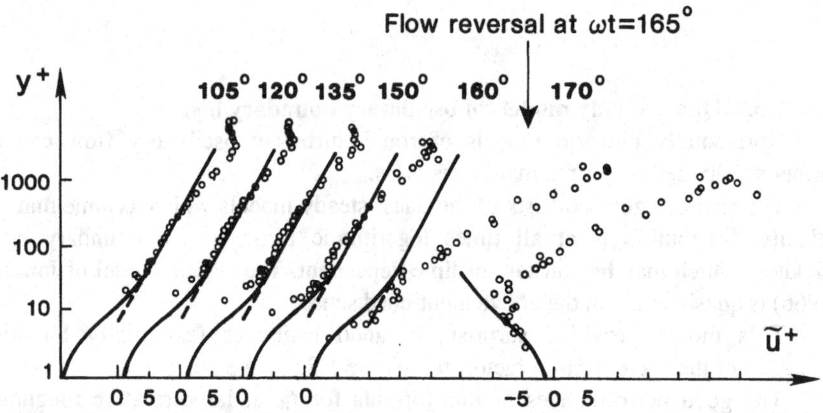

Figure 1.3.1: Dimensionless velocity profiles for smooth turbulent oscillatory flow at different phases of the free stream velocity $A\omega \sin \omega t$; $A = 3.1m$, $T = 9.72\,s$, $y^+ = z\,\hat{u}_* / \nu$, $\tilde{u}^+ = \tilde{u}(z,t)/\tilde{u}_*(t)$. After Jensen (1989).

1.3.3 Velocity distribution models for oscillatory boundary layers

The second category of horizontally uniform models seeks empirical expressions for the velocity distribution $u(z,t)$ in more or less close analogy with smooth, laminar oscillatory flow. Kalkanis (1957, 1964), Sleath (1970) and Nielsen (1985) all followed this line of approach.

Drawing on analogies with smooth, laminar oscillatory flow, which is simple harmonic everywhere, these models are only designed to describe the fundamental harmonic mode of the flow.

The physical similarities between turbulent flows (rough or smooth) and the smooth, laminar solution become particularly striking when the data are plotted in the appropriate way for the purpose. Hence we take a closer look at the mathematical structure of the smooth laminar solution

$$u(z,t) = A\omega\,[1 - D_1(z)]\,e^{i\omega t} \tag{1.2.13}$$

$$D_1(z) = \exp\left[-(1+i)\frac{z}{\sqrt{2\nu/\omega}}\right] \tag{1.3.1}$$

From this expression we note that $\ln|D_1|$ and $\operatorname{Arg} D_1$ are identical quantities and both proportional to the distance from the bed

$$\ln|D_1| = \operatorname{Re}\{\ln D_1\} = -\frac{z}{\sqrt{2\nu/\omega}} \tag{1.3.2}$$

$$\operatorname{Arg} D_1 = \operatorname{Im}\{\ln D_1\} = -\frac{z}{\sqrt{2\nu/\omega}} \tag{1.3.3}$$

The vertical scale of the velocity distribution is thus $\delta_s = \sqrt{2\nu/\omega}$ which is called the *Stokes length.*

These mathematical relationships prompted Nielsen (1985) to study the quantities $\ln|D_1|$ and $\operatorname{Arg}D_1$ (subscript 1 referring to the fundamental harmonic component) of measured velocities and to plot them as function of the elevation z. Examples of such plots are shown in Figures 1.2.17 and 1.3.2 .

Obviously the two quantities $\ln|D_1|$ and $\operatorname{Arg} D_1$ are all but identical almost throughout the boundary layer. Hence, it seems reasonable to apply a description of the form (1.2.13) with

$$D_1 = \exp[-(1+i)\,F(z)] \tag{1.3.4}$$

where $F(z)$ is a real valued function of z. However, $F(z)$ is not necessarily a linear function as in the smooth laminar case. If one accepts a straight line in Figure 1.3.2 as a reasonable approximation, we get an expression of the form

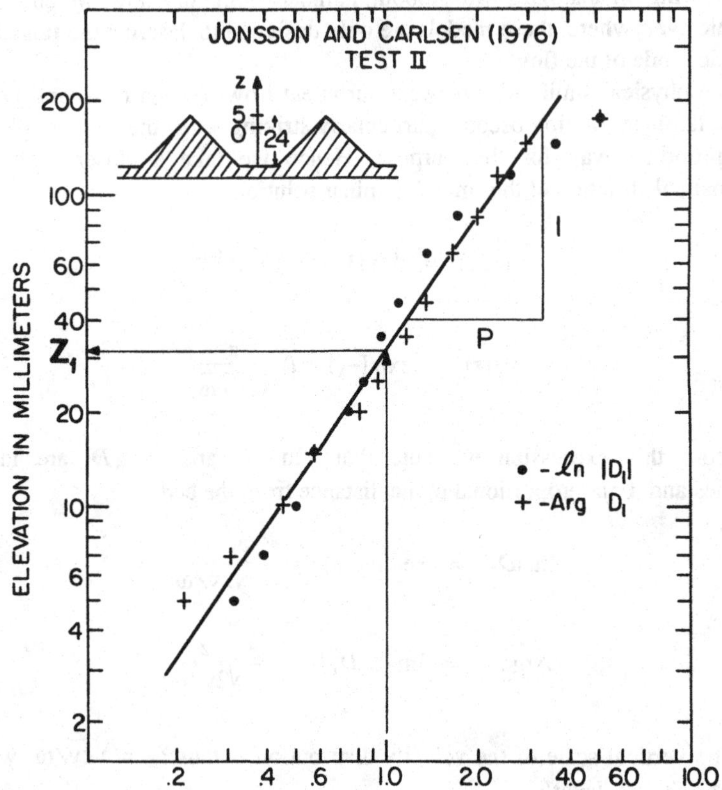

Figure 1.3.2: The analogies between smooth, laminar and rough, turbulent oscillatory flows become visible when real and imaginary parts of $\ln D_1$ are plotted together against z. Data from Jonsson & Carlsen (1976) *Test 2*, $(A,T,r) = (1.79m, 7.2s, 63mm)$.

$$u_1(z,t) \;=\; A\omega\,[1 - D_1(z)]\,e^{i\omega t} \tag{1.2.13}$$

with

$$D_1 \;=\; \exp[-(1+i)\,(\frac{z}{z_1})^p\,] \tag{1.3.6}$$

where the two parameters z_1 and p are derived as shown in the figure above.

From presently available data it appears that Equations (1.2.13) and (1.3.6) describe $u_1(z,t)$ reasonably well for turbulent and transitional oscillatory boundary layers with relative roughness greater than about 0.01. We may note here, that all the available measurements of friction factors over beds of loose sand by Carstens et al (1969), and Lofquist (1986) indicate relative roughness values well within this range, see Section 3.6.

For smooth and almost smooth turbulent flows, like those investigated by Jensen (1989), there is no longer an identity between $\text{Arg}\, D_1$ and $\ln |D_1|$, see Figure 1.3.3.

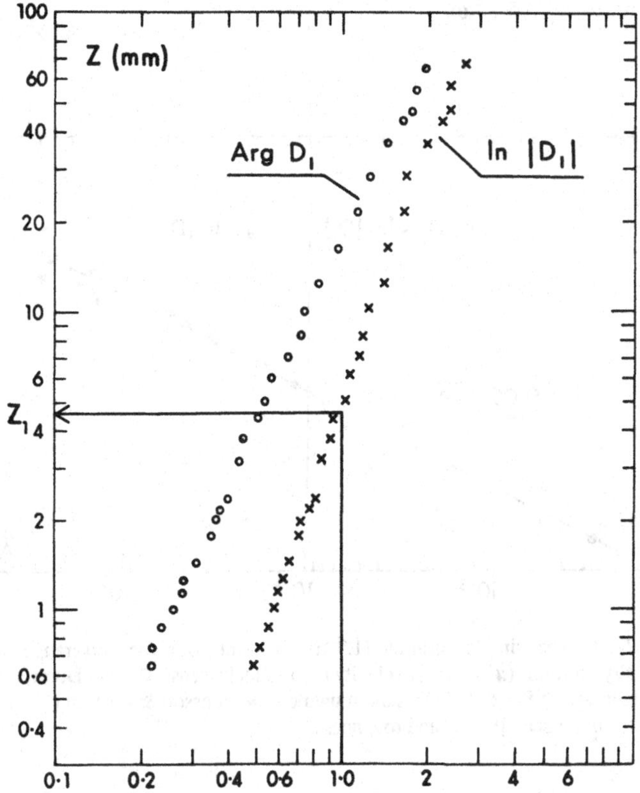

Figure 1.3.3: $\text{Arg}\, D_1$ and $\ln |D_1|$ derived from the measurements of Jensen (1989) Test 13, $A = 3.1m$, $T = 9.7s$, $r = 0.84mm$. Even for these, almost smooth flow conditions the formula (1.3.11) predicts the defect magnitude $|D_1|$ with great accuracy. For thes experiments : $0.09\sqrt{rA} = 0.0046m$.

For very low relative roughness values *Arg* D_1 is smaller than $\ln |D_1|$ down to 40%, but the two quantities do behave in a very similar fashion throughout the boundary layer.

As mentioned in connection with Figure 1.2.17, there is a very close analogy between $u_1(z,t)$ from fairly rough turbulent flows $(r/A > 0.06)$ and the smooth laminar solution. For such flows one finds $p = 1$ and hence

$$D_1 = \exp[-(1+i)\frac{z}{z_1}] \tag{1.3.7}$$

and as mentioned in Section 1.2.9 this flow structure corresponds to a constant eddy viscosity $\nu_t = 0.5\, z_1^2\, \omega$.

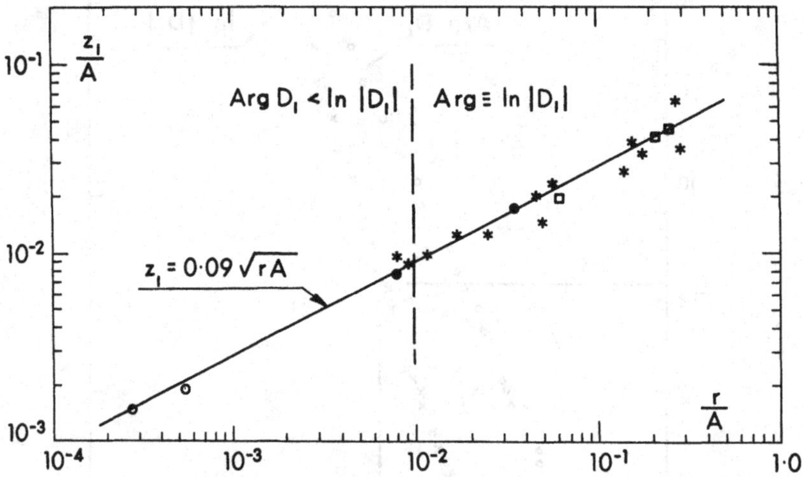

Figure 1.3.4: The simple formula (1.3.10) is valid over the full range where horizontally uniform ($u = u(z,t)$) velocity models make sense. □ : van Doorn (1981), strip roughness, *: Sleath (1987), sand roughness, •: Jonsson & Carlsen (1976) strip roughness, o: Jensen (1989), sand roughness.

According to the dimensional analysis of Jonsson (1966) it should be possible to prescribe the two parameters z_1 and p as functions of the relative bed roughness and the Reynolds number i e

$$z_1 = z_1\left(\frac{r}{A}, \frac{A^2\omega}{\nu}\right) \tag{1.3.8}$$

$$p = p\left(\frac{r}{A}, \frac{A^2\omega}{\nu}\right) \tag{1.3.9}$$

where the dependence on $A^2\omega/\nu$ is expected to be weak or absent for most practical cases. This was attempted by Nielsen (1985) who found that z_1 can be predicted by

$$z_1 = 0.09\sqrt{rA} \tag{1.3.10}$$

for the complete range $0.01 < r/A < 0.5$ where Equation (1.3.6) is applicable, see Figure 1.3.4.

In fact, this.formula can also be used for prediction of the magnitude of the velocity defect for almost smooth turbulent flows where Arg D_1 and ln $|D_1|$ are no longer identical. Thus, for such flows we still have

$$|D_1| = \exp\left(-\frac{z}{0.09\sqrt{rA}}\right)^p \tag{1.3.11}$$

see Figure 1.3.3.

For smooth turbulent oscillatory flows, the data of Jensen (1989) indicate

$$|D_1| = \exp\left(-\frac{z}{\sqrt{2\nu/\omega}}\right)^p, \qquad p \approx 1/3 \tag{1.3.12}$$

i e , the parameter z_1 maintains its "laminar value", the Stokes length, for smooth oscillatory flows, even when they become fully turbulent.

With respect to the second parameter p the message from the experimental data is not quite as clear, see Figure 1.3.5.

However, it can be seen that the power p varies from unity for very rough flows to about *1/3* for smooth turbulent flows. It was noted by Nielsen (1985) that the *p-value* of approximately *1/3* for smooth turbulent flows corresponds to maximum energy dissipation for fixed z_1.

The fact that $p \approx 1$ for $r/A \gtrsim 0.06$ indicates similarity with laminar flow, and hence that the assumption of a constant, real-valued eddy viscosity ($\nu_1 = 0.5\,\omega\, z_1^2$) provides a reasonable model in this range. See also page 40.

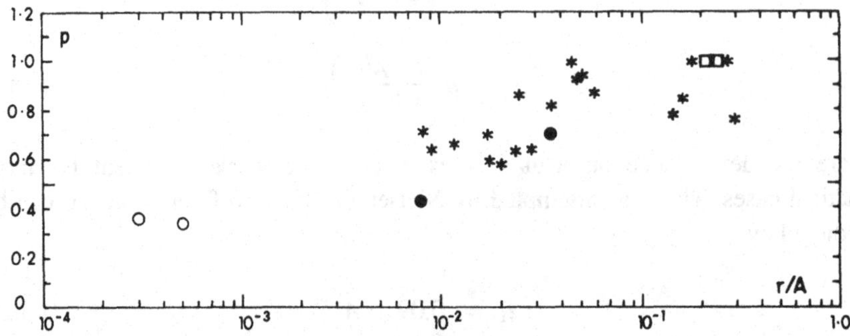

Figure 1.3.5: Best fit p – values (Equation 1.3.11) from experiments with different roughness types. Legend as for Figure 1.3.4.

1.3.4 Eddy viscosity based models of oscillatory boundary layers

The third category of horizontally uniform models contain the eddy viscosity based models which have all so far been based on the equation of motion (1.1.10) in the form

$$\frac{\partial}{\partial t}(u - u_\infty) = \frac{\partial}{\partial z}(\nu_t \frac{\partial u}{\partial z}) \tag{1.3.13}$$

and, with the exception of the model by Trowbridge and Madsen (1984) all of the existing eddy viscosity models for oscillatory boundary layers assume that the eddy viscosity is a function of z but not of t.

Kajiura (1968) developed a very detailed model with very few oscillatory flow data at hand and consequently had to rely very heavily on experimental information from steady flows. He suggested the three layered eddy viscosity distribution which is shown in Figure 1.3.6.

The mathematics involved in deriving the velocity distribution from Kajiura's model are somewhat overwhelming and various attempts have been made to simplify it. Thus, Brevik (1981) omitted the inner layer while Grant and Madsen (1979) assumed $\nu_t = \kappa \hat{u}_* z$ for the complete range $0 < z < \infty$.

The latter assumption can at best be valid for almost smooth conditions and the continued growth of ν_t for $z \to \infty$ seems particularly unrealistic since it is well known that there is virtually no turbulence above a few roughness heights from the bed.

Nevertheless, the resulting velocity distributions look quite reasonable even with this unrealistic eddy viscosity distribution. This is due to the very forgiving nature of the governing equation (1.3.13) with the boundary conditions $u(r/30, t) = 0$ and $\partial u/\partial z \to 0$ for $z \to \infty$.

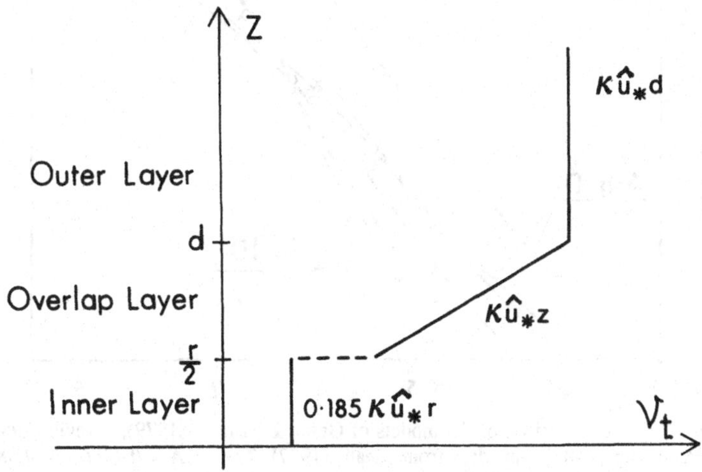

Figure 1.3.6: The eddy viscosity distribution applied by Kajiura (1968). The overlap layer, which is analogous to the logarithmic part of a steady boundary layer, was only expected to exist for almost smooth beds. Kajiura suggested that it would probably disappear completely for $r/A > 1/30$. The friction velocity is defined by $\hat{u}_* = \sqrt{f_w/2}\, A\omega$, and the thickness d of the wall layer is given by $d = 0.05\, \hat{u}_* / \omega$.

In order to be satisfactory, an oscillatory boundary layer model must be able to predict both magnitude and phase of the simple harmonic velocity $u_1(z,t)$.

An efficient way of testing this ability is to plot the complex velocity defect function $D_1(z)$ in terms of $\ln|D_1|$ and $\operatorname{Arg} D_1$ against z together with corresponding data. The defect function is defined by Equations 1.3.2 and 1.3.3. This has been done in Figure 1.3.7 to compare the models of Grant & Madsen (1979), Brevik (1981) and Myrhaug (1982) to the measurements of Sleath (1987) Test 4.

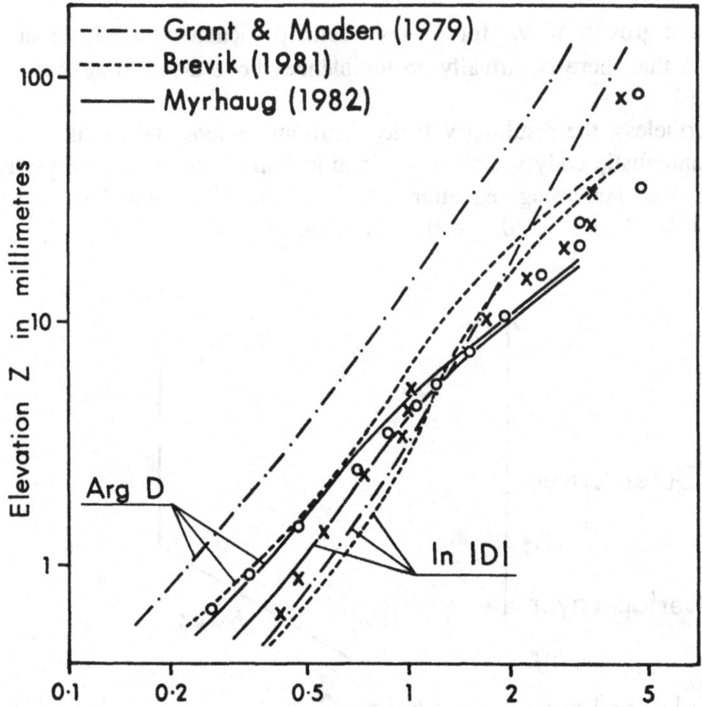

Figure 1.3.7: Comparison of the models of Grant & Madsen (1979), Brevik (1981) and Myrhaug (1982) with data from Sleath (1987), Test 4. $A = 0.45m$, $T = 4.58s$, $r = 2d = 3.26\,mm$. *x:* ln $|D|$ from measurements, o: Arg D from measurements.

The comparison in Figure 1.3.7 shows that all three models predict the magnitude of the velocity defect, and hence ln $|D|$ quite well, but only Myrhaug's model performs well with respect to the phase (Arg D). The performance generally improves with increasing model complexity.

The fact that Myrhaug's model fares best among the ones tested in Figure 1.3.7 corresponds to the fact that his assumed eddy viscosity distribution corresponds rather closely to the empirical values shown in Figure 1.2.16 for the relevant Reynolds number. Sleath's Test 4 has a similar Reynolds Number to Jensen's test 12. Myrhaug's ν_t-distribution may be less suited to model some of the other cases shown in Figure 1.2.16.

It is not surprising that the agreement between these theories and measurements is fairly unimpressive. They all assume ν_t to be a real-valued function of z only, while most data show a general tendency for the local shear stresses to be out of phase with the local velocity gradients, see Figure 1.2.11. These phase shifts make it necessary for ν_t to be either a complex function of z or to be strongly time-dependent, see Sections 1.2.8 and 1.2.9.

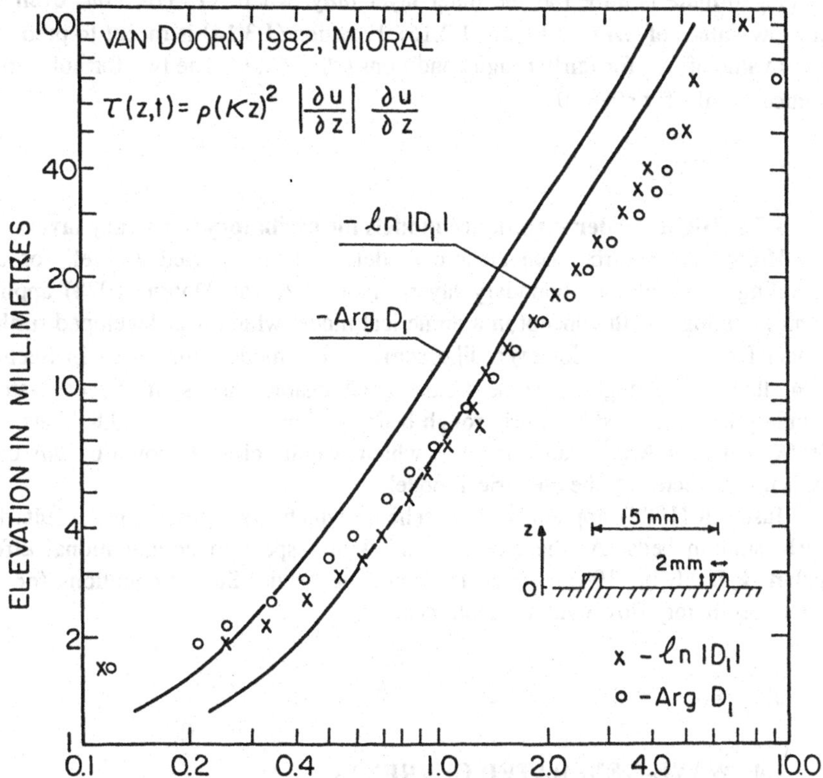

Figure 1.3.8: **Comparison of the Bakker & van Doorn mixing length model with data from van Doorn (1982), Test M10 RAL. We see that the model predicts fairly reasonable values of $\ln|D_1|$ while the predicted values of $\text{Arg}\, D_1$ are clearly different and further removed from the experimental values. The identity between $\text{Arg}\, D_1$ and $\ln|D_1|$ is very clearly shown for these conditions, *(T, A, r) = (2s, 0.33m, 0.021m)*.**

The only case where it seems well justified to assume that ν_t is real-valued and independent of time is for fairly rough beds, $r/A > 0.06$ where ν_t is independent of z as well and given by

$$\nu_t \;=\; 0.5\omega z_1^2 \;\approx\; 0.5\,\omega\,(0.09\sqrt{rA}\,)^2 \;=\; 0.004\,\omega\,r\,A \tag{1.3.14}$$

see Section 1.2.8 and Figure 1.2.16. It will be noted that this formula is not in general agreement with Equation (1.2.44) which was suggested by Sleath (1991). The reason is that the two formulae are not really predicting the same thing. Sleath's formula is aimed at the outer layer only, where $\nu_t \approx constant$ even for very low values of r/A (see Figure 1.2.16). Equation (1.3.14) is meant to predict a global value of ν_t for fairly rough conditions ($r/A > 0.06$). The two formulae give identical results for $r/A \approx 0.4$.

1.3.5 Higher order turbulence models for oscillatory boundary layers

Higher order turbulence closure models have been tried as well for the modelling of oscillatory boundary layers. Bakker & van Doorn (1978) applied Prantl's mixing length concept in a numerical model which was developed further by van Doorn (1983). However, like some of the models discussed in Section 1.3.4, this model neglects some fundamental characteristics of the oscillatory boundary layer, at least for fairly rough beds, as shown in Figure 1.3.8. Thus, the identity between $\text{Arg}\, D_1$ and $\ln |D_1|$, which is quite clearly shown by this data set, is not predicted by the numerical model.

Justesen (1988) applied a κ–ε scheme which gives reasonable results for nearly smooth beds. At the extreme end with respect to computational effort Spalart & Baldwin (1987) solved the complete Navier-Stokes equations for the case of oscillatory flow over a smooth bed.

1.4 WAVE-GENERATED CURRENTS

1.4.1 Introduction

Wave motion is often thought of as purely oscillatory. However, measurements of the velocity field under real waves show the existence of time-averaged velocity components almost everywhere.

The magnitude of these steady flow components is generally much smaller than that of the oscillatory components. However, because their effect is cumulative, their contribution to the net sediment transport may well be significant.

1.4.2 Eulerian drift in Stokes waves

Imagine a velocity probe positioned just above the mean water level of a sine wave. This probe will only be wet during part of each wave period but for all of this time, it will be recording positive horizontal velocities. Hence, the time-averaged Eulerian velocity at this point will be non-zero and directed in the direction of wave propagation. Similarly, a probe anywhere between the MWL and the wave trough will also record a positive mean velocity.

The Eulerian net flow between the crest and the trough amounts to $gH^2/8c$, and this is the total, because a probe positioned anywhere below the trough level will measure zero average velocity under a sine wave.

Now, in most cases, the presence of a beach or the end of the wave flume prevent net transport of water in the shoreward direction. The zero net transport thus required, is most simply obtained by superposition of a uniform, negative, steady velocity

$$\bar{u}_{stokes} = -\frac{g H^2}{8 c D} \qquad (1.4.1)$$

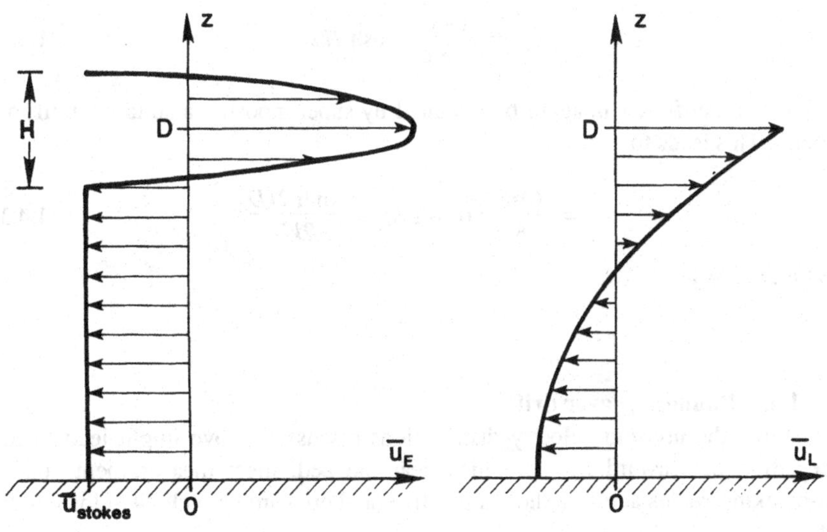

Figure 1.4.1: Eulerian (left) and Lagrangian (right) net flow velocities resulting from superimposing a uniform drift on a sine wave in order to obtain zero flow of water. These velocities are based solely on continuity considerations. That is, the forces required to set up such a flow pattern in a real fluid are not considered.

onto the sine wave velocity field. This uniform, seaward drift velocity is generally referred to as the *Stokes drift*. The considerations behind it are based on continuity only, i e nothing is said about the forces which would be needed to drive it with a real (viscous) fluid.

1.4.3 Lagrangian drift in Stokes waves

Similarly to the Eulerian net velocity described above, a pure sine wave will also result in a positive, Lagrangian mean velocity at all levels. The Lagrangian net velocity is qualitatively due to two properties of the sine wave velocity field.

Firstly, a fluid particle in a sine wave will move with larger forward velocities at the top of its orbit than the backward velocities at the bottom.

Secondly, the particle moves with the wave during its forward motion and against it during its backward motion, and will therefore spend more time moving forewards than backwards.

The Lagrangian net velocity thus obtained is often referred to as the mass transport velocity and its distribution over the depth is given by

$$\bar{u}_L = \frac{(A\omega)^2}{8\,c} \cosh 2kz \tag{1.4.2}$$

A zero net flow can again be obtained by superimposing a suitable, uniform velocity. This leads to

$$\overline{u_L} = \frac{(A\omega)^2}{8\,c} \left(\cosh 2kz - \frac{\sinh 2kD}{2kD}\right) \tag{1.4.3}$$

See Figure 1.4.1

1.4.4 Boundary layer drift

While the inviscid velocity distributions discussed above might lead to the expectation of seaward net velocities near the bed, most measurements under non-breaking waves actually show a positive or shoreward net velocity close to the bed.

An explanation for this forward boundary layer drift was given by Longuet-Higgins (1953, 1956). It occurs because the horizontal and vertical velocities in a wave motion with a viscous bottom boundary layer are not exactly ninety degrees out of phase, as they would be in a perfectly inviscid wave motion.

This gives rise to finite time-averaged stress terms of the form $-\rho\,\overline{\tilde{u}\tilde{w}}$ which are analogous to the familiar Reynolds stresses $-\rho\,\overline{u'w'}$.

These stress terms grow from zero at the bed (where $\tilde{w} = 0$), towards an asymptotic value of

$$-\rho\,(\overline{\tilde{u}\tilde{w}})_\infty \;=\; \rho\,(A\omega)^2 k\,\delta_s/4 \qquad (1.4.4)$$

where k is the wave number and δ_s is the Stokes length $\sqrt{2\nu/\omega}$. See Figure 1.4.2. For the case of constant eddy viscosity as well as for laminar flow, the distribution of $-\rho\,\overline{\tilde{u}\tilde{w}}$ is given by

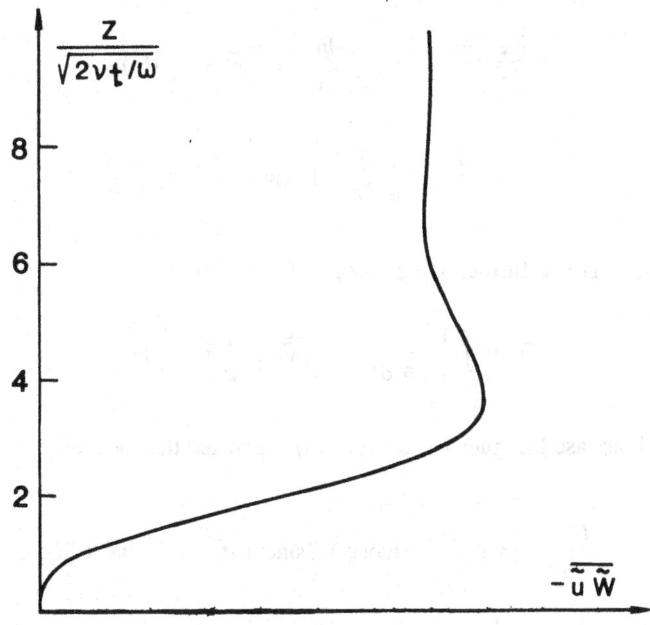

Figure 1.4.2: Distribution of $-\rho\,\overline{\tilde{u}\tilde{w}}$ in a wave boundary layer with constant eddy viscosity. By comparing the expression (1.4.4) with the expression $\hat{\tau} = \rho\sqrt{\omega\nu}\,A\omega$ for the peak bed shear stress due to a purely oscillatory flow, we see that the asymptotic value of $-\rho\,\overline{\tilde{u}\tilde{w}}$ amounts to $\frac{kA}{\sqrt{8}}\hat{\tau}$.

$$-\rho\,(\overline{\tilde{u}\tilde{w}}) \;=\; \rho\frac{1}{4}(A\omega)^2 k\,\delta_s\,[1 - e^{-\xi}\,(\,2\cos\xi - e^{-\xi} + 2\xi\sin\xi)\,] \qquad (1.4.5)$$

where $\xi = z/\sqrt{2\nu/\omega}$. This distribution is shown in Figure 1.4.2.

These "Reynolds like" stresses influence the $\overline{u}$-distribution through the equation of motion (1.1.21) which for a two dimensional case with uniform flow conditions $(\frac{\partial}{\partial x} = 0)$ can be written

$$\frac{\partial}{\partial z}(\,\nu\frac{\partial\overline{u}}{\partial z} - \overline{u'w'}\,) \;=\; \frac{1}{\rho}\frac{\partial\overline{p}}{\partial x} + \frac{\partial}{\partial z}\overline{\tilde{u}\tilde{w}} \qquad (1.4.6)$$

which may be integrated once to yield

$$\nu\frac{\partial\overline{u}}{\partial z} - \overline{u'w'} \;=\; \frac{1}{\rho}\frac{\partial\overline{p}}{\partial x}z + \overline{\tilde{u}\tilde{w}} + \frac{1}{\rho}\overline{\tau}(o) \qquad (1.4.7)$$

or

$$\nu_t\frac{\partial\overline{u}}{\partial z} \;=\; \frac{1}{\rho}\frac{\partial\overline{p}}{\partial x}z + \overline{\tilde{u}\tilde{w}} + \frac{1}{\rho}\overline{\tau}(o) \qquad (1.4.8)$$

This can be integrated further using, $\overline{u}(o) = 0$, to give

$$\overline{u} \;=\; \int_o^z \frac{1}{\nu_t}\left(\frac{1}{\rho}\frac{\partial\overline{p}}{\partial x}z + \overline{\tilde{u}\tilde{w}} + \frac{1}{\rho}\overline{\tau}(o)\right)dz \qquad (1.4.9)$$

For the laminar case Longuet-Higgins (1956) suggested the solution

$$\overline{u_E} \;=\; \frac{(A\omega)^2}{4\,c}\,[3 + e^{-\xi}\,(-4\cos\xi + 2\sin\xi + e^{-\xi} - 2\xi\sin\xi + 2\xi\cos\xi)] \qquad (1.4.10)$$

which is based on the assumptions of $\frac{\partial\overline{p}}{\partial x} = 0$ and $\overline{\tau}(o) = -\rho\,(\overline{\tilde{u}\,\tilde{w}})_\infty$.

We see that according to this solution, the Eulerian mean velocity tends towards an asymptotic value of magnitude $3(A\omega)^2/4c$ for $z \gg \delta_s$. The corresponding mass transport velocity, i e the asymptotic time-averaged,

Lagrangian velocity is 67% larger at $5(A\omega)^2/4c$.

The velocity distribution (1.4.10) corresponds to laminar flow or to a turbulent flow with constant eddy viscosity, but Longuet-Higgins showed that the asymptotic velocities above the boundary layer are the same for any distribution of the eddy viscosity, as long as it is time-independent, $\nu_t = \nu_t(z)$.

It should be kept in mind however, that the solution (1.4.10), for the drift velocity, relies on the condition

$$\bar{\tau}(o) \;=\; \tau_W - \rho\, g\, D\frac{d\bar{\eta}}{dx} - \frac{dS_{xx}}{dx} \;=\; -\rho(\overline{\tilde{u}\tilde{w}})_\infty\,.$$

If that is changed, the velocity distribution changes accordingly. See Figure 1.4.3.

The behaviour of $-\rho\,\overline{\tilde{u}\tilde{w}}$ and hence of $\overline{u_E}$, in turbulent boundary layers in general, is at present neither well documented nor well understood. However, the results in Section 1.3.3 concerning the structure of turbulent oscillatory boundary layers indicate that the laminar solution (with ν replaced by ν_t) may have fairly wide applicability. It was found in Section 1.3.3 that turbulent oscillatory boundary layers in the roughness range $0.06 < r/A < 0.5$ seem to correspond to a constant eddy viscosity with $\sqrt{2\nu_t/\omega} = z_1 = 0.09\sqrt{rA}$. Hence, the stress distribution (1.4.5) with $\delta = 0.09\sqrt{rA}$ may provide reasonable estimates for these flows.

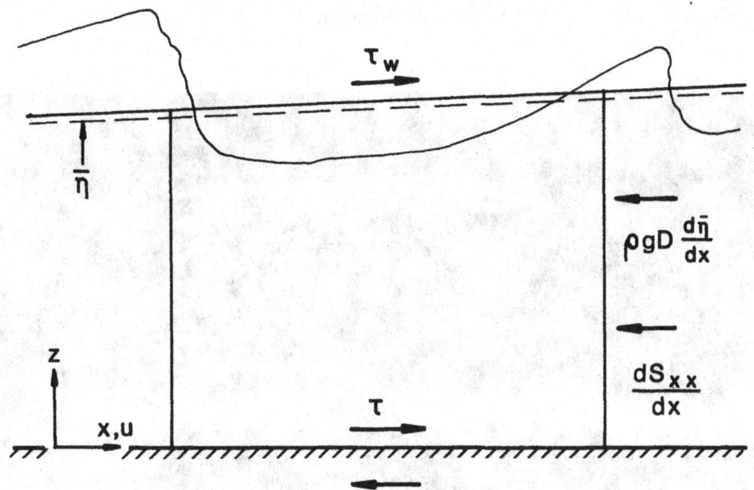

Figure 1.4.3: The bed shear stress must in general balance the sum of wind stress on the surface, the radiation stress gradient and the pressure force due to mean surface slope.

The driving mechanism behind the velocity distribution (1.4.10) is also present under standing waves and partially standing waves. In that case however, the wave conditions vary significantly in the x-direction so, several more terms from Equation (1.1.21) must be retained. In that case, the result is a series of counter rotating cells with the velocities at the bed directed from the surface nodes towards the surface antinodes.

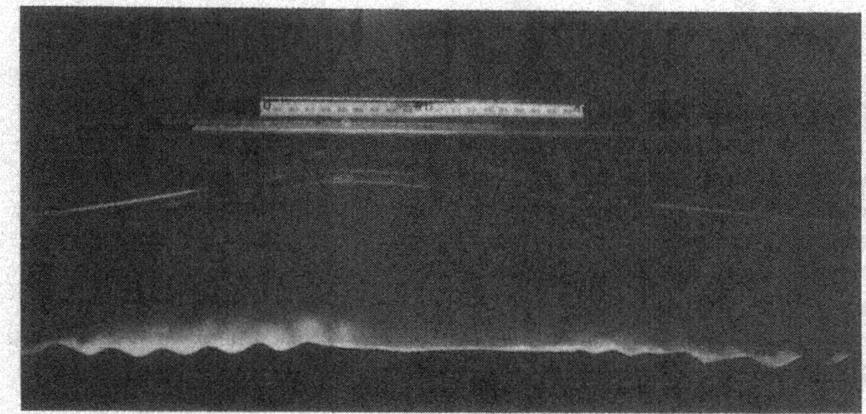

Figure 1.4.4a: The wave maker is turned on and sand is entrained under the surface nodes where the near-bed velocities are strongest.

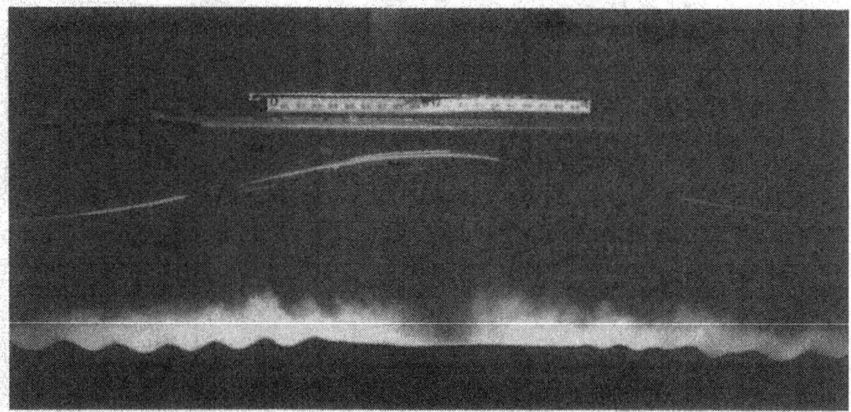

Figure 1.4.4b: The suspended sediment is convected by the boundary layer drift towards the surface antinode.

These cells generate a clearly observable sediment transport pattern resulting in bar formation, as pointed out by Carter et al (1973), and illustrated by the photographs in Figures 1.4.4 a through d. The first three (a through c) show how the sediment suspension is re-established when the wavemaker is turned on after a period of quiescence. Sand is picked up only under the surface nodes and then convected by $\overline{u}$ towards the surface antinodes and then towards the surface.

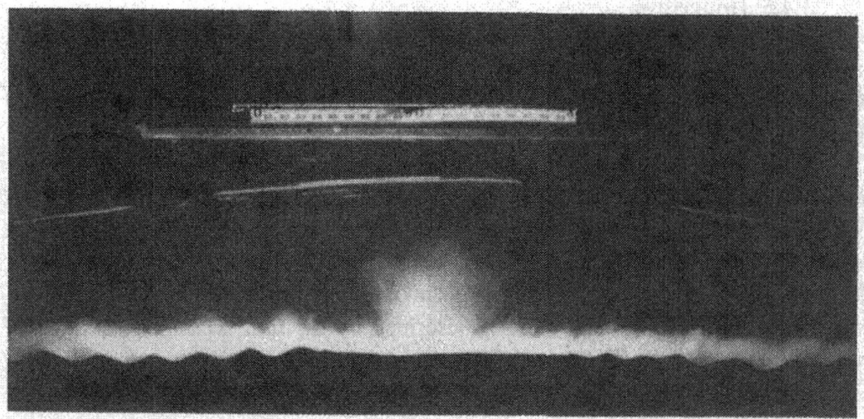

Figure 1.4.4c: Some sand is rising towards the surface under the antinode.

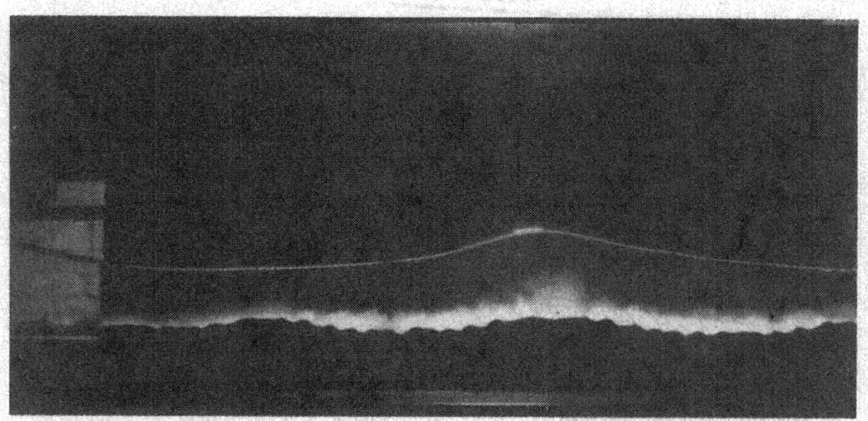

Figure 1.4.4d: Eventually, the sand which settles under the antinodes forms bars.

Figure 1.4.4d shows the resulting bar pattern.

The water depth in these experiments was *0.12m*, the wave period was *0.5s* and the median sand size was *0.08mm* .

For a detailed discussion of drift patterns over flat and rippled beds under standing waves as well as progressive waves see Sleath (1984).

1.4.5 Undertow

The shoreward boundary layer drift indicated by the solution (1.4.10) is usually not observed in the surf zone. In the surf zone, the $\bar{u}$-picture tends to be dominated by the so-called undertow, a seaward mean velocity between the bed and the wave trough level, see Figure 1.4.5.

The undertow is a gravity driven current related to the phenomenon of wave setup as illustrated very lucidly by the experiments of Longuet-Higgins (1983). It occurs because the radiation stress gradient dS_{xx}/dx is not uniform over the depth under breaking waves while the opposing pressure gradient $d\bar{p}/dx = \rho g \, d\bar{\eta}/dx$ from the wave setup is (nearly) uniform, and therefore dominates near the bed.

Quantitatively, the undertow is described by the time averaged equation of motion (1.1.21) which for the two dimensional case can be written

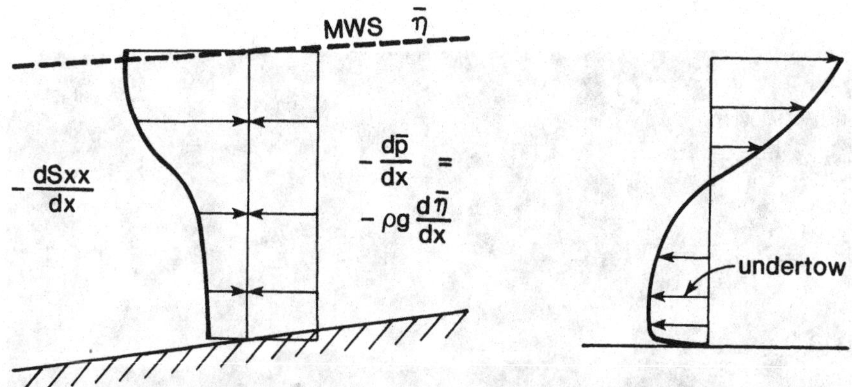

Figure 1.4.5: The undertow occurs in the surf zone because the radiation stress gradient is concentrated near the surface, while the opposing pressure gradient due to the setup slope is essentially uniform over the depth.

$$\frac{\partial}{\partial z}\left(\nu \frac{\partial \bar{u}}{\partial z} - \overline{u'w'} \right) = \frac{1}{\rho}\frac{\partial \bar{p}}{\partial x} + \frac{\partial}{\partial x}\bar{u}^2 + \frac{\partial}{\partial x}\overline{\tilde{u}^2} + \frac{\partial}{\partial x}\overline{u'^2} + \frac{\partial}{\partial z}\overline{\tilde{u}\tilde{w}} \qquad (1.4.11)$$

If we introduce the eddy viscosity

$$\nu_{c1} = \nu + \frac{-\overline{u'w'}}{\frac{\partial \bar{u}}{\partial z}}$$

and assume hydrostatic mean pressure ($\frac{\partial \bar{p}}{\partial x} \equiv \rho g \frac{d\bar{\eta}}{dx}$), this can be rewritten as

$$\frac{\partial}{\partial z}\left(\nu_{c1} \frac{\partial \bar{u}}{\partial z} \right) = g\frac{\partial \bar{\eta}}{\partial x} + \frac{\partial}{\partial x}\bar{u}^2 + \frac{\partial}{\partial x}\overline{\tilde{u}^2} + \frac{\partial}{\partial x}\overline{u'^2} + \frac{\partial}{\partial z}\overline{\tilde{u}\tilde{w}} \qquad (1.4.12)$$

which can be integrated once to give

$$\nu_{c1}\frac{\partial \bar{u}}{\partial z} = g\frac{d\bar{\eta}}{dx}z + \int_o^z \left(\frac{\partial}{\partial x}\bar{u}^2 + \frac{\partial}{\partial x}\overline{\tilde{u}^2} + \frac{\partial}{\partial x}\overline{u'^2} \right) dz + \overline{\tilde{u}\tilde{w}} + \frac{\bar{\tau}(o)}{\rho} \qquad (1.4.13)$$

where the boundary condition $\nu_{c1} \frac{\partial \bar{u}}{\partial z}\Big|_{z=o} = \bar{\tau}(o)$ has been used.

The bed shear stress $\bar{\tau}(o)$ may include a wind stress as well as the radiation stress contribution $-\frac{dS_{xx}}{dx}$ and the setup contribution $-\rho g D \frac{d\bar{\eta}}{dx}$. See Figure 1.4.3.

To obtain the undertow distribution, Equation (1.4.13) is integrated once more with the use of the non-slip boundary condition $\bar{u}(o) = 0$. This leads to

$$\bar{u}(z) = \int_o^z \frac{1}{\nu_{c1}} \left(\int_o^z \left(\frac{\partial}{\partial x}\bar{u}^2 + \frac{\partial}{\partial x}\overline{\tilde{u}^2} + \frac{\partial}{\partial x}\overline{u'^2} \right) dz + \overline{\tilde{u}\tilde{w}} + g\frac{d\bar{\eta}}{dx}z + \frac{\bar{\tau}(o)}{\rho} \right) dz \qquad (1.4.14)$$

From this expression it is obvious that in order to model the undertow, one must consider the complete system of turbulence from wave breaking (to get the eddy viscosity ν_{c1}), the wave setup $\bar{\eta}$, and the process of wave transformation and decay across the surf zone. For examples of recent models see Svendsen et al (1987), Roelvink & Stive (1989) and Deigaard et al (1991).

1.5 WAVE-CURRENT BOUNDARY LAYER INTERACTION

1.5.1 Introduction

The hydraulic conditions which cause problems of siltation or erosion on the coast are nearly always mixed in the sense that the velocity field has both a steady component $\bar{u}$ and a periodic component $\tilde{u}$ which may have very different relative magnitudes and different directions.

We are, in the present context, mainly interested in the sediment transport which results from these combined flows and hence we shall concentrate on the flow structure near the bed where most of the sediment transport occurs. The changes in wave height and direction which occur when waves travel through a variable current field are considered outside the scope of this text. For a recent review of these phenomena see Jonsson (1990).

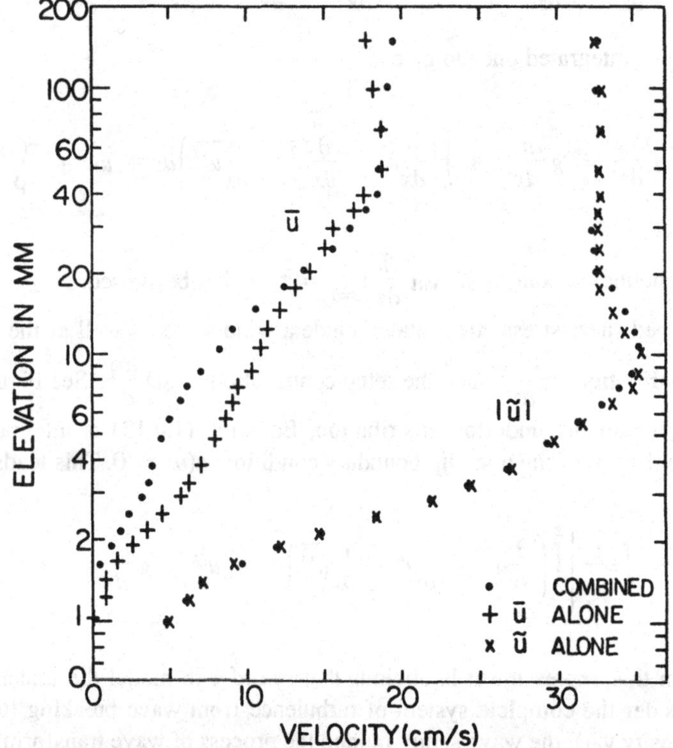

Figure 1.5.1: Profile of a pure current $\bar{u}$ (+), the velocity amplitude $|\tilde{u}|$ in a pure oscillatory flow (x), and both $\bar{u}$ and $|\tilde{u}|$ (•) from the combination of the two flows. Measurements in an oscillating water tunnel by van Doorn (1982).

An example of velocity profiles from corresponding pure current, pure wave and combined wave-current flows are shown in Figure 1.5.1.

These data show that while the structure of the oscillatory flow is unchanged by the addition of the current, the addition of the waves changes the current profile considerably.

In essence the effect of the waves is to suppress the current gradients and in turn the current strength inside the wave boundary layer, an effect which is generally attributed to increased wave-induced mixing near the bed.

The current profile $\bar{u}(z)$ in a combined flow is commonly split conceptually into three parts which are more obvious in Figure 1.5.2. than in Figure 1.5.1.

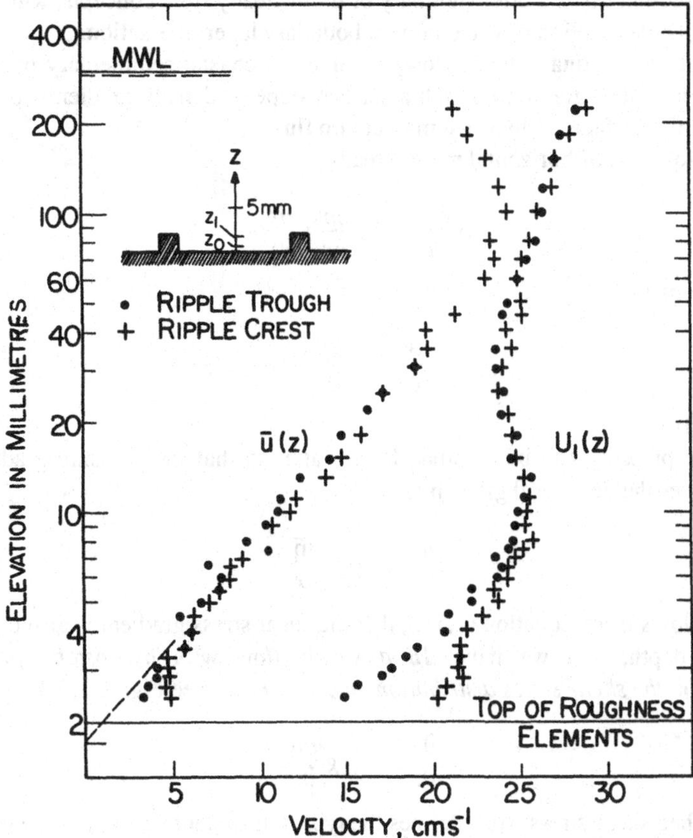

Figure 1.5.2: Measurements of the steady component $\bar{u}$ and the amplitude U_1 of the primary oscillatory component from a combined flow (following current) in a wave flume. Data from van Doorn (1981) Test V20 RA+RB.

There is the wave-dominated layer near the bed. Then there is a logarithmic layer and finally an upper layer which covers sixty to eighty percent of the water depth.

Since most of the sediment transport normally occurs in the two lower layers i e, the wave-dominated layer and the logarithmic layer, the following sections are focussed on those.

1.5.2. Resume of steady boundary layer concepts

This section gives a brief summary of those steady flow concepts, which will be used in the description of wave-current boundary layer interaction.

The steady or quasi-steady flows involved in coastal sedimentary processes are driven by either gravity, through a surface slope or density gradients, by wind stresses on the surface, or by wave momentum flux.

The equation of horizontal motion reads

$$\rho \frac{\partial u}{\partial t} = -\frac{\partial p}{\partial x} + \frac{\partial \tau}{\partial z} \tag{1.5.1}$$

which becomes

$$\frac{\partial \bar{\tau}}{\partial z} = \frac{\partial \bar{p}}{\partial x} \tag{1.5.2}$$

for steady flow.

If the pressure can be assumed hydrostatic, so that the pressure gradient is constant over the depth and given by

$$\frac{\partial \bar{p}}{\partial x} \equiv \rho g \frac{\partial \bar{\eta}}{\partial x} \tag{1.5.3}$$

then, it follows from Equation (1.5.2), that the shear stress gradient is also constant over the depth, and we have: *In a steady flow with hydrostatic pressure distribution, the shear stress distribution must be linear.* See Figure 1.5.3.

$$\frac{\partial \bar{\tau}}{\partial z} \equiv \rho g \frac{\partial \bar{\eta}}{\partial x} \tag{1.5.4}$$

The bed shear stress $\bar{\tau}(o)$ defines the typical turbulent velocity at the bed

$$\overline{u_*} = \sqrt{\bar{\tau}(o)/\rho} \tag{1.5.5}$$

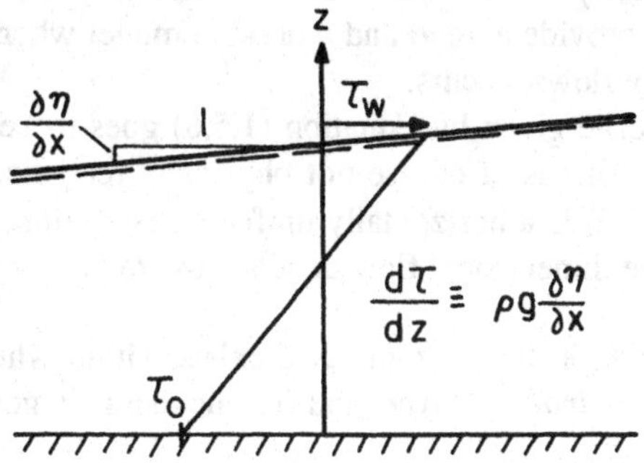

Figure 1.5.3: In a steady flow with hydrostatic pressure distribution, the shear stress is linearly distributed. The bed shear stress $\overline{\tau}(o)$ defines the typical turbulent velocity at the bed: $\overline{u_*} = \sqrt{\overline{\tau}(o)/\rho}$.

which is called the friction velocity.

It is well established experimentally that the velocity profile in a steady, uniform, turbulent boundary layer can be described by

$$\frac{\overline{u}(z)}{\overline{u_*}} = \frac{1}{\kappa} \ln \frac{z}{z_o} \tag{1.5.6}$$

where κ is called von Karman's constant and has a value of approximately *0.4*, see Figure 1.5.4.

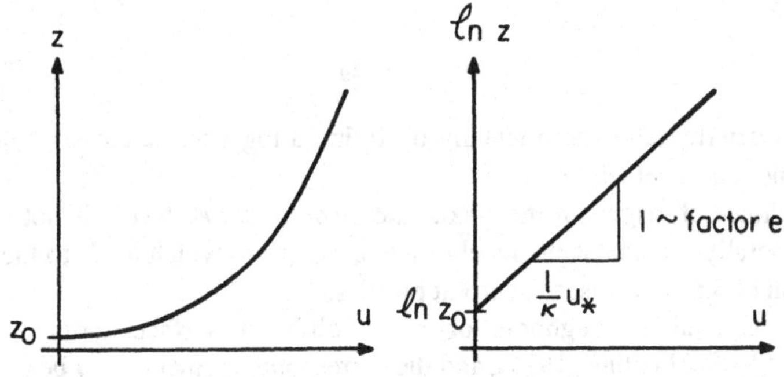

Figure 1.5.4: Geometric properties of the logarithmic velocity distribution.

This logarithmic velocity law which is also called the "*law of the wall*" is not a good approximation to the velocity distribution near the water surface, see for example Figure 1.5.2, but it does provide a good and workable model where most of the sediment transport, in steady flows, occurs.

The logarithmic velocity profile given by Equation (1.5.6) goes to zero at a finite elevation z_0 above the bed. This is of course not physically realistic, but it reflects the fact that Equation (1.5.6) is a horizontally uniform description, which does not attempt to model the three dimensional flow details close to the roughness elements. See Figure 1.1.5.

If the bed is perfectly smooth, a thin bottom layer exists within which the flow is laminar. It is called the *laminar sublayer,* and its thickness is generally taken to be

$$\delta_{lam} = 11.6\,\nu/\overline{u_*} \tag{1.5.7}$$

where ν is the laminar viscosity.

If the irregularities of the bed surface are larger than δ_{lam}, so that they penetrate the laminar sublayer, we say that the flow is *rough.*

Nikuradse (1933) found that the logarithmic velocity profile over a bed of closely packed spheres of diameter d goes through zero at

$$z_0 = \frac{1}{30}d \tag{1.5.8}$$

when the flow is fully turbulent, and this has been observed to be the case for $r\,\overline{u_*}/\nu > 70$. Based on this observation we define the *equivalent Nikuradse roughness* of other surface geometries by

$$r = 30\,z_o \tag{1.5.9}$$

where z_o is determined experimentally by fitting a logarithmic curve of the form (1.5.6) to measured velocities.

The choice of origin for the z-axis: the *theoretical bed level,* is not obvious, but it is generally defined as the level of the origin (z=o) which leads to the best fit by Equation (1.5.6) to measured current profiles.

The equivalent roughness of many different surface types has been investigated by Schlichting (1979), and the corresponding, theoretical bed level has been discussed in detail by Jackson (1981).

In analogy with Newton's equation

$$\tau = \rho \nu \frac{\partial u}{\partial z} \tag{1.5.10}$$

for shear stresses in laminar flow, we can define a turbulent *eddy viscosity* by

$$\tau = \rho \nu_t \frac{\partial u}{\partial z} \tag{1.5.11}$$

Then, from the logarithmic velocity distribution (1.5.6) and the corresponding shear stress distribution

$$\overline{\tau}(z) = \overline{\tau}(o)\,(1 - z/D) \tag{1.5.12}$$

where D is the water depth, we find that the eddy viscosity in steady, uniform flow with no surface stress is parabolic

$$\nu_t(z) = \kappa\,\overline{u_*}\,z\,(1 - z/D) \tag{1.5.13}$$

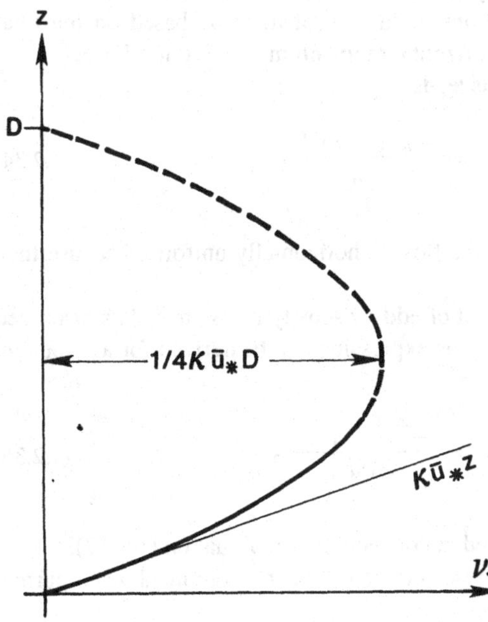

Figure 1.5.5: The eddy viscosity distribution which corresponds to the logarithmic velocity profile is parabolic. However, for sediment transport calculations, the linear approximation near the bed is often sufficient.

This is a simple and useful formula, but since it is based on the law of the wall, Equation (1.5.6), it is not reliable near the free surface.

If the layer of interest is very thin compared to the total flow depth, we often apply the so-called *constant stress assumption.* That is, the last terms in the expressions for shear stress, Equation (1.5.12) and for eddy viscosity, Equation (1.5.13), are left out. Thus, in the *constant stress layer* we have $\overline{\tau} \equiv \overline{\tau}(o)$ and $\nu_t = \kappa \overline{u_*} z$.

The velocity distribution in the constant stress layer is still logarithmic and given by Equation (1.5.6), and the eddy viscosity corresponds to the straight line in Figure 1.5.5.

1.5.3 The eddy viscosity concept for combined flows

The simplest way of defining the relationship between the velocity field and the shear stresses in a turbulent fluid is by using the eddy viscosity concept. However, we found in Section 1.2.8 that the eddy viscosity for a purely oscillatory flow is much more complicated than that of a steady flow, and this holds even more so for combined wave-current flows.

In order for the eddy viscosity to be useful in the sense that it enables the greatest simplification of the equations of motion, it must be based on the total transfer in the vertical direction of horizontal momentum, see Section 1.1.5.

For a steady, turbulent flow this leads to

$$\nu_t = \frac{\overline{\tau}/\rho}{\frac{d\overline{u}}{dz}} = \frac{-\overline{u}\,\overline{w} - \overline{u'w'}}{\frac{d\overline{u}}{dz}} + \nu \tag{1.2.34}$$

where the term $-\overline{u}\,\overline{w}$ vanishes if the flow is horizontally uniform, because then $\overline{w} = 0$ by continuity.

For oscillatory flows the concept of eddy viscosity is less trivial as discussed in Section 1.2.8. However, an analogous expression can be derived for a turbulent, purely oscillatory flow ($\overline{u} = 0$)

$$\nu_w = \frac{\tilde{\tau}/\rho}{\frac{\partial\tilde{u}}{\partial z}} = \frac{-\widetilde{\tilde{u}\tilde{w}} - \widetilde{u'w'}}{\frac{\partial\tilde{u}}{\partial z}} + \nu \tag{1.2.35}$$

where the meaning of $\widetilde{\tilde{u}\tilde{w}}$ is defined in connection with Equation (1.1.17).

Let us now consider the eddy viscosity concept for a combined wave-current motion.

Based on the expressions (1.1.22) and (1.1.25), for the steady and the periodic components respectively of the total momentum transfer, we obtain the following eddy viscosities

$$\nu_c = \frac{\bar{\tau}/\rho}{\frac{d\bar{u}}{dz}} = \frac{-\bar{u}\,\bar{w} - \overline{\tilde{u}\tilde{w}} - \overline{u'w'}}{\frac{d\bar{u}}{dz}} + \nu \tag{1.5.14}$$

and

$$\nu_w = \frac{\tilde{\tau}/\rho}{\frac{\partial \tilde{u}}{\partial z}} = \frac{-\bar{u}\,\tilde{w} - \tilde{u}\,\bar{w} - \widetilde{\tilde{u}\,\tilde{w}} - \widetilde{u'w'}}{\frac{\partial \tilde{u}}{\partial z}} + \nu \tag{1.5.15}$$

where ν_c is the eddy viscosity felt by the steady flow component $\bar{u}(z)$ and ν_w is the one felt by the periodic component $\tilde{u}(z,t)$. The interesting thing about these two expressions is that they are not similar at all. Judging from the appearance of these expressions, it would be quite surprising if ν_c and ν_w turned out to be identical or just similar.

From (1.5.14) it is clear that ν_c must be constant in time, but ν_w may well be a function of time as well as of z. As for the pure wave case, which was discussed in Sections 1.2.8 and 1.2.9, it is quite possible for the expression (1.5.15) to take both negative and infinite values.

Coffey & Nielsen (1984, 1986) showed examples of corresponding values of ν_c and ν_w from wave flume data (van Doorn 1981 Tests V10 and V20) and from tunnel data (van Doorn 1982, Test S10). They found that in both cases ν_c was typically three to four times greater than ν_w.

A similar data set derived from the velocity data of van Doorn (1982) Test M10 is shown in Figure 1.5.6. The plotted values of ν_w are calculated as $|\nu_1|$ from the first harmonic of $\tilde{u}(z,t)$ in accordance with Equation (1.2.42) from Section 1.2.9.

The plotted values of ν_c are derived from

$$\nu_c = \frac{\bar{\tau}(z)/\rho}{\frac{d\bar{u}}{dz}} = \frac{\overline{u_*}^2\,(1 - z/D)}{\frac{d\bar{u}}{dz}} \tag{1.5.16}$$

with $\overline{u_*}$ determined from the logarithmic part of the $\bar{u}$-distribution.

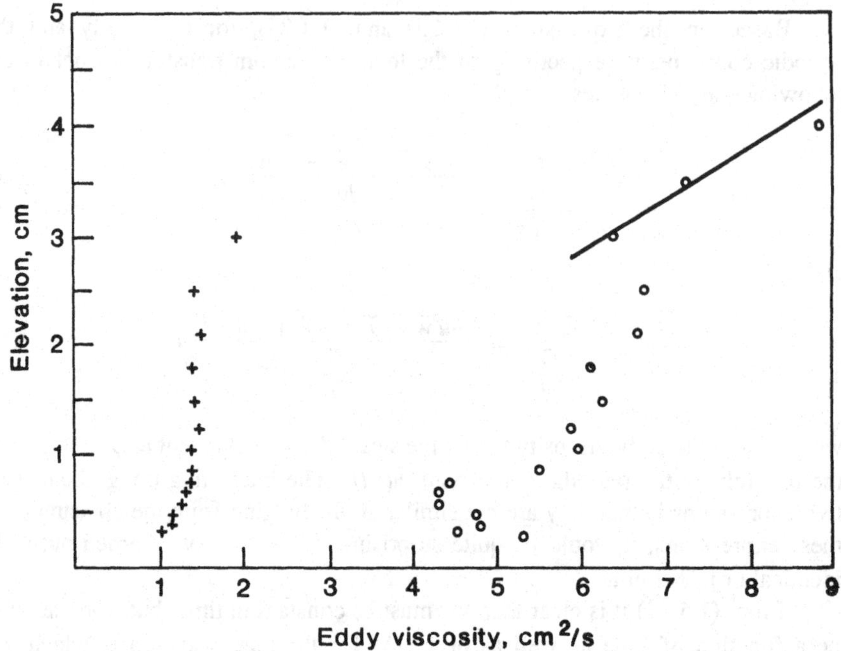

Figure 1.5.6: The magnitudes of the eddy viscosities ν_c (o) and ν_w (+), based on Equations (1.5.14) and (1.5.15), may well be very different. Data from oscillating water tunnel, van Doorn (1982) Tests M20, $(A, T, r, <\bar{u}>) = (0.33m, 2s, 0.0021m, 0.2m/s)$. The line corresponds to $\nu_c = 0.4\,\overline{u_*}\,z$, see Figure 1.5.5.

Despite the scatter in the values of ν_c it is quite clear that the magnitudes of ν_c and ν_w are quite different. Typically, ν_c is three to four times greater than ν_w.

While this conclusion is perfectly in line with those previously drawn by Coffey & Nielsen (1984, 1986), it is acknowledged, that the determination of $\overline{u_*}$ through a log-curve fit introduces an element of arbitrariness when the extent of the "logarithmic layer" is poorly defined.

The definitions (1.5.14) and (1.5.15), of ν_c and ν_w are the ones applied, with more or less explicit acknowledgement, in previous eddy viscosity based models of wave current boundary layer interaction. There are however, other options which may turn out to be more appropriate.

It may for example be reasonable to give the dominating term $-\overline{\tilde{u}\tilde{w}}$ of the total momentum transfer expression (1.1.22) explicit consideration in the equation of motion for $\bar{u}(z)$ instead of hiding it under the collective label of $\bar{\tau}(z)$. This approach is outlined in Section 1.5.9 page 91, and in Section 7.2 page 293.

With respect to the eddy viscosity concept, the advantage of this approach is that the expression for the current eddy viscosity ν_C then becomes analogous to that from steady turbulent flow as the term $-\overline{\tilde{u}\tilde{w}}$ is removed. Compare the formulae (1.2.34) and (1.5.14). The only difference will be due to the difference in turbulence intensity.

With the definition (1.5.14) however, ν_C is a function of the angle between the current and the direction of wave propagation. The term $-\overline{\tilde{u}\tilde{w}}$, which is positive or zero (see Figure 1.4.2) makes a positive contribution to ν_C for a following current and a negative contribution for an opposing current.

Sleath (1987) found, for purely oscillatory flow in an oscillating water tunnel, that $\overline{\tilde{u}\tilde{w}}$ was typically ten times greater than $\overline{u'w'}$. It is therefore natural to expect $-\overline{\tilde{u}\tilde{w}}$ to be the dominant term in ν_C defined by Equation (1.5.14). The magnitude of $-\overline{\tilde{u}\tilde{w}}$ will be greater under real progressive waves than in a tunnel where the contribution described by Longuet-Higgins Equation (1.4.5) is absent. The corresponding difference between $\bar{u}$-profiles in wave flumes and in tunnels can be seen by comparing Figures 1.5.1 and 1.5.2. For more details, see Section 7.2.

1.5.4 Turbulence intensity in combined wave-current flows

The majority of existing models for wave-current boundary layer interaction are based on the philosophy that the interaction and the observed changes to the current profiles in particular are due to wave-induced changes in the turbulence intensity rather than to terms like $\overline{\tilde{u}\tilde{w}}$. We shall, therefore, take a look at some measurements of the turbulence intensity in combined flows.

As a measure of the turbulence intensity some measured values of $w'_{rms} = (\overline{w'^2})^{0.5}$ are shown in Figure 1.5.7. This data was recorded in an oscillating water tunnel under five different sets of conditions: a pure wave motion $(A,T) = (0.10m, 2.0s)$; two cases of pure current with a depth average of $0.1m/s$ and $0.2m/s$ respectively; and for the wave motion superimposed on each of the currents. The strip-roughness bed had a hydraulic roughness of $21mm$.

The pure wave case is the one for which the boundary layer structure data is shown in Figure 1.2.17. From that figure it may be inferred that this flow is analogous to a smooth laminar oscillatory flow with $z_1 = \sqrt{2\nu_W/\omega} = 4.2mm$ corresponding to $\nu_W = 2.8 \cdot 10^{-5} m^2/s$.

The data in Figure 1.5.7 show a clear difference between the cases with waves and those without.

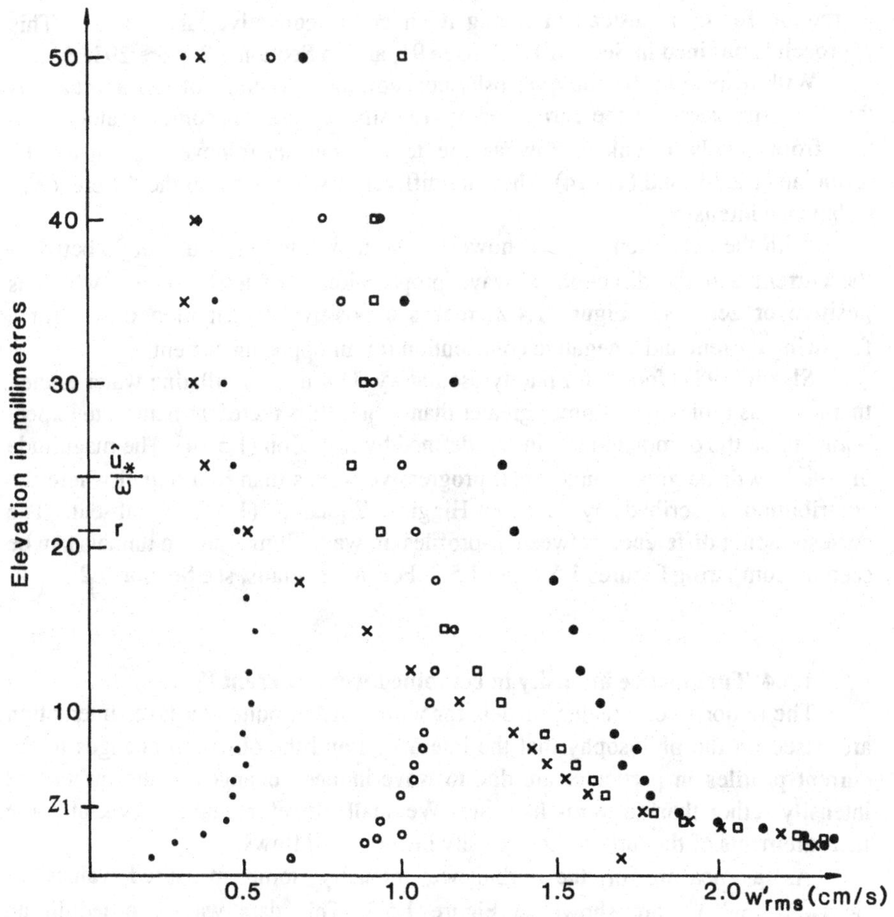

Figure 1.5.7: Root mean square vertical velocity fluctuations in five cases; pure current *0.1m/s* (•) and *0.2m/s* (o) depth average, a pure wave motion *(A, T) = (0.10m, 2.0s)* (x), and the waves superimposed on each of the currents (▪) and (o). Data from oscillating water tunnel, van Doorn (1982).

For all cases with waves, w'_{rms} peaks at the top of the roughness elements ($z = 2mm$) with a maximum value roughly equal to $0.5\, \hat{u}_*$.

For $z < z_1$ there is no apparent increase in turbulence intensity due to the addition of currents, but for $z > z_1$ the data do not contradict the simple superposition rule suggested by Sleath (1991). He suggested that the turbulence intensity of combined flows could be estimated simply by adding the values for the corresponding pure cases : $w'_{rms,cw} = w'_{rms,w} + w'_{rms,c}$.

For the pure current cases, the turbulence intensity is almost independent of z and proportional to the current strength, $w'_{rms,c} \approx 0.5\, \overline{u_*}$.

We note that the thickness of the layer with considerable wave-generated turbulence is, in this case, of the order $5\, z_1 \approx 5 \cdot 0.09 \sqrt{r A} \approx 0.5 \sqrt{r A}$. For this experiment that coincides roughly with the magnitude of $\hat{u}_* / \omega$ and of r.

1.5.5 The influence of currents on the wave boundary layer

The present section deals with the possible effects of a superimposed current on the wave boundary layer structure. More precisely, the question is as follows. Assume that the wave motion just above the boundary layer is unchanged and given by

$$u_\infty (t) = A\, \omega\, e^{i\omega t} \tag{1.5.17}$$

How will the wave boundary layer structure then change if a steady current flow is superimposed?

Most of the existing theoretical models, e g Grant & Madsen (1979), Christoffersen & Jonsson (1985) and Myrhaug & Slaattelid (1989), suggest that the addition of $\overline{u}$ will increase the turbulence intensity and the eddy viscosity applicable to the wave motion. Hence the wave boundary layer thickness should increase as $\sqrt{\nu_w/\omega}$. However, the experimental evidence indicates that the effect is somewhat weaker than the models predict. For example, the data in Figure 1.5.8 show no change to the wave boundary layer structure, and also the Figures 6 and 7 of Myrhaug & Slaattelid (1989) show that their model over-estimates the wave boundary layer thickness compared to the measurements. That is, their predicted increase of ν_w seems to be exaggerated.

This may be understood by considering the expression (1.5.15) for ν_w

$$\nu_w = \frac{\tilde{\tau}/\rho}{\dfrac{\partial \tilde{u}}{\partial z}} = \frac{-\overline{u}\,\tilde{w} - \tilde{u}\,\overline{w} - \widetilde{\tilde{u}\,\tilde{w}} - \widetilde{u'w'}}{\dfrac{\partial \tilde{u}}{\partial z}} + \nu \tag{1.5.15}$$

and remembering that Sleath (1987) found, for purely oscillatory flow, that the term $-\widetilde{\tilde{u}\tilde{w}}$ totally over-shadowed the turbulence term $-\widetilde{u'w'}$. If the turbulent contribution $-\widetilde{u'w'}$ to ν_w is generally insignificant it is unlikely that current-induced turbulence will have much impact on ν_w in a combined wave-current flow. The lack of sensitivity to the superposition of currents is also indicated by the turbulence intensity measurements shown in Figure 1.5.7, in particular for $z < z_1$.

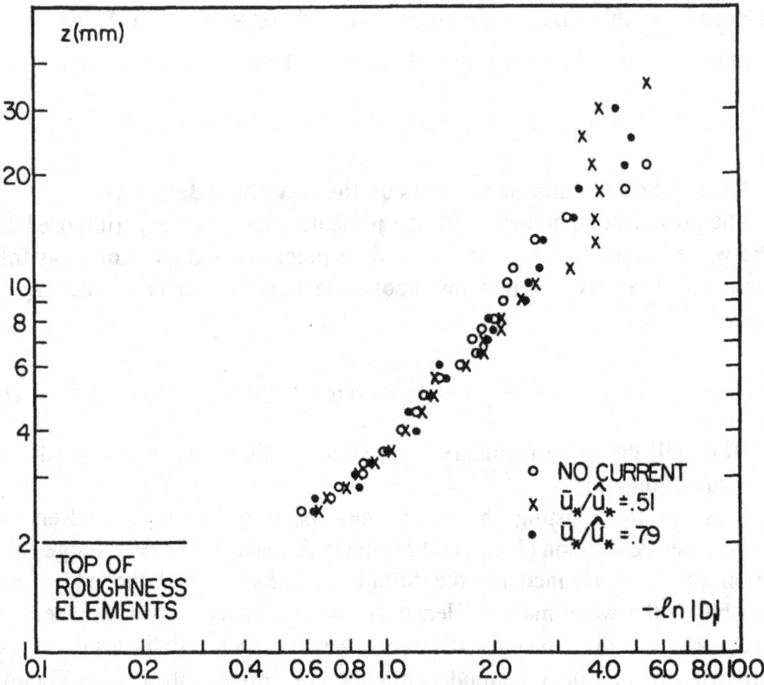

Figure 1.5.8: Measured values of the dimensionless velocity defect $D_1(z)$ for tests V00RA, V10RA, and V20RA from van Doorn (1981). Even for superimposed currents as strong as $\bar{u}_*/\hat{u}_* = 0.79$ there is no evidence of change to the wave boundary layer structure.

The absence of experimental evidence for current-induced changes to the wave boundary layer structure prompted Coffey & Nielsen (1986) to put forward a simpler wave-current interaction model which ignored current-induced changes to the wave boundary layer. It is a clear practical advantage to be able to ignore the effect because then, the wave component of a combined boundary layer flow can be calculated as if the waves were there alone.

It must of course, be expected that with very strong currents, such as some rip currents, the wave boundary layer structure can no longer be assumed unchanged. However, the presently available data show no significant effect for the parameter range $0 < \overline{u_*}/\hat{u}_* < 0.79$, see Figure 1.5.8.

The data in Figure 1.5.8 correspond to colinear waves and currents, but recent measurements of the perpendicular case by Sleath (1990) also show little or no change to the wave boundary layer structure due to superimposed currents of moderate strengths, $\overline{u_*}/A\,\omega < 0.08$.

1.5.6 Energy dissipation by waves superimposed on a current

Several authors (Kemp & Simons 1983, Asano et al 1986 and Simons et al 1988) have observed that waves travelling along a wave flume will lose height more rapidly if they travel on an opposing current and more slowly if travelling on a following current. It has been suggested that this is due to large changes of the wave boundary layer structure and a resulting change in energy dissipation rate. This is however, an interpretation which depends on the definitions applied.

In the following, it will be demonstrated that the above mentioned effect can be attributed mainly to the change in wave group velocity (which together with the wave height determines the wave energy flux), rather than to a change of the wave boundary layer structure. This agrees with the measurements by Sleath (1990) which showed very little change of the wave friction factor due to superimposed currents of moderate strengths ($\overline{u_*}/A\,\omega < 0.08$).

By applying the definitions for energy dissipation and shear stresses suggested by Christoffersen & Jonsson (1985) and by assuming that the waves and the current have separate energy budgets the interpretation of the wave height attenuation data comes into line with the evidence presented in the previous section. That is, the data indicate that the influence from a current on the wave boundary layer structure and hence on the wave energy dissipation rate is weak.

When experiments with extremely large relative roughness ($r/A > 10$) are excluded, the presently available data does not indicate significant changes to the effective wave energy dissipation rate or to the wave boundary layer structure for relative current strengths ($\overline{u_*}/\hat{u}_*$) up to about *10*. See Figure 1.5.9.

The energy dissipation factors in Figure 1.5.9 are based on the following definitions. It is assumed that the waves and the current have separate energy budgets so that the energy balance for waves in a flume with uniform cross section reads

$$\frac{dE_f}{dx} = -D_E \tag{1.5.18}$$

where E_f is the wave energy flux per unit length of wave crest and D_E is rate of wave energy dissipation per unit bed area, see Section 1.2.6. The energy flux is - according to linear wave theory - given by

$$E_f = \frac{1}{8}\rho g H^2 (C_{gr} + U) \tag{1.5.19}$$

where U is the depth-averaged current velocity and C_{gr} is the group velocity of the waves in the frame of reference which moves with the current. This group velocity can be calculated from the usual formula

$$C_{gr} = \frac{gT_r}{4\pi} \tanh k D \, (1 + \frac{2kD}{\sinh 2kD}) \tag{1.5.20}$$

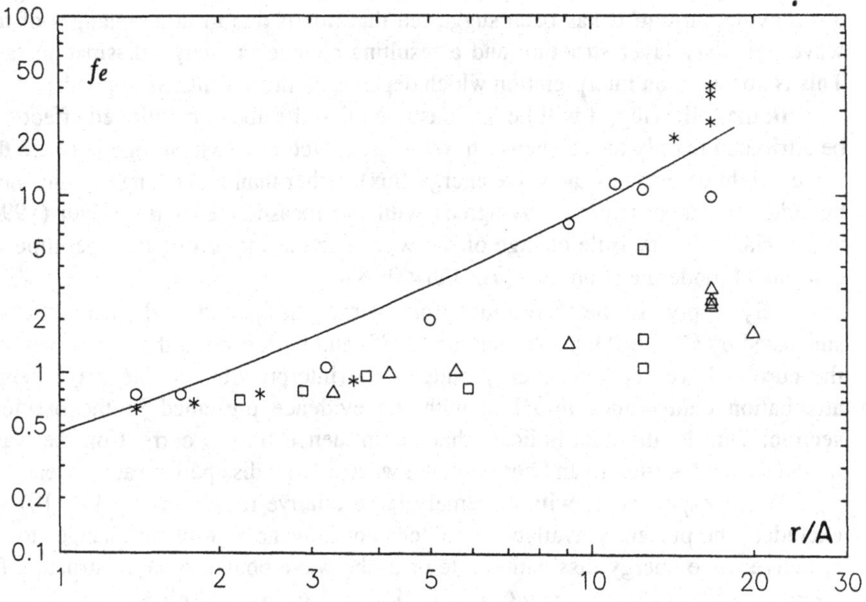

Figure 1.5.9: Wave energy dissipation factors derived, through Equation (1.5.29), from the wave height decays measured by Simons et al (1988). The symbols *, o, □, Δ correspond respectively to the following relative current strengths ($\overline{u_*}/\hat{u}_*$): *0, 1.3-2.1, 4.1-9.8, 5.5-10.5.* The curve corresponds to Swart's formula, Equation (1.2.22).

applied in the frame of reference which travels with speed U. That is, T_r is the wave period in this moving frame of reference and the wave number must be determined from the dispersion relation

$$kD \tanh kD = \frac{4\pi^2}{gT_r^2} D = \frac{D}{g}\left(\frac{2\pi}{T_a} - kU\right)^2 \tag{1.5.21}$$

where T_a is the wave period seen from the laboratory frame of reference.

For the wave energy dissipation we apply Jonsson's (1966) definition

$$D_E = \frac{2}{3\pi} \rho f_e \,(A\omega_a)^3 \tag{1.5.22}$$

so we have

$$\frac{d}{dx}\left(\frac{1}{8}\rho g H^2 (C_{gr} + U)\right) = -\frac{2}{3\pi} \rho f_e \,(A\omega_a)^3 \tag{1.5.23}$$

where $C_{gr} + U$ is a constant when the depth is fixed and the near bed velocity amplitude is given by

$$A\omega_a = \frac{\pi H}{T_a \sinh kD} \tag{1.5.24}$$

Then the energy dissipation equation can be reduced to

$$\frac{d}{dx}\left(\frac{1}{H}\right) = \frac{8\pi^2 f_e}{3gT_a^3 (C_{gr} + U) \sinh^3 kD} \tag{1.5.25}$$

which, with the starting conditions $H(x_o) = H_o$ has the solution

$$H(x) = \frac{H_0}{1 + H_0 \dfrac{8\pi^2 f_e}{3gT_a^3 (C_{gr} + U) \sinh^3 kD} (x - x_0)} \tag{1.5.26}$$

In most previous studies, the observed wave height variation has been fitted by an exponential which corresponds to viscous or laminar energy dissipation, i e

$$H(x) = H(x_o)\, e^{-\alpha(x - x_o)} \tag{1.5.27}$$

and the measurements have been discussed in terms of the dissipation factor

$$\alpha = -\frac{1}{H}\frac{dH}{dx} \tag{1.5.28}$$

For the real wave height variation under turbulent conditions, which is given by Equation (1.5.26) there is no constant corresponding to α but for small energy dissipation rates we have

$$\alpha \approx H_0 \frac{8\pi^2}{3\, g\, T_a^3\, (C_{gr} + U) \sinh^3 kD} f_e \tag{1.5.29}$$

This shows how, for fixed f_e, the observed α-values will decrease for increasing U and vice versa. This effect of the mean current on dH/dx was also noted by Simons et al (1988).

The f_e-values in Figure 1.5.9 which are based on measured α-values and on the analysis above show that f_e and presumably the general structure of the wave boundary layer are practically unchanged by the superposition of currents. This agrees with the friction factor measurements of Sleath (1990) and with the findings related to Figure 1.5.8 about the wave velocity defect function.

Hence, the data indicate that, for practical purposes, the wave boundary layer structure and the wave energy dissipation factor for a combined flow can be calculated as if the current was not there. That is of course, after the near bed velocity amplitude $A\omega$ has been calculated with due respect to the current by using the dispersion relation (1.5.21). The conclusion above should not be applied uncritically to conditions outside the range of conditions represented by the data in Figures 1.5.8 and 1.5.9.

1.5.7 The influence of waves on current profiles

In relation to the modelling of coastal processes the most important aspect of wave-current boundary layer interaction is the change of current velocity distributions due to the superposition of waves.

The scene can be set by either of the following two questions

a) Assuming that the average bed shear stress and hence $\overline{u_*}$ is kept unchanged, what is the change to the current profile due to the superposition of waves? See Figure 1.5.10, left-hand side.

b) What are the changes to the current profile when waves are added in such a way that the reference current velocity $\overline{u_r} = \overline{u}(z_r)$ is kept unchanged? See Figure 1.5.10, right.

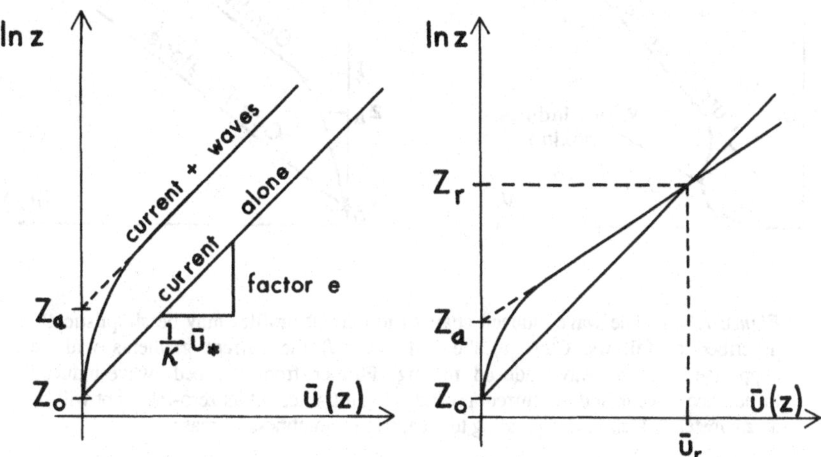

Figure 1.5.10: Two ways of considering the problem of wave-induced changes to current profiles. Left) The friction velocity $\overline{u_*}$ is considered known and fixed. right) The velocity at a certain level z_r is considered known and fixed.

The first approach which was taken by Lundgren (1972) is the most rational because it leads to logically straight model building, while the latter approach leads to iterative thinking, as applied in the model of Grant & Madsen (1979).

In the following we shall follow Lundgren's line of thought where $\overline{u_*}$ is taken to be known and fixed, - for example determined by a fixed mean surface slope

$$\overline{u_*} = (\overline{\tau}(o)/\rho)^{0.5} = (-gD\, d\overline{\eta}/dx\,)^{0.5}$$

Then, the wave-induced changes to $\overline{u}(z)$ are as outlined by Figure 1.5.11.

Let the undisturbed current profile be given by

$$\overline{u}(z) = \frac{\overline{u_*}}{\kappa} \ln\frac{z}{z_o} \tag{1.5.30}$$

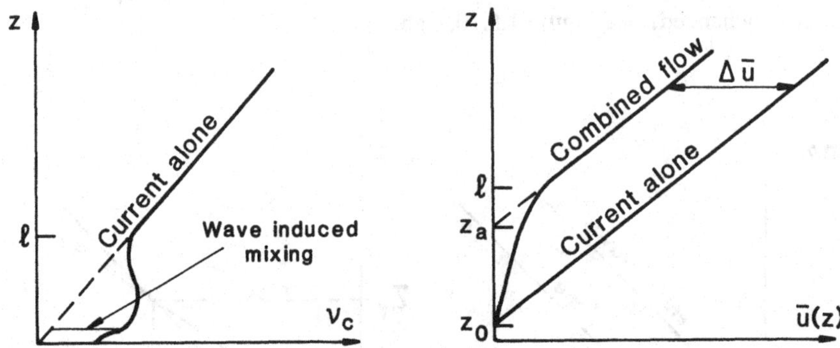

Figure 1.5.11: The wave-induced changes to current profiles may be simplistically described as follows. Close to the bed $(z < l)$ the current gradients will be suppressed due to wave-induced mixing. Further from the bed, wave-induced turbulence is weak and the current profile is logarithmic, but its zero-intercept will be at z_a instead of z_o corresponding to an apparent roughness increase

where $z_o \approx r/30$ as usual.

With waves superimposed, there will be a layer of thickness l close to the bed with direct wave influence on $\bar{u}(z)$. That is, inside this layer the current gradient $d\bar{u}/dz$ is suppressed due to wave-induced mixing, see Figures 1.5.1 and 1.5.7.

Judging from the turbulence measurements shown in Figure 1.5.7, the thickness l of the layer with predominantly wave-generated turbulence should, in that particular case, be of the order

$$l \approx 5\,z_1 = 5 \cdot 0.09\sqrt{r\,A} \approx 0.5\sqrt{r\,A} \qquad (1.5.31)$$

see Equation (1.3.10), or alternatively,

$$l \approx \hat{u}_* / \omega \qquad (1.5.32)$$

However, the relationship between l and the various boundary layer thickness values (Section 1.2.7) may depend on the relative roughness r/A, the relative current strength $\overline{u_*}/\hat{u}_*$ and possibly on the angle φ between the current and the direction of wave propagation

$$\frac{l}{z_o} = F_l(\frac{r}{A}, \frac{\overline{u_*}}{A\omega}, \varphi). \qquad (1.5.33)$$

The shape of the current profile for $z < l$ seems to be as much like a linear function of z as any other simple shape including the logarithmic shape suggested by Grant & Madsen (1979), Fredsoe (1984), and by Christoffersen & Jonsson (1985), at least for moderate values of r/A, see Figure 1.5.12.

Outside this layer, the wave-generated mixing is relatively weak, so the law of the wall may be assumed to apply in which case the current distribution must be logarithmic and can be written as

$$\overline{u}(z) = \frac{\overline{u_*}}{\kappa}\ln\frac{z}{z_a} \quad \text{for } z > l \qquad (1.5.34)$$

where z_a is analogous to, but larger than $z_o = r/30$.

The existence of a logarithmic layer corresponding to Equation (1.5.34) is not always equally obvious in the experimental data, compare Figures 1.5.1 and 1.5.2, and in some cases the fitting of a logarithmic curve is somewhat arbitrary. However, the description above, which is essentially due to Lundgren (1972), is reasonably adequate considering the presently available experimental details.

The apparent roughness increase z_a/z_o and the corresponding velocity reduction

$$\Delta\overline{u} = \frac{\overline{u_*}}{\kappa}\ln\frac{z_a}{z_o} \qquad (1.5.35)$$

can be expected to depend on the relative current strength, the relative roughness, and on the angle between the current and the direction of wave propagation

$$\frac{z_a}{z_o} = F(\frac{A\,\omega}{\overline{u_*}}, \frac{r}{A}, \varphi) \qquad (1.5.36)$$

The presently available data indicate that, the influence of φ on z_a/z_o is strongest under real waves where it stems from the term $-\overline{\tilde{u}\tilde{w}}$, as discussed in Section 1.5.9. In tunnels and on oscillating plates where the main part of this momentum transfer term is absent, the angle φ seems to have little impact on z_a/z_o. Thus, the "$\varphi = 90^o$data" of Sleath (1990) measured over an oscillating plate, follow the same trend as the tunnel data of van Doorn (1982), see Figure 1.5.13.

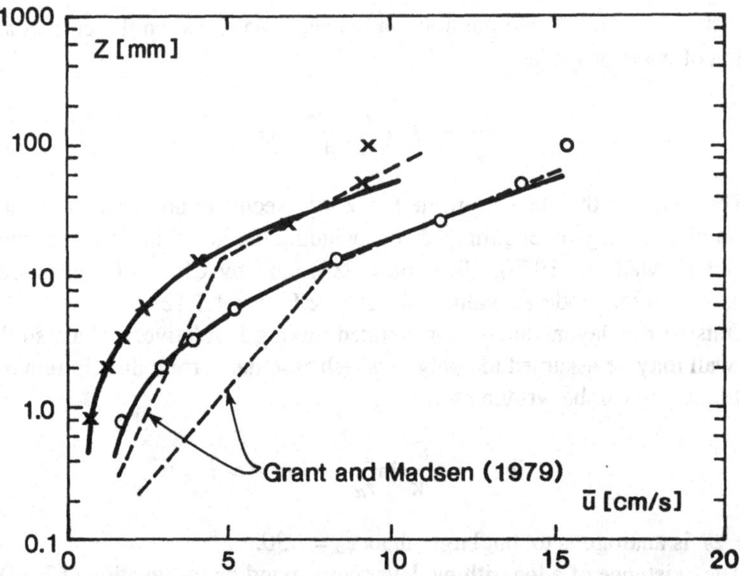

Figure 1.5.12: Even for fairly small relative roughness (r/A), the current profile in the inner, wave-dominated layer is not logarithmic. Data from Sleath (1990), A = $0.141m$, $T = 1.76s$, $r = 1.49mm$, $\overline{u_*}$ = $1.9cm/s$ (x) or $2.3cm/s$ (o).

There is some evidence however, which suggests that opposing currents ($\varphi = 180^o$) are affected in a different way than following currents ($\varphi = 0^o$). Thus the "opposing current data" of Kemp & Simons (1983) plot somewhat higher than the "following current data" from 1982 in Figure 1.5.13. That is, the apparent roughness increase is greater for opposing currents than for following currents.

This can be explained by considering explicitly the $\overline{\tilde{u}\tilde{w}}$-terms in the equation of motion (1.1.21) instead of making them part of $\overline{\tau}$ as done in the present, simplistic approach. See Section 1.5.9.

Based on the data, which was then available, Coffey & Nielsen (1986) suggested that it might be possible to express the apparent roughness increase as function of a single parameter, namely the friction velocity ratio $\hat{u}_*/\overline{u_*}$. They suggested the formula :

$$\frac{z_a}{z_o} = 1 + 0.06\left(\frac{\hat{u}_*}{\overline{u_*}}\right)^3 \tag{1.5.37}$$

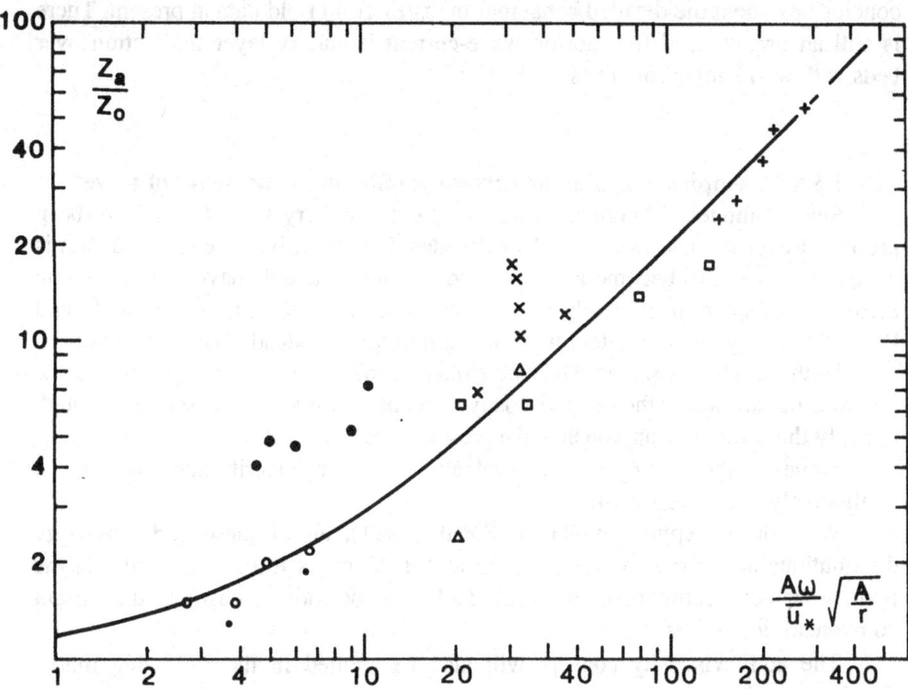

Figure 1.5.13: Apparent roughness increase as function of relative current strength and relative roughness. The curve corresponds to the theory of Sleath (1991). o: Kemp & Simons 1982 flume $\varphi = 0$, ●: Kemp & Simons 1983 flume $\varphi = 180^o$, •: Asano et al 1986 flume $\varphi = 0^o$, Δ : van Doorn 1981 flume $\varphi = 0^o$, □: van Doorn 1982 tunnel, x,+ : Sleath 1990 $\varphi = 90^o$

However, Sleath (1990) found that this formula seemed inadequate for data with smaller relative roughness ($r/A < 0.1$). Subsequently Sleath (1991) developed a model which leads to the expression

$$\frac{z_a}{z_o} = 1 + 0.19 \frac{A\omega}{u_*}\sqrt{\frac{A}{r}} \tag{1.5.38}$$

which is shown together with the presently available data in Figure 1.5.13.

Several field studies, e g Cacchione & Drake (1982), Grant et al (1983), Coffey (1987), Lambrakos et al (1988) and Slaattelid et al (1990), provide useful data on z_a. However, since the corresponding values of the hydraulic roughness r (and hence of $z_0=r/30$) are very hard to estimate, it is impossible to make firm conclusions about the detailed behaviour of z_a/z_0 from field data at present. There is still an urgent need for data on wave-current boundary layer interaction over beds with well known roughness.

1.5.8 An empirical model for current profiles in the presence of waves

Several models of combined wave-current boundary layer flows have been presented over the last two and a half decades. However, because detailed data is only now starting to become available, most of these models have been based on theoretical adaptation of steady flow concepts such as *the law of the wall* and Prandtl's mixing length model for momentum transfer in steady boundary layers.

With the exception of those of Bijker (1967) and Fredsoe (1984) all the previous models apply the eddy viscosity concept, and it seems reasonable enough to apply this simplistic approach at the present state of the art.

Some of the previously assumed eddy viscosity distributions are shown qualitatively in Figure 1.5.14.

With the exception of that of Sleath (1991), all of these eddy viscosity distributions are discontinuous at $z=l$, and in all cases it was assumed that the periodic velocity component $\tilde{u}$ would feel the same eddy viscosity as the current component $\overline{u}$, $\nu_c \equiv \nu_w$.

The eddy viscosity concept will also be applied in the following model development but with one major difference from previous models. We shall not assume that the wave component $\tilde{u}(z,t)$ feels the same eddy viscosity as the steady component $\overline{u}(z)$. Instead, we shall acknowledge the empirical fact that the eddy viscosity ν_c which is felt by $\overline{u}(z)$ is generally about four times larger than ν_w which is felt by $\tilde{u}(z,t)$

$$\nu_c \approx 4\nu_w \tag{1.5.39}$$

see the discussion in Section 1.5.3, page 70.

For the sake of simplicity, it shall also be assumed that $\tilde{u}(z,t)$ can be calculated from (A, ω, r) alone, as if the current was not there. This approach seems justified by the available experimental data, which indicate that the wave boundary layer structure, represented by $\tilde{u}(z,t)/A\,\omega$, changes very little even with fairly strong currents superimposed (see Sections 1.5.4 and 1.5.5). The velocity defect data in Figure 1.5.8 show no effect from superimposed currents with

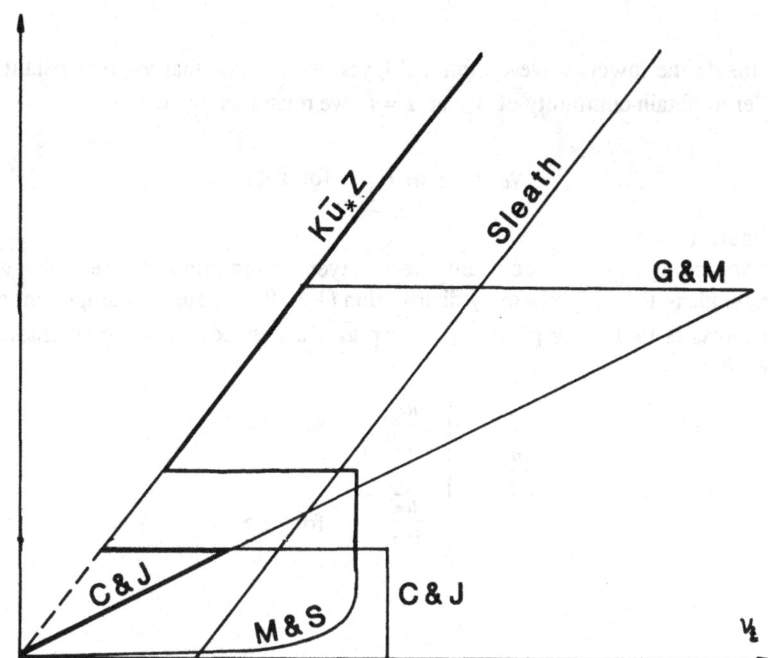

Figure 1.5.14: Some of the eddy viscosity distributions assumed by previous authors for combined wave-current flow (shown qualitatively). G&M: Grant & Madsen (1979), C&J: Christoffersen & Jonsson (1985), M&S: Myrhaug & Slaattelid (1989), Sleath (1991). In all of these studies, it was assumed that the waves felt the same eddy viscosity as the current.

strength $\overline{u_*}/\hat{u}_*$ up to *0.79*. In addition, the wave energy dissipation data in Figure 1.5.9, page 76, indicate no change to the wave friction factor for much higher values of the relative current strength $\overline{u_*}/\hat{u}_*$.

The essence of the problem is then to specify the distribution of the eddy viscosity ν_c which governs the distribution of $\overline{u}(z)$.

Following the line of thought of Lundgren (1972) we assume that an outer layer exists, where the eddy viscosity felt by the current $\overline{u}(z)$ is given by

$$\nu_c = \kappa \overline{u_*} z \qquad \text{for } z > l$$

just as in a pure current boundary layer which is thin compared with the flow depth (a constant stress layer), see page 68.

Inside the lower, wave-dominated layer, we assume that ν_c is constant and, in order to obtain continuity of ν_c at $z = l$, we must then have

$$\nu_c \;=\; \kappa\,\overline{u_*}\,l \qquad \text{for } z < l \tag{1.5.40}$$

see Figure 1.5.15.

According to the constant-stress layer assumption (page 68) which corresponds to the eddy viscosity distribution (1.5.39), the steady component of the shear stress is uniformly given by $\overline{\tau} \equiv \rho\,\overline{u_*}^2$ and hence the current gradient is given by

$$\frac{d\overline{u}}{dz} \;=\; \begin{cases} \dfrac{\overline{u_*}}{\kappa\,l} & \text{for } z < l \\[2ex] \dfrac{\overline{u_*}}{\kappa\,z} & \text{for } z > l \end{cases} \tag{1.5.41}$$

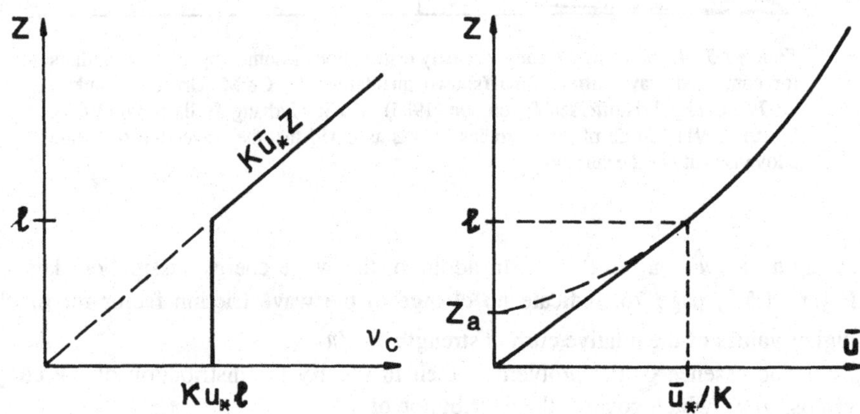

Figure 1.5.15, left: The assumed distribution of the eddy viscosity ν_c which is felt by the steady component $\overline{u}(z)$ of a combined wave-current boundary layer flow. *Right:* The resulting velocity distribution.

Starting with $\overline{u} = 0$ at $z = 0$ we get the linear current distribution

$$\bar{u}(z) = \frac{\overline{u_*}}{\kappa}\frac{z}{l} \qquad \text{for} \quad z < l \tag{1.5.42}$$

in the inner layer.

In the outer layer the current distribution is logarithmic and may be written

$$\bar{u}(z) = \frac{\overline{u_*}}{\kappa}\ln\frac{z}{z_a} \qquad \text{for} \quad z > l \tag{1.5.43}$$

where matching with the lower-layer solution is obtained when

$$z_a = e^{-1}\, l \tag{1.5.44}$$

At the present state of the art, it does not seem possible to derive a general expression for l which is valid for all flow conditions (see the discussion related to Equation 1.5.36). As more experimental information becomes available, for example in the form of Equation (1.5.36), it will be straightforward to make use of it using the framework of Figure 1.5.15 with l and z_a determined from

$$e^{-1}\frac{l}{z_0} = \frac{z_a}{z_0} = F\left(\frac{A\omega}{\overline{u_*}}, \frac{r}{A}, \varphi\right) \tag{1.5.45}$$

***Example 1.5.1*: A simple estimate of F for fairly rough conditions**

It seems futile at present to speculate too much about the detailed form of the function $F\left(\frac{A\omega}{\overline{u_*}}, \frac{r}{A}, \varphi\right)$ since so many aspects of the underlying physics are unknown. However, for fairly large relative roughness, $0.06 < r/A < 0.5$, where the structure of the wave boundary layer corresponds to a constant, real-valued eddy viscosity (see Section 1.2.9, page 40), a simple estimate of F can be obtained if the influence of φ is neglected.

For such wave boundary layers, the eddy viscosity can be estimated by

$$\nu_w = 0.5\,\omega\, z_1^2 \approx 0.5\,\omega\,(0.09\sqrt{rA})^2 = 0.004\,\omega\, r A \tag{1.3.14}$$

see page 52. With this estimate of ν_w and with the observation that $\nu_c \approx 4\,\nu_w$ (see Section 1.5.3, page 70), the following estimate for ν_c is obtained

$$\nu_c \approx 0.016\,\omega A r \qquad \text{for} \quad z < l \tag{1.5.46}$$

Then, since the current eddy viscosity in the lower layer is written as $\nu_c = \kappa \overline{u_*} l$ (see Figure 1.5.15), this gives the following expression for l

$$l \approx \frac{0.016}{\kappa} \frac{A\,\omega}{\overline{u_*}} r \tag{1.5.47}$$

and hence with $z_a = e^{-1} l$, $\kappa \approx 0.4$, and $z_0 = r/30$

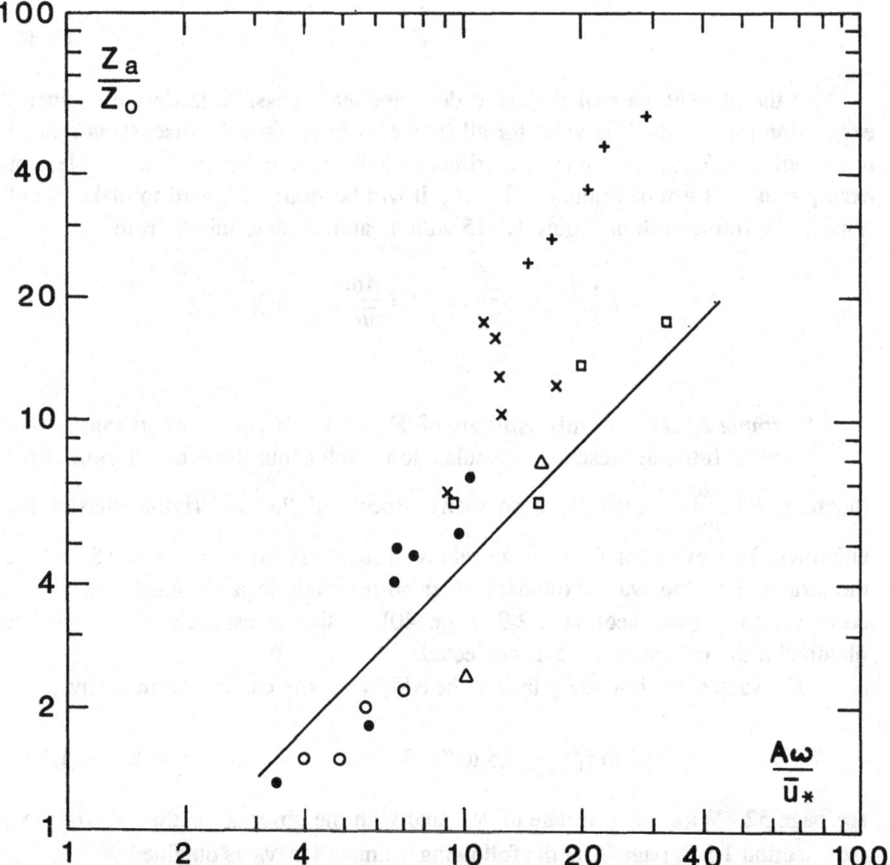

Figure 1.5.16: Apparent roughness increase as function of the relative current strength for the same data which was shown in Figure 1.3.13. The straight line corresponds to Equation (1.5.48).

$$F = \frac{z_a}{z_o} \approx 0.44 \frac{A\,\omega}{\overline{u_*}} \qquad (1.5.48)$$

This expression for F corresponds to the line in Figure 1.3.16. The agreement is fair, except for the data of Sleath (1991) (+), for which z_a/z_o is systematically underpredicted. These data are, however, outside the validity range of the assumed eddy viscosity formula (1.3.14), page 52, so this is not surprising. The validity range of Equation (1.3.14) is approximately $0.06 < r/A < 0.5$, while the mentioned data correspond to $r/A = 0.011$ or $r/A = 0.020$.

Although the model in the example above was developed following Lundgren (1972) i e, the time-averaged friction velocity was assumed known and fixed, it can still be applied to problems where current friction velocity $\overline{u_*}$ is not given.

Consider the situation where data from a current meter at $z=z_r$ above the wave boundary layer are available, so that we know the values of $[A, \omega, \overline{u}(z_r), \varphi]$. Assume further that the bed roughness and hence z_o is somehow known.

Information about current distribution in general and about the time averaged friction velocity in particular can then be derived in the following way.

We assume that z_r is within the logarithmic part of the current profile. Hence, we have, in accordance with Equation (1.5.34)

$$\overline{u}(z_r) = \frac{\overline{u_*}}{\kappa} \ln \frac{z_r}{z_a} = \frac{\overline{u_*}}{\kappa} \ln \frac{z_r}{z_o F\left(\frac{A\,\omega}{\overline{u_*}}, \frac{r}{A}, \varphi\right)} \qquad (1.5.49)$$

where the unknown is the time-averaged friction velocity $\overline{u_*}$. The form of the function F is assumed known, for example, given by Sleath's (1991) expression (1.5.38) or by Equation (1.5.48).

For the purpose of iterative solution, it will generally be convenient to rewrite this Equation (1.5.49) in the form

$$\overline{u_*} = \frac{\kappa\, \overline{u}(z_r)}{\ln \frac{z_r}{z_o} - \ln F\left(\frac{A\,\omega}{\overline{u_*}}, \frac{r}{A}, \varphi\right)} \qquad (1.5.50)$$

which will normally converge rather rapidly.

When the current friction velocity $\overline{u_*}$ has been found from the iteration, the value of F can be calculated and so can $z_a = z_0 F$ and $l = e^1 z_0 F$. All components are then available for drawing the velocity distribution

$$\overline{u}(z) = \begin{cases} \dfrac{\overline{u_*}}{\kappa} \dfrac{z}{l} & \text{for } z < l \\ \dfrac{\overline{u_*}}{\kappa} \ln \dfrac{z}{z_a} & \text{for } z > l \end{cases} \tag{1.5.51}$$

Example 1.5.2: Current profile from wave data and reference current

To illustrate the use of the model above for deriving details of the current distribution $\overline{u}(z)$ from the current velocity at a single point plus the wave data and the bed roughness, consider the numerical example $[A, T, r, \overline{u}(1.0m)]$ = $[0.9m, 8s, 6cm, 0.45m/s]$.

Inserting these data into Equation (1.5.50) with F given by Equation (1.5.48) gives

$$\overline{u_*} = \frac{0.4 * 0.45}{\ln \dfrac{1.0}{0.06/30} - \ln (0.44 \dfrac{0.9 * 2\pi/8}{\overline{u_*}})} = \frac{0.18}{7.383 + \ln \overline{u_*}} \tag{1.5.52}$$

which converges rapidly to the friction velocity value $\overline{u_*} = 0.0425m/s$. This corresponds to $F = 0.44 \dfrac{A\omega}{\overline{u_*}} = 7.32$, and hence $z_a = z_0 F = 0.0146m$, and $l = z_a e = 0.040m$. The corresponding velocity distribution, given by Equation (1.5.51), is shown in Figure 1.5.17.

Alternatively, if Sleath's (1991) expression (1.5.38) is used for F in Equation (1.5.50), we get

$$\overline{u_*} = \frac{0.4 \cdot 0.45}{\ln \dfrac{1.0}{0.06/30} - \ln (1 + 0.19 \dfrac{0.9 \cdot 2\pi/8}{\overline{u_*}} \sqrt{\dfrac{0.9}{0.06}})} = \frac{0.18}{6.22 - \ln (1 + \dfrac{0.52}{\overline{u_*}})} \tag{1.5.53}$$

which equally rapidly converges to a friction velocity of $\overline{u_*} = 0.048m/s$. With this value, the model of Sleath (1991) gives the velocity distribution which is shown by the broken line in Figure 1.5.17 .

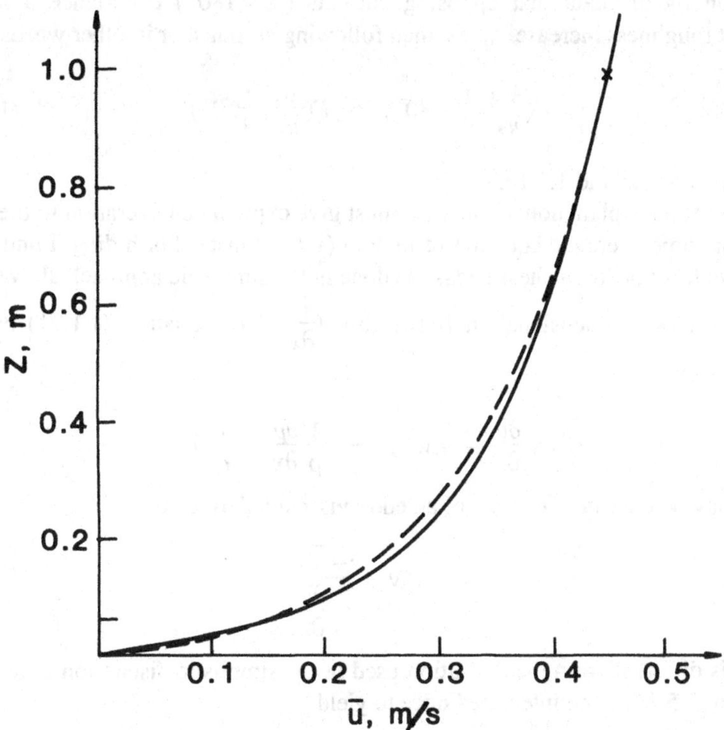

Figure 1.5.17: Velocity profiles obtained from the data set $[A, T, r, \bar{u}(1.0m)]$ = $[0.9m, 8s, 6cm, 0.45m/s]$, with two different wave-current boundary layer models. The full line corresponds to the model defined in Figure 1.5.15 with $F = z_a/z_o = 0.44\ A\omega/\overline{u_*}$, while the broken line corresponds to the model of Sleath (1991).

1.5.9 Wave-current interaction models with explicit consideration of $\widetilde{u}\widetilde{w}$

As mentioned above, a detailed theoretical prediction of the function $z_a/z_o = F(\frac{A\omega}{\overline{u_*}}, \frac{r}{A}, \varphi)$ will not be attempted in the present text. However, one particular aspect of the available experimental data, which is not addressed directly by the simplistic approach above, is easily addressed in theoretical terms. That is the indication by the data that opposing currents ($\varphi = 180^o$) experience a larger, apparent roughness increase z_a/z_o than following currents, or in other words,

$$F(\frac{A\omega}{\overline{u_*}}, \frac{r}{A}, 180^o) \ > \ F(\frac{A\omega}{\overline{u_*}}, \frac{r}{A}, 0^o) \qquad (1.5.54)$$

see Figures 1.5.13 and 1.5.16.

To get an explanation for this we must give explicit consideration to the term $\overline{\tilde{u}\tilde{w}}$ in the time-averaged equation of motion (1.1.21) instead of hiding it under the collective label of "total shear stress" as done in the simplistic approach above.

For a two dimensional, uniform case $(\frac{\partial}{\partial x} = 0)$ Equation (1.1.21) can be written

$$\frac{\partial}{\partial z}(\nu \frac{\partial \bar{u}}{\partial z} - \overline{u'w'}) \ = \ \frac{1}{\rho}\frac{\partial \bar{p}}{\partial x} + \frac{\partial}{\partial z}\overline{\tilde{u}\tilde{w}} \qquad (1.4.6)$$

which may be rewritten in terms of an eddy viscosity defined by

$$\nu_{c1} \ = \ \nu + \frac{-\overline{u'w'}}{\frac{\partial \bar{u}}{\partial z}} \qquad (1.5.55)$$

(which is different from the definition used in the simplistic discussion above, see Equation (1.5.14)), and integrated once to yield

$$\nu_{c1}\frac{\partial \bar{u}}{\partial z} \ = \ \frac{1}{\rho}\frac{\partial \bar{p}}{\partial x} z + \overline{\tilde{u}\tilde{w}} + \frac{\bar{\tau}(o)}{\rho} \qquad (1.5.56)$$

For a hydrostatic mean pressure gradient due to the mean surface slope $\frac{d\bar{\eta}}{dx}$, where the mean bed shear stress equals $-\rho\, g\, D \frac{d\bar{\eta}}{dx}$, Equation (1.5.56) can be rewritten as

$$\nu_{c1}\frac{\partial \bar{u}}{\partial z} \ = \ g\,(z{-}D)\frac{\partial \bar{\eta}}{\partial x} + \overline{\tilde{u}\tilde{w}} \ = \ -\,g\,D\,(1 - \frac{z}{D})\frac{\partial \bar{\eta}}{\partial x} + \overline{\tilde{u}\tilde{w}} \qquad (1.5.57)$$

which, for a thin bottom boundary layer where $z \ll D$, can be conveniently approximated by

$$\nu_{c1}\frac{\partial \bar{u}}{\partial z} \ = \ \frac{\bar{\tau}(o)}{\rho} + \overline{\tilde{u}\tilde{w}} \ = \ \overline{|u_*|\,u_*} + \overline{\tilde{u}\tilde{w}} \qquad (1.5.58)$$

where the friction velocity is defined by

$$|\overline{u_*}|\,\overline{u_*} = \frac{\overline{\tau}(o)}{\rho} = -g\,D\frac{d\overline{\eta}}{dx}$$

The current distribution is then obtained by integration and use of the non-slip boundary condition $\overline{u}(o) = 0$

$$\overline{u}(z) = \int_o^z \left(\frac{|\overline{u_*}|\,\overline{u_*} + \overline{\widetilde{u}\widetilde{w}}}{\nu_{c1}} \right) dz \tag{1.5.59}$$

With u positive in the direction of wave propagation, the term $\overline{\widetilde{u}\widetilde{w}}$ is likely to be always negative, see page 55. The effect of $\overline{\widetilde{u}\widetilde{w}}$ on the current distribution (with fixed $\overline{u_*}$) is therefore to reduce the magnitude of $\overline{u}(z)$ for following currents and increase it for opposing currents.

However, when z_a is found for each profile by extrapolation of the upper, logarithmic part of the current profile, one finds larger z_a/z_o for the opposing currents than for the following currents, see Figure 1.5.18. This agrees with the data of Kemp & Simons (1982, 1983), see Figure 1.5.13, page 83.

In addition, Equation (1.5.57) shows that due to $\overline{\widetilde{u}\widetilde{w}}$ being negative, the current gradient $(d\overline{u}/dz)$, for a following current, will become negative close to the free surface. This was observed by Kemp & Simons (1982) and can be seen from the wave flume data of van Doorn (1981) in Figure 1.5.2, page 63.

For perpendicular currents and for currents in an oscillating water tunnel $\overline{\widetilde{u}\widetilde{w}}$ is much smaller because the the part of it, which is equivalent to the laminar expression (1.4.5), page 57, is absent. Thus, the logarithmic part of the $\overline{u}$-profile shows the "correct" slope, and $d\overline{u}/dz$ does not turn negative, see Figure 1.5.1.

The situation is demonstrated qualitatively by the example in Figure 1.5.18 where the current distribution has been calculated from Equation (1.5.59) with the eddy viscosity distribution

$$\nu_{c1} = \begin{cases} |\overline{u_*}|\,\delta_s & \text{for } z/\delta_s < 4 \\ 0.25\,|\overline{u_*}|\,z & \text{for } z/\delta_s \geq 4 \end{cases} \tag{1.5.60}$$

and with $\overline{\widetilde{u}\widetilde{w}}$ based on a constant eddy viscosity of $\nu_w = 0.5\,\omega\,\delta_s^2$.

The magnitude of $|\,(\overline{\widetilde{u}\widetilde{w}})_\infty\,|$ under progressive waves, relative to the maximum wave-induced bed shear stress $\hat{\tau}$ may be estimated roughly on the basis of their respective values for boundary layers with constant eddy viscosity:

$$\frac{-(\overline{\widetilde{u}\widetilde{w}})_\infty}{\hat{\tau}/\rho} = \frac{\frac{1}{4}(A\omega)^2\,k\,\delta_s}{\frac{1}{2}f_w\,(A\omega)^2} = \frac{1}{2\sqrt{2}}\,k\,A \tag{1.5.61}$$

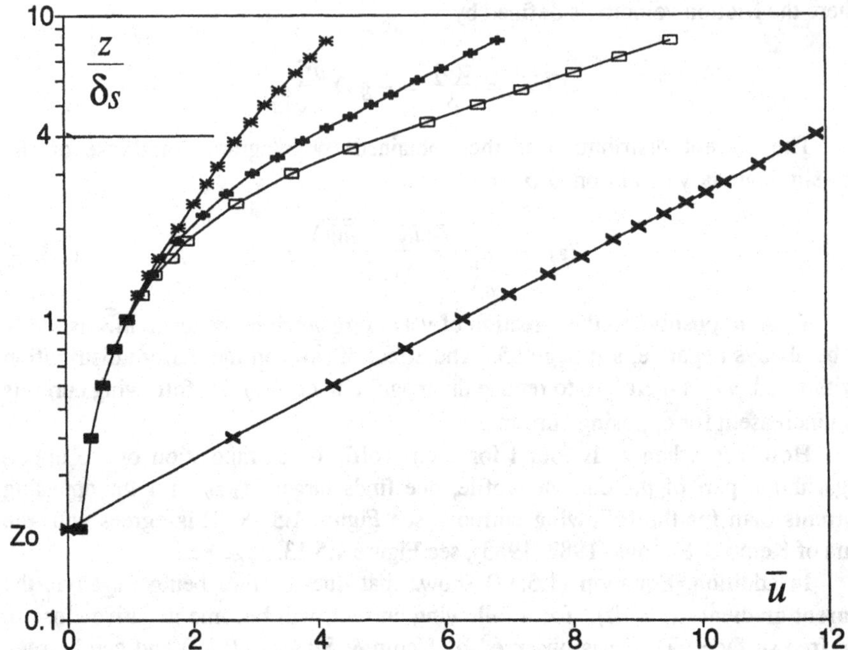

Figure 1.5.18: Current distributions for a pure current and for a current with following, opposing and perpendicular waves superimposed according to Equation (1.5.59) with the eddy viscosity distribution (1.5.60). The relative "drift strength" $|(\overline{\tilde{u}\tilde{w}})_\infty/\overline{u_*}^2|$ is, for both following and opposing waves, equal to *0.5*. Legend: x: pure current, *: following waves , +: perpendicular waves or tunnel, *rectangle* : opposing waves.

since the wave friction factor for constant eddy viscosity amounts to $2/\sqrt{\text{Re}} = 2\sqrt{\nu/(A^2\omega)} = \sqrt{2}\,\delta_s/A$, see Equations (1.2.16) and (1.4.4).

It seems rather natural to incorporate $\overline{\tilde{u}\tilde{w}}$ into new models of wave-current boundary layer interaction, along the lines suggested above. For further incentive see also Section 7.2, page 293. However, some questions need to be addressed regarding the general nature of $\overline{\tilde{u}\tilde{w}}$ in turbulent boundary layers over rough beds.

Longuet-Higgins' expression (1.4.5) was derived for laminar flow over a plane bed and the only vertical velocities involved are due to the thickening and thinning of the boundary layer under a progressive wave. Over a rough bed however, there will be additional periodic, vertical velocities caused by the bed geometry, and the significance of their contribution to $\overline{\tilde{u}\tilde{w}}$ must be considered.

CHAPTER 2

SEDIMENT MOBILITY, BED-LOAD AND SHEET-FLOW

2.1 FORCES ON SEDIMENT PARTICLES

2.1.1 Introduction

For the purpose of sediment transport modelling, it is necessary to consider three types of forces which govern the behaviour of cohesionless sediment particles whether they are resting at the bed or moving around in a slurry or a thin suspension. These are: the gravity force $F_g = Mg$; intergranular forces related to collisions or continuous contact; and the fluid forces which may be due to surface drag or fluid pressures.

2.1.2 Intergranular forces

The intergranular forces are well understood as far as resting (non-shearing) grains are concerned.

For resting grains the static angle of repose φ_s is determined by the frictional coefficient, i e the ratio between the effective normal stress σ_e and the maximum sustainable shear stress τ_{max} , see Figure 2.1.1.

The angle φ_s applies to dry sand or to sand which is entirely under water. Sand which is wet but not saturated may stand at a much steeper angle because the negative porewater pressure increases the effective normal stress σ_e.

For most sandy materials the static angle of repose is between 26 and 34 degrees, being greatest when the material is most densely packed. For moving (shearing) materials of near maximum concentration, the frictional coefficient. Hence the dynamical angle of repose, φ_d is of similar magnitude to φ_s.

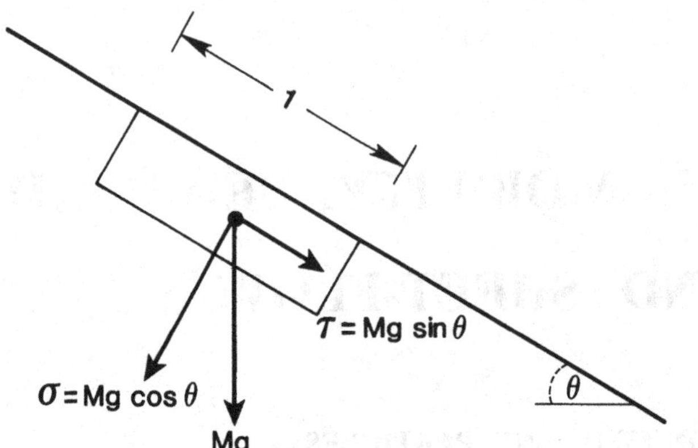

Figure 2.1.1: The frictional coefficient $(\tau/\sigma_e)_{max}$ equals tan φ_s, where φ_s is the static angle of repose.

Hanes & Inman(1985b) suggested a typical value of *31* degrees for beach sand.

When a horizontal sand bed is exposed to a fast, steady flow, a finite top layer of sand will start to move with the flow, partly as bed-load and partly in suspension. The fact that the moving layer is of finite thickness is significant although seemingly trivial, because it shows that the moving sand has increased the strength of the sand below.

Since the shear stress is not decreasing downward, the top layer of immobile sand is able to withstand the shear stress which eroded the top layers when the flow was started. This is due to the fact that the moving sand is transferring at least part of its weight to the bed as effective stresses and thereby increasing the effective normal stress in the bed. The effective nomal stress transferred by the moving sand is generally referred to as the dispersive stress.

Bagnold (1954,1956) studied the normal and tangential stresses in granular flows and suggested that they be given as functions of the shear rate du/dz and of the linear sediment concentration λ which is the relative surface proximity between sediment particles, see Figure 2.1.2. The linear sediment concentration is related to the volumetric concentration c by

$$\lambda = \frac{1}{(c_{max}/c)^{1/3} - 1} \tag{2.1.1}$$

where c_{max} is the maximum concentration corresponding to grain contact. The linear concentration λ increases drastically as c approaches c_{max}, see Figure 2.1.2.

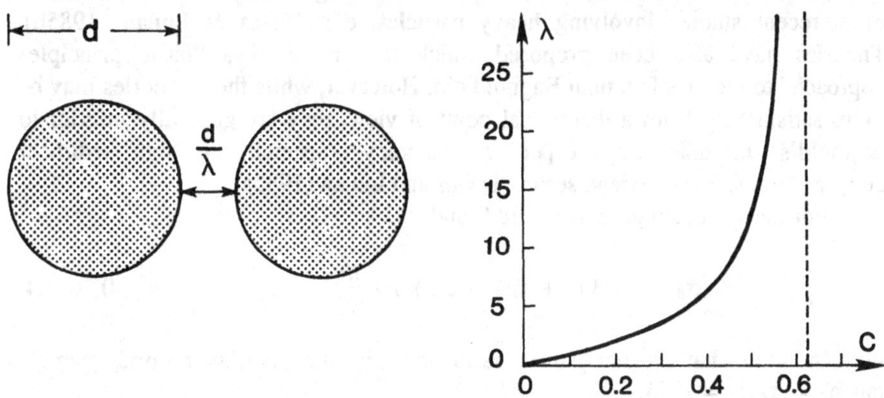

Figure 2.1.2: Bagnold's linear sediment concentration λ becomes infinite as the volumetric concentration c approaches its maximum value c_{max} which is generally of the order *0.65*.

Bagnold considered two different regimes in which different types of interactions dominate the behaviour of the fluid-grain-mixture. For small, light grains in a very viscous fluid the interactions are dominated by viscosity and Bagnold termed this the macro-viscous regime. For large, dense particles at high shear rates the interactions are dominated by particle collisions and this is called the inertial regime. The dimensionless parameter which separates the different regimes is

$$B = \frac{s\, d^2 \sqrt{\lambda}\, \frac{du}{dz}}{\nu} \tag{2.1.2}$$

which has been termed the *Bagnold number*. A purely inertial regime is found for $B > 450$, and a purely macro-viscous regime corresponds to $B < 40$.

In the inertial regime the stresses are proportional to the linear sediment concentration and to the shear rate squared. Bagnold obtained the following semi-empirical formula

$$\frac{\tau}{\tan\varphi_d} = \sigma_e = \frac{1}{25}\,\rho\, s\,(\lambda\, d\, \frac{du}{dz})^2 \qquad \text{for } B > 450 \tag{2.1.3}$$

where the dynamical angle of repose was found to be about 18 degrees corresponding to $\tan\varphi_d = \tau/\sigma_e = 0.32$. Bagnold's experimental results, which were obtained with neutrally bouyant particles have generally been confirmed by more recent studies involving heavy particles, e g Hanes & Inman (1985b). Theories have also been proposed which take more of a "basic principles approach" to the problem than Bagnold did. However, while these theories may be more satisfactory from a theoretical point of view, they are generally inferior to Bagnold's formulae with respect to accuracy, especially at high sediment concentrations. For a review, see e g Hanes and Inman (1985a).

For the viscous regime Bagnold found

$$\frac{\tau}{\tan\varphi_d} = \sigma_e = 1.3\,(1+\lambda)(1+\lambda/2)\,\rho\,\nu\,\frac{du}{dz} \qquad \text{for } B < 40 \quad (2.1.4)$$

with a somewhat greater stress ratio than in the inertial regime, namely $\tan\varphi_d = \tau/\sigma_e = 0.75$.

With typical values (s, d, λ, du/dz, ν) = $(2.65, 2{\cdot}10^{-4} m, 1, 100\ s^{-1}, 10^{-6}\ m^2/s)$, corresponding to $B = 10.6$, sheet flow conditions under waves will generally be in the macro viscous regime.

2.1.3 Fluid forces

The fluid forces on sediment particles are of two kinds. Namely surface drag forces and pressure forces resulting from pressure gradients in the fluid.

The total pressure force which is determined as the surface integral of the pressure is by Green's theorem equal to minus the volume integral of the pressure gradient, $\nabla p = (\frac{\partial p}{\partial x}, \frac{\partial p}{\partial z})$. Hence, for the situation in Figure 2.1.3 where the pressure along a vertical is hydrostatic and there is a constant horizontal pressure gradient, the total pressure force on the body is

$$F_p = \begin{pmatrix} -V\dfrac{\partial p}{\partial x} \\ -V\dfrac{\partial p}{\partial z} \end{pmatrix} = \begin{pmatrix} -V\dfrac{\partial p}{\partial x} \\ \rho g V \end{pmatrix} \qquad (2.1.5)$$

where the vertical component is the familiar bouyancy force corresponding to $\partial p/\partial z = -\rho g$. V is the particle volume.

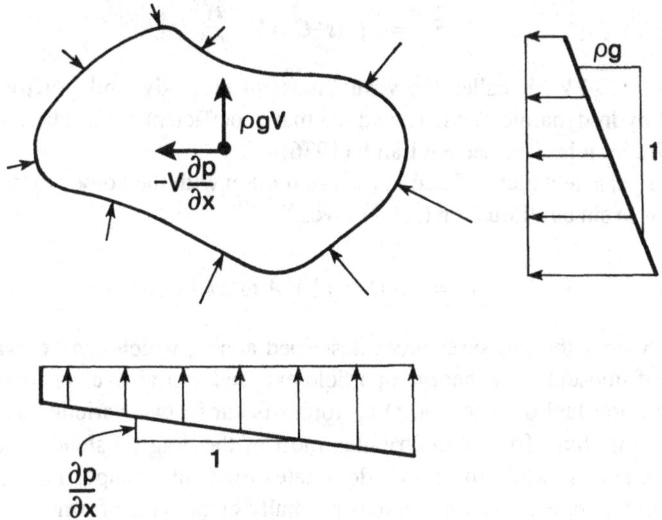

Figure 2.1.3: The total pressure force is, by Green's theorem, equal to the volume times the pressure gradient when the pressure gradient can be considered constant over the body.

If the shear stresses in the fluid are relatively small (and in particular for inviscid fluids), the equation of motion (1.1.10) may be used to write the horizontal pressure force in terms of the horizontal fluid acceleration

$$-V\frac{\partial p}{\partial x} = \rho V\frac{du}{dt} \tag{2.1.6}$$

When calculating the pressure force F_p on a body which is held fixed while the fluid is accelerating past it, an extra mass $\rho C_M V$ must be added corresponding to the volume of surrounding fluid which the body keeps from accelerating, so for a fixed body in a horizontally accelerated fluid we get

$$F_{p,x} = \rho\,(1+C_M)\,V\frac{du}{dt} \tag{2.1.7}$$

Correspondingly, the force required to give a sediment particle with volume V and relative density s the acceleration $\frac{du_s}{dt}$ through a resting fluid is

$$F = \rho\,(s+C_M)\,V\frac{du_s}{dt} \qquad (2.1.8)$$

where $\rho\,(s+C_M)\,V$ is called the virtual mass of the body, and $\rho C_M V$ is called the added hydrodynamic mass. The added mass coefficient for a sphere is *0.5*, for a long cylinder it is *1.0*, see e g Lamb (1936).

For a particle which is fixed in a wave motion with the homogeneous velocity field $u = A\,\omega \sin \omega t$ Equation (2.1.7) gives

$$F_P = \rho\,(1+C_M)\,V A\,\omega^2 \cos \omega t \qquad (2.1.9)$$

Apart from the pressure forces described above, which can be evaluated on the basis of inviscid flow theory, a particle exposed to a viscous or turbulent flow will in addition feel drag forces. Drag forces occur in two varieties: skin friction and form drag. Skin friction contributes most of the drag on slender, streamlined bodies like kajaks, while form drag dominates for plump shapes like spheres and most sediment grains. The drag force is normally given on the form

$$\boldsymbol{F_D} = \tfrac{1}{2}\rho\,A\,C_D\,|\boldsymbol{u}|\,\boldsymbol{u} \qquad (2.1.10)$$

where $A\ (\approx \pi\,d^2/4)$ is the cross sectional area facing the flow, and C_D is the drag coefficient which depends on the sediment shape and on the Reynolds number, $d\,|\boldsymbol{u}|/\nu$. When the flow becomes laminar the drag force is actually proportional to the flow velocity and, for small ($d\,|\boldsymbol{u}|/\nu < 1$) spherical particles, the drag force is given by Stokes' law

$$\boldsymbol{F_D} = 3\,\pi\,\rho\,\nu\,d\,\boldsymbol{u} \qquad (2.1.11)$$

which corresponds to a drag coefficient of

$$C_D = \frac{24}{d\,|\boldsymbol{u}|/\nu} \qquad (2.1.12)$$

Drag coefficients for spherical particles are plotted in Figure 4.2.1, page 164.

By comparing the expressions (2.1.9) and (2.1.10) for the pressure force and the drag force respectively on a spherical particle in a wave motion, we see that the force amplitude ratio

$$\frac{F_{P,\max}}{F_{D,\max}} = \frac{\rho\frac{\pi}{6}d^3(1+C_M)A\omega^2}{\frac{1}{2}\rho\frac{\pi}{4}d^2 C_D A^2\omega^2} = \frac{4}{3}\frac{1+C_M}{C_D}\frac{d}{A} \tag{2.1.13}$$

is proportional to d/A which is called the Keulegan Carpenter number.

A sand particle in a flat sand bed will on the average have to carry a drag force of magnitude $\tau_o d^2$ but the instantaneous force experienced by individual particles will be highly variable because no two particles will be equally exposed to the flow and because the flow at any point of the bed constantly changes with the formation of high velocity streaks and low velocity streaks, see Kline et al (1967) or Lian (1990).

A particle on the bed will also experience a lift force F_L which is due to the curvature of the stream lines in the flow over the top of it. Similarly to the force on a stone on a string which is Mu^2/r, the lift force is also proportional to the square of the fluid velocity and inversely proportional to the orbit radius which in this case is of the order of d. The force on the sediment particle with volume of the order d^3 then is

$$F_L = \rho C_L \frac{u^2}{d} d^3 = \rho C_L u^2 d^2 \tag{2.1.14}$$

Seepage or infiltration, i e a flow perpendicular to the sand surface may have a stabilising or destabilising effect on the sand because the vertical fluid drag changes the effective normal stress.

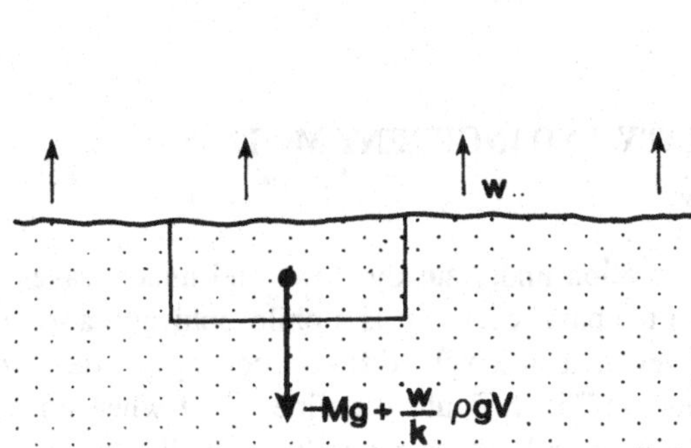

Figure 2.1.4: An outflow velocity w corresponds to a lift force of $\frac{w}{K}\rho g$ per unit volume, where K is the permeability.

This is the mechanism which causes the formation of quicksand. Ground water flows through sandy materials are generally described in terms of Darcy's Law

$$\mathbf{u} \;=\; -K\left(\frac{\nabla p - \rho \mathbf{g}}{\rho g}\right) \tag{2.1.15}$$

where K is the permeability which has the dimension of velocity, or

$$\begin{pmatrix} u \\ w \end{pmatrix} \;=\; -K\begin{pmatrix} \dfrac{1}{\rho g}\dfrac{\partial p}{\partial x} \\ \dfrac{1}{\rho g}\dfrac{\partial p}{\partial z}+1 \end{pmatrix} \tag{2.1.16}$$

The water velocity $\mathbf{u} = (u, w)$ is in this definition the equivalent clear water velocity, i e, the flow rate per unit area. The actual velocities of the water in the pores are somewhat greater.

For the situation in Figure 2.1.4, where the seepage rate corresponds to the vertical, equivalent clear water velocity w, the total vertical pressure gradient in the pore water must be

$$\frac{\partial p}{\partial z} \;=\; -\rho\, g\,(1+\frac{w}{K}) \tag{2.1.17}$$

corresponding to a bouyancy force of $\rho\, g\,(1+\frac{w}{K})$ per unit volume. Thus, in order to lift a sediment particle with density $s\,\rho$ a vertical outflow velocity (flow rate per unit area) of magnitude $(s - 1)K$ is required.

2.2 SEDIMENT MOBILITY AND INCIPIENT MOTION

2.2.1 Introduction

The initiation of sediment motion under steady flows and under waves has attracted considerable interest in the past because it is a philosophically appealing concept. In practical terms however, it is a very difficult concept to deal with. Firstly, because "initiation of motion" is difficult to define - is it when one in a thousand grains moves or when one in a hundred moves? Secondly, because the complicating variables in a natural situation is very large. For example, the sand bed is never left perfectly smooth from previous events. Relict bed forms will be present and initiation of motion will occur sooner near the crest of these bedforms due to local enhancement of the bed shear stresses.

In addition, biological activity will also complicate the micro-topography and excretions from animals may tend to glue the sand particles together. Nevertheless, we shall consider a few classical approaches to the description of incipient sediment motion in the following.

2.2.2 The mobility number

A simple, yet useful, dimensionless measure of the fluid forces on a sediment particle under waves is the mobility number ψ which will be defined in the following.

For sand size particles (*d~0.2 mm*) under waves with typical semi-excursions, A of the order $0.1m$ - $2m$, the Keulegan Carpenter number d/A is very small and hence the drag force will tend to dominate over the pressure force, see Equation (2.1.13), page 101.

Hence, the total disturbing force on a sand particle at the bed is approximately proportional to the square of the velocity amplitude $A\omega$, and the ratio between this disturbing force and the stabilising force due to gravity is reasonably described by the mobility number

$$\psi = \frac{(A\,\omega)^2}{(s-1)\,g\,d} \tag{2.2.1}$$

2.2.3 The Shields parameter

A different measure of the balance between disturbing and stabilising forces on sand grains at the bed was suggested by Shields (1936) in a study of the incipient sediment motion in steady flow,

$$\theta = \frac{\tau(o)}{\rho\,(s-1)\,g\,d} = \frac{u_*^{\,2}}{(s-1)\,g\,d} \tag{2.2.2}$$

Accordingly, θ is known as the Shields parameter.

This parameter is particularly convenient to use in connection with steady flow because there, the steady bed shear stress, $\overline{\tau}(o)$ and hence the friction velocity $\overline{u_*}$ are quantities which are easily measured, $\overline{\tau}(o) = \rho\, g\, D\, I$, where D is the flow depth and I is the hydraulic gradient.

In connection with wave motion, the Shields parameter (corresponding to total stress) is generally defined in terms of the peak bed shear stress $\hat{\tau}$

$$\theta = \frac{\hat{\tau}}{\rho\,(s-1)\,g\,d} = \frac{\frac{1}{2} f_w\,(A\,\omega)^2}{(s-1)\,g\,d} = \frac{1}{2} f_w\,\psi \qquad (2.2.3)$$

where f_w is the wave friction factor, defined on page 23, and this is the notation which will be used in the following.

2.2.4 Skin friction

The total bed shear stress τ may be seen as consisting of two contributions namely, the form drag τ'' and the skin friction τ'. The significance of each of these for the sediment transport is quite different as described by Engelund & Hansen (1972).

The form drag is generated by the difference in pressures between the upstream and the downstream sides of bedforms, and it does not directly affect the stability of individual surface sediment particles. The main disturbing influence to the surface grains is generally considered to come from the skin friction τ'. The corresponding skin friction Shields parameter,

$$\theta' = \frac{\tau'}{\rho\,(s-1)\,g\,d} \qquad (2.2.4)$$

is therefore frequently used to predict initiation of motion and the magnitude of moving sediment concentrations.

Correspondingly, τ' is often referred to as the *effective stress* in connection with sediment transport.

A comprehensive discussion of the concept of effective stress in steady flows is given by Engelund & Hansen (1972).

If the bed is flat, the form drag is absent so, from that point of view, $\tau' = \tau$ and $\theta' = \theta$.

There is, however, some indication that the total shear stresses on flat sand beds under waves may not be totally effective with respect to transporting sediment, see Section 2.4.4, page 121.

2.2.5 The grain roughness Shields parameter.

Flat beds of loose sand under waves as well as under steady flow may offer considerably more resistance to the flow than sand paper with the same grain size.

This is a consequence of the momentum transfer by moving sand from the flow to the bed, see Section 3.6, page 145.

It is, however, difficult to estimate this momentum transfer and the amount of related data is very limited. So a generally accepted method for calculating the skin friction on a bed of highly mobile sand under waves is not yet available (an empirical formula will be presented in Section 3.6.6, p 155 ff).

On the other hand, the mobility number ψ is not a totally adequate measure of the sediment mobility because it neglects the dependence on the ratio d/A of the force exerted by waves on sediment particles, see for example Figure 3.4.6, page 141.

It was therefore suggested by Madsen & Grant (1976) that the sediment mobility be estimated in terms of a grain roughness Shields parameter, and this approach has since been quite popular.

Following Engelund & Hansen (1972) and Nielsen (1979), we shall adopt the value $2.5d_{50}$ for the *grain roughness* of a flat bed of sand with median size d_{50}, and correspondingly operate with a grain roughness Shields parameter $\theta_{2.5}$ defined by

$$\theta_{2.5} = \frac{\frac{1}{2} f_{2.5}\, \rho\, (A\,\omega)^2}{\rho\,(s-1)\, g\, d} = \frac{1}{2} f_{2.5}\, \psi \tag{2.2.5}$$

where the special grain roughness friction factor, $f_{2.5}$ is based Swart's (1974) formula (1.2.22) and a roughness of $2.5d_{50}$

$$f_{2.5} = \exp\left[5.213 \left(\frac{2.5d_{50}}{A}\right)^{0.194} - 5.977\right] \tag{2.2.6}$$

The relationship between $\theta_{2.5}$ and θ for both rippled and flat sand beds in oscillatory flows is illustrated by the data of Carstens et al (1969) and of Lofquist (1986) in Figure 2.2.1.

The data show that for rippled beds θ is generally an order of magnitude greater than $\theta_{2.5}$ with no systematic trend between different grain sizes.

For flat beds, θ (= θ') is also considerably larger than $\theta_{2.5}$, by a factor five or so, when the activity level is high ($\theta_{2.5} \gtrsim 0.3$). For a flat bed at low activity level, one would expect to find $\theta = \theta' \approx \theta_{2.5}$.

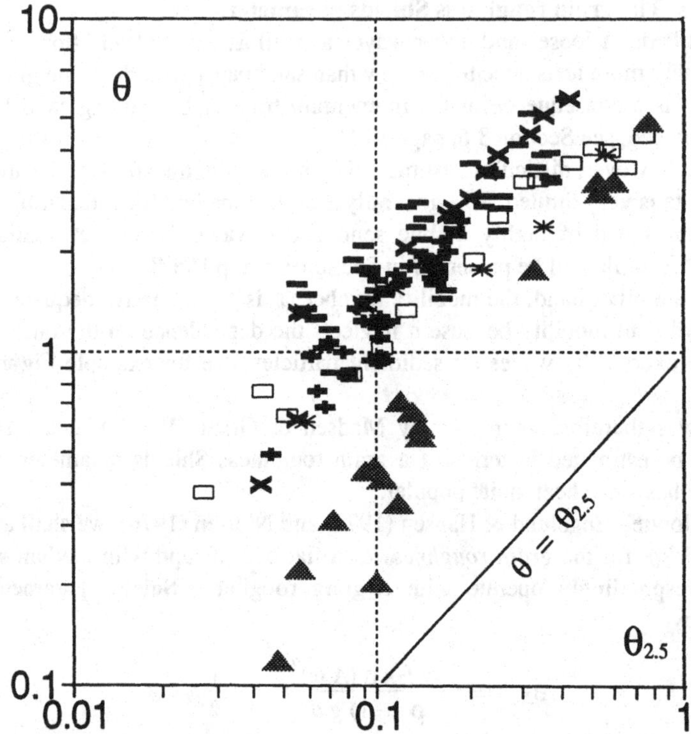

Figure 2.2.1: **Relationship between the "total Shields parameter" θ and the grain roughness Shields parameter $\theta_{2.5}$ for the presently available oscillatory flow data. Legend *bar* : Lofquist *0.55mm* sand, + : Lofquist *0.18mm*, * : Carstens et al *0.19mm*, *rectangle* : Carstens et al 0.30mm, X : Carstens et al *0.59mm*. All of the above correspond to rippled beds, while the *triangles* correspond to flat beds, Carstens et al *0.19mm* and *0.30mm*.**

While the data of Carstens et al and of Lofquist consistently show high energy dissipation rates over flat sand beds, these values do not immediately agree with the available information on corresponding sediment transport rates.

Thus, while the available friction/energy-dissipation data indicate that the total bed friction and the energy dissipation on a flat sand bed correspond to a stress five times as large as $\tau_{2.5}$, a different picture is presented by the sediment transport data examined in Section 2.4, page 116. They indicate that the effective stress for moving sediment over a flat bed under waves is of the order $\tau_{2.5}$, rather than $5\tau_{2.5}$. The writer knows of no satisfactory explanation for this at the present time. Wave friction factors and the corresponding hydraulic roughness of sand beds under waves are discussed in detail in Section 3.6, page 145.

2.2.6 The critical Shields parameter and the Shields diagram

The *critical Shields parameter* θ_c is the effective Shields parameter (θ') at which sediment movement starts.

The value of θ_c is a function of the sediment size and density, of the fluid density and viscosity, and of the flow structure. Typical θ_c-values for sand in water are of the order *0.05*. For sand in air, they are somewhat lower, usually in the range $0.01<\theta_c<0.02$, see Allen (1982). In both air and water, θ_c becomes much larger in the silt range of grain sizes ($d<0.063mm$).

To obtain a simple description of the behaviour of θ_c, Shields (1936) noted that both the drag force F_D and the lift force F_L, on a bed sediment particle, are proportional to u_*^2, and to functions of the *grain Reynolds number*, $u_* d/\nu$. He therefore plotted observed values of θ_c, against the the grain Reynolds number. The resulting diagram is called the *Shields diagram.*

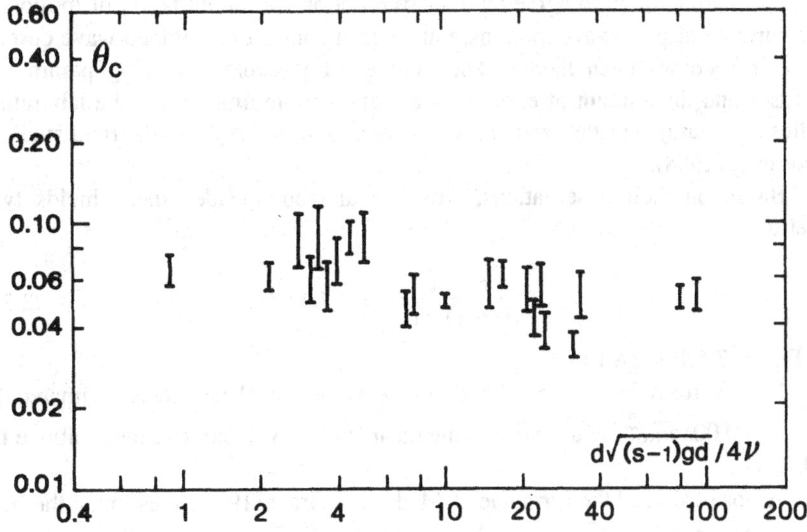

Figure 2.2.2: The Madsen-Grant diagram for the initiation of sediment motion in oscillatory flow and on oscillating trays. Data from Manohar (1955) and from Carstens et al (1969).

A slightly different diagram was suggested by Madsen & Grant (1976). Instead of the grain Reynolds number they used $d\sqrt{(s-1)gd}/4\nu$ as the abscissa. An example of such a diagram is shown in Figure 2.2.2.

For typical beach sand with $d = 0.2mm$ and $s = 2.65$, the value of $d\sqrt{(s-1)gd}/4\nu$ is approximately 11. Around this value the data show no significant trend. Hence, the use of an all round value of

$$\theta_c \approx 0.05 \tag{2.2.7}$$

for the critical Shields parameter seems justified in most practical cases.

The effective shear stress τ_c which corresponds to θ_c $[\tau_c = \rho\,(s-1)\,g\,d\,\theta_c]$ is called the *critical shear stress* for the particular sediment.

The *Shields criterion* $\theta_c = \theta_c(u_*d/\nu)$ is but one of many criteria for the initiation of sediment motion. Many others have been suggested. For a comprehensive review, see Hallermeier (1980).

2.2.7 Initiation of motion in combined wave-current flows.

It is hard enough to agree on a sharp criterion for the initiation of motion in pure currents or pure wave motions, but when it comes to combined wave current flows, it becomes even harder. The number of possible governing parameters increases and the amount of experimental data is more limited, e g the laboratory studies of Natarajan (1969) and Hammond & Collins (1979), and the field study of Amos et al (1988).

Based on their observations, Amos et al recommended the "Shields type criterion"

$$\frac{\overline{\tau} + \hat{\tau}}{\rho\,(s-1)\,g\,d} = 0.04 \tag{2.2.8}$$

see Figure 3.5.1, page 144.

In this formula $\overline{\tau}$ is the time-averaged bed shear stress estimated by $\overline{\tau} = \frac{1}{2}\rho\, 0.003\, \overline{u_{100}}^2$ and $\overline{u_{100}}$ is the mean current velocity one metre above the bed.

Amos et al used the formulae of Madsen & Grant (1976) to estimate the peak wave-induced bed shear stress $\hat{\tau}$, but very similar values are obtained from Equations (1.2.18) and (1.2.22) with $r = 2.5d_{50}$.

2.2.8 The depth of closure

A related concept to the initiation of motion is the depth of closure on a beach profile. The depth of closure is the depth beyond which sand level changes between seasonal surveys become unmeasurable or insignificant.

The depth of closure is a prerequisite for the use of the Bruun rule (Bruun 1962) which provides a simple estimate of the shoreline recession in response to sea level rise under a number of simplifying assumptions, see e g Bruun (1983, 1990).

Practical estimates of the depth of closure are of the order *3.5* times the annual maximum significant wave height. For a review of the concept and related formulae see Hallermeier (1981).

2.3 STEADY BED-LOAD AND SHEET-FLOW

2.3.1 Introduction

The following section summarises current experimental facts about steady bed-load transport, mainly in the light of the theory of Bagnold (1956), and with emphasis on concepts and formulae which are transferable to oscillatory flows and combined wave current flows.

Most of the steady bed-load formulae are of the form $\Phi = \Phi(\theta')$. That is, they state direct relationships between effective bed shear stresses and dimensionless transport rates without considering underlying details such as, the amount of moving sediment, and the typical speed with which that sediment is moving.

It is necessary, however, to know about these details in order to adapt steady flow models to unsteady flow situations. Sections 2.3.3 and 2.3.5 therefore attempt to extract information about the amount of bed-load and about the typical speed with which the bed-load moves in steady flows.

2.3.2 What is bed-load ?

The total load of moving sediment is generally seen as composed of three parts: the wash-load, the suspended load and the bed-load.

The bed-load has been defined in different ways depending on the context. In relation to measurements, it is often defined as that part of the total load which travels below a certain level or (very pragmatically) as the part which gets caught in bed-load traps. For modelling purposes however, it is more convenient to apply the definition of Bagnold (1956).

Bagnold defines the bed-load as that part of the total load which is supported by intergranular forces. The rest, i e, the suspended load and the wash load are supported by fluid drag.

Obviously a given grain may well be supported partly by intergranular forces and partly by fluid drag and hence contribute to both the suspended load and the bed-load. This makes the bed-load practically unmeasureable in situations where suspension is present as well, and this is of course of some concern. We shall however stay with Bagnold's definition in order to make use of the advantages it gives with respect to rational discussion and modelling.

2.3.3 The amount of bed-load

Making use of Bagnold's definition of bed-load, it is fairly easy to estimate the weight of material which will be moved as bed-load under a certain effective stress τ'.

The bed-load must, due to its immersed weight, deliver an effective normal stress

$$\sigma_e = \rho\,(s-1)\,g \int_0^\infty c_B(z)\,dz \qquad (2.3.1)$$

onto the top-most layer of the immobile bed; $c_B(z)$ is the volumetric concentration of bed-load.

Assuming then, that the yield criterion for the top layer of immobile grains is

$$\tau_{max} = \tau_c + \sigma_e \tan\varphi_s \qquad (2.3.2)$$

see Figure 2.3.1, we see that the amount of bed-load which is in equilibrium with τ' is given by

$$\int_0^\infty c_B(z)\,dz = \frac{\tau' - \tau_c}{\rho\,(s-1)\,g\,\tan\varphi_s} \qquad (2.3.3)$$

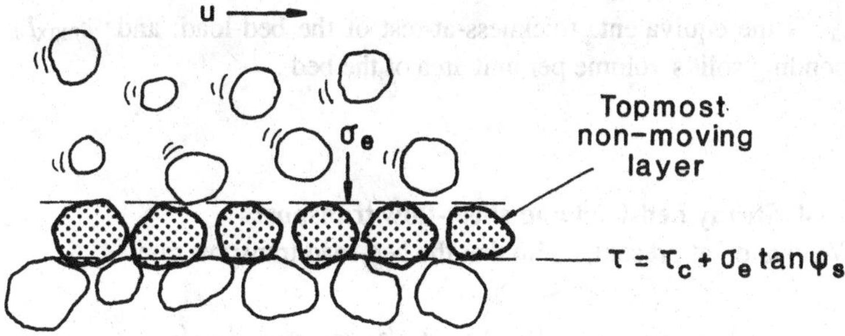

Figure 2.3.1: For the equilibrium bed-load transport rate, the dispersive stress σ_e must satisfy the yield criterion $\tau = \tau_c + \sigma_e \tan\varphi_s$.

Here it is convenient to introduce the maximum concentration c_{max}, which is the volumetric concentration of solid sediment in the immobile bed. In terms of c_{max}, the vertical scale of the bed-load distribution is then defined by

$$L_B = \frac{1}{c_{max}} \int_0^\infty c_B(z)\, dz \qquad (2.3.4)$$

Introducing this expression for L_B into Equation (2.3.3) we see that the vertical distribution scale measured in grain diameters is

$$\frac{L_B}{d} = \frac{\theta' - \theta_c}{c_{max}\ \tan\varphi_s} \qquad (2.3.5)$$

Bagnold (1956) gave $\tan\varphi_s = 0.63$ as a typical value for fairly rounded grains corresponding to a maximum concentration of the same value, i e $c_{max} = 0.63$ [vol/vol], and he noted that the product $c_{max}\ \tan\varphi_s$ is fairly constant at about 0.4 for different grain shapes.

Hence, as rules of thumb we have

$$L_B = 2.5\,(\theta' - \theta_c)\, d \qquad (2.3.6)$$

and

$$c_{max} L_B = 2.5\,(\theta' - \theta_c)\, d\, c_{max} \tag{2.3.7}$$

L_B is the equivalent thickness-at-rest of the bed-load, and $c_{max}L_B$ is the corresponding solids volume per unit area of the bed .

2.3.4 Steady bed-load and sheet-flow transport

We cannot, at present, claim that the bed-load transport rate

$$Q_B = \int_0^{\infty} c_B(z)\, u_S(z)\, dz = c_{max}\, L_B\, U_B \tag{2.3.8}$$

is well understood because neither the sediment velocity distribution $u_S(z)$ nor the concentration distribution $c_B(z)$ through "the bed-load layer" are well understood. We can, however, still predict Q_B empirically with reasonable confidence for steady flow because it has been measured directly in a large number of experiments. Most of the data support transport formulae of the form

$$\Phi_B \propto (\theta' - \theta_c)\sqrt{\theta'} \tag{2.3.9}$$

see Figure 2.3.2. The dimensionless flux Φ_B is defined by

$$\Phi_B = \frac{Q_B}{d\sqrt{(s-1)\, g\, d}} \tag{2.3.10}$$

The plotted Φ-values correspond to the total, measured sediment transport, but the contribution from suspended load amounted only to a small fraction. Engelund (1981) estimated at most 20%, even for Wilson's high-stress data. The general trend of the data is closely matched by the classical bed-load formula

$$\Phi_B = 8\,(\theta' - \theta_c)^{1.5} \tag{2.3.11}$$

of Meyer-Peter & Muller (1948), except that for high stress values ($\theta > 1$), the numerical constant *8* is too small and a value of about *12* seems more appropriate. The slightly different formula

$$\Phi = 12\,(\theta' - \theta_c)\sqrt{\theta'} \tag{2.3.12}$$

also represents a reasonable approximation to the data.

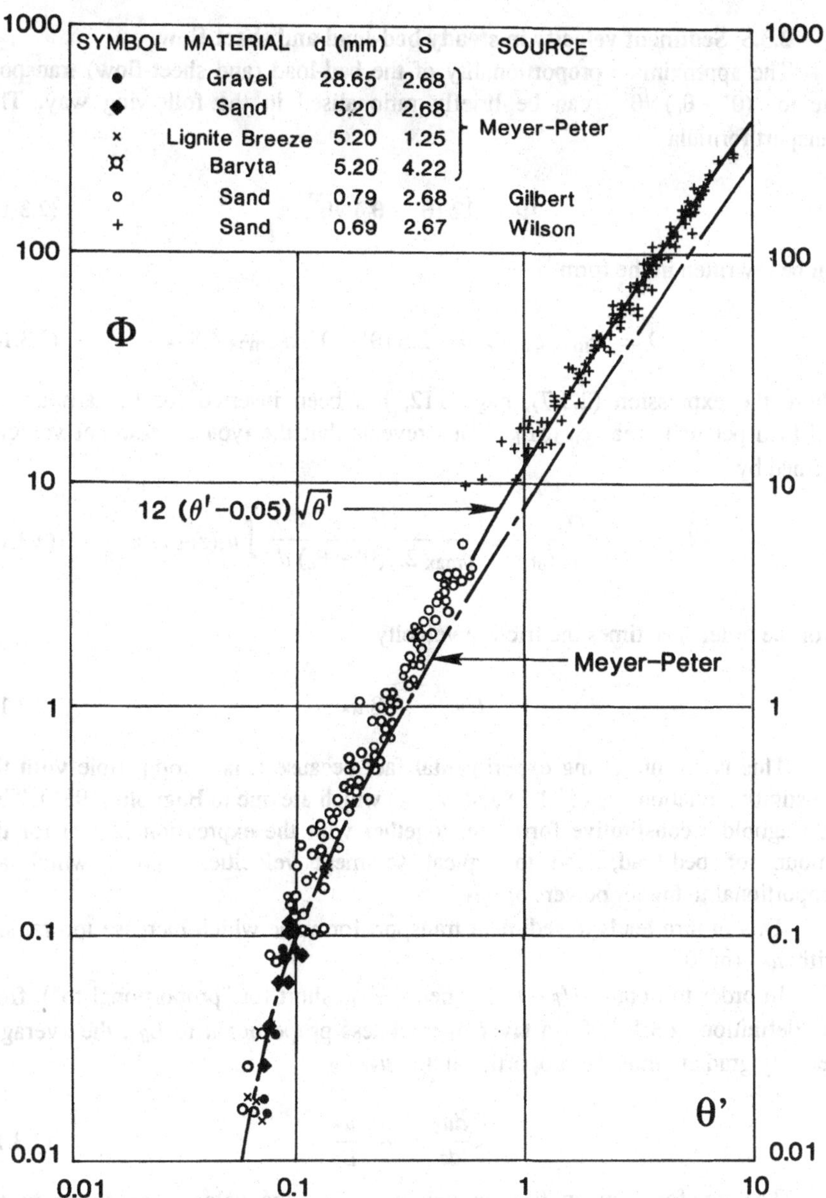

Figure 2.3.2: Total sediment transport rate under steady flows over flat beds. Due to the fact that the sediment was fairly coarse (0.7mm), suspended load contributed little to the transport rates for the shown "high stress data" ($\theta' > 0.5$).

2.3.5 Sediment velocity in steady bed-load and sheet-flow

The approximate proportionality of the bed-load (and sheet-flow) transport rate to $(\theta' - \theta_c)\sqrt{\theta'}$ can be briefly rationalised in the following way. The transport formula

$$\Phi = 12\,(\theta' - \theta_c)\sqrt{\theta'} \qquad (2.3.12)$$

can be rewritten in the form

$$Q = c_{max}\, L_B\, U_B \approx 2.5\,(\theta' - \theta_c)\, d\, c_{max}\, 4.8 u_* \qquad (2.3.13)$$

where the expression (2.3.7), page 112, has been inserted for the amount of bed-load per unit area $c_{max} L_B$. This reveals that the typical sediment velocity defined by

$$U_B = \frac{Q_B}{c_{max}\, L_B} = \frac{1}{c_{max}\, 2.5(\theta' - \theta_c)\, d} \int_0^\infty u_s(z)\, c_B\, dz \qquad (2.3.14)$$

is of the order *4.8* times the friction velocity

$$U_B \approx 4.8\, u_* \qquad (2.3.15)$$

This is an interesting experimental fact because it is incompatible with the constitutive relationships (2.1.3) and (2.1.4) which are due to Bagnold (1956). That is, Bagnold's constitutive formulae, together with the expression (2.3.7) for the amount of bed-load, lead to typical sediment velocities U_B which are proportional to higher powers of u_*.

This in turn leads to sediment transport formulae which increase too rapidly with u_* (or θ').

In order to obtain $U_B \sim u_*$ (where "~" is short for "proportional to"), from the definition (2.3.14), for a layer of thickness proportional to L_B , the averaged velocity gradient must be proportional to u_*/L_B :

$$\left\langle \frac{du_s}{dz} \right\rangle \sim \frac{u_*}{L_B} \qquad (2.3.16)$$

That is, for a given friction velocity u_*, the velocity gradients in the bed-load layer are inversely proportional to the weight of bed-load. This weight being proportional to L_B.

From the form of Equation (2.3.16), it can be seen that, if the constitutive relationship for the granular flow is written in terms of an effective viscosity, ν_e

$$\tau/\rho \;=\; u_*^2 \;=\; \nu_e < \frac{du_s}{dz} >$$

this viscosity must be proportional to $u_* L_B$

$$\nu_e \;\sim\; u_* L_B \qquad (2.3.17)$$

As mentioned above the result (2.3.15) for the typical sediment velocity and the corresponding result (2.3.16) for the averaged velocity gradient disagree with Bagnold's constitutive relationships from Section 2.1.2. Thus, Bagnold's formula

$$\tau/\rho \;=\; u_*^2 \;=\; 0.013\, s\, (\lambda\, d \frac{du}{dz})^2 \qquad \text{for } \; B > 450 \qquad (2.3.18)$$

obviously corresponds to

$$< \frac{du_s}{dz} > \;\sim\; u_* \;\sim\; \theta'^{\frac{1}{2}}$$

which is in conflict with Equation (2.3.16), and with $L_B \sim \theta'$ it leads to

$$Q_B \;\sim\; c_{\max} L_B U_B \;=\; c_{\max} L_B (L_B < \frac{du_s}{dz} >) \;\sim\; \theta'^{\frac{5}{2}} \qquad (2.3.19)$$

A model of this type was developed and discussed in detail by Hanes & Bowen (1985).

If Bagnold's formula (2.1.4) for the hyper-viscous regime is used instead of (2.3.18) above, the result for the transport rate is

$$Q_B \;\sim\; \theta'^3 \qquad (2.3.20)$$

which is even further from the experimental trend of the data in Figure 2.3.2.

If data with fine sand at high effective stresses ($\theta' \gtrsim 1$) are included in a plot like Figure 2.3.2, they may indicate a more rapid increase of the total transport rate than $\theta'^{1.5}$. This is however, most likely due to the rapidly increasing rate of suspended transport, and that is another matter.

The discussion above is only related to the constitutive relationships for the bed-load layer.

2.4 BED-LOAD AND SHEET-FLOW UNDER WAVES

2.4.1 Introduction

The following section describes the processes of bed-load and sheet-flow sediment transport under waves. Or in other words, the quasi-steady processes of sediment transport which occur under waves over effectively flat sand beds.

Under sheet-flow conditions, the sediment transport rate is not entirely due to bed-load transport in Bagnold's sense, see Section 2.3.2. However, the assumption of quasi-steadiness may be applied to the total transport rate under certain conditions. That is, when the thickness of the sediment transporting layer is small compared to the typical distance $w_o T$ settled by a suspended sediment particle during one wave period.

Such conditions can be observed at moderate flow intensities ($\theta_{2.5} \lesssim 1$) over artificially flattened beds before vortex ripples form, and for high flow intensities where vortex ripples are naturally absent, see Section 3.4, page 135 ff.

The approach taken is essentially to adapt the steady flow sediment transport models from the previous section to oscillatory flows and to combined wave-current flows through quasi-steady considerations.

The quasi-steady approach is used because, it seems fairly reasonable to consider the processes of bed-load and sheet-flow sediment transport as quasi steady. There are however two main problems with this transfer of technology from steady flows to oscillatory and combined flows.

Firstly, steady flow transport formulae are generally of the form $\Phi = \Phi(\theta')$, i e, they assume knowledge of the effective (skin friction) stress θ', and our knowledge about the effective Shields parameter θ' for waves over sand beds is sparse. See Figure 2.2.1, page 106, and Section 3.6, page 145 ff.

The most commonly attempted way around this problem is to replace θ' by the more "calculable" $\theta_{2.5}$ and we shall see that this approach is quite justifiable in the sense that it leads to good predictions.

It should be kept in mind, however, that $\theta_{2.5}$ and θ' are conceptually different as discussed in Sections 2.2.4 and 2.2.5.

The second main problem with transferring steady flow sediment transport formulae to coastal conditions, is to deal with those effects of unsteadiness which cannot be ignored.

This problem is addressed in detail in Section 2.4.4, page 121, and the conclusion which is reached is in essence as follows.

It may well be possible to model bed-load transport over flat beds under waves by quasi-steady expressions of the form $Q(t) = Q(\tau[t])$, but it is not, in general, possible to use expressions of the form $Q(t) = Q(u_\infty[t])$.

The reason is that the bed shear stress which occurs at a certain time is not just a function of the instantaneous free stream velocity, but depends strongly on whether this u_∞ has been achieved through rapid or very gradual acceleration.

2.4.2 The amount of bed-load under waves

The amount of bed-load, quantified by the maximum value $L_{B,\max}$ of the equivalent thickness at rest L_B, was measured by Sawamoto & Yamashita (1986) in an oscillating water tunnel. Their data are plotted in Figure 2.4.1 versus the grain roughness Shields parameter $\theta_{2.5}$.

The line corresponds to the steady flow formula (2.3.6) with θ' replaced by $\theta_{2.5}$:

$$L_{B,\max} = 2.5\,(\theta_{2.5} - 0.05)\,d \tag{2.4.1}$$

The remarkably close agreement with the directly adapted steady flow formula is unexpected because the data in Figure 2.2.1, page 106, indicate that the total Shields parameter on a flat sand bed is generally much greater than $\theta_{2.5}$. If the data in Figure 2.2.1 are to be trusted, in the sense that measured energy dissipation rates are true indications of the effective sediment transporting stress, this effective stress should be of the order $5\,\theta_{2.5}$ when $\theta_{2.5} \approx 1$. Hence the expected magnitude of $L_{B,\max}$, based on Equation (2.3.7) and assuming "perfect quasi-steadiness", would be $L_{B,\max} \approx 2.5\,\theta' d = 2.5\,(5\,\theta_{2.5})\,d = 12.5\,\theta_{2.5}\,d$. That is however, about five times more than what was observed.

The writer can offer no clear explanation for this at present, but the observed amounts can of course still be reasonably predicted by Equation (2.4.1), or maybe, a fine-tuned version of it, using a power of about *3/4* rather than *1* as suggested by Sawamoto and Yamashita.

2.4.3 Bed-load and sheet-flow under sine waves

We saw in the previous section that the amount of bed-load moving under a sine wave can be predicted by the directly adapted steady flow formula Equation (2.4.1). That is, by a steady flow formula with θ' replaced b $\theta_{2.5}$ and, possibly,

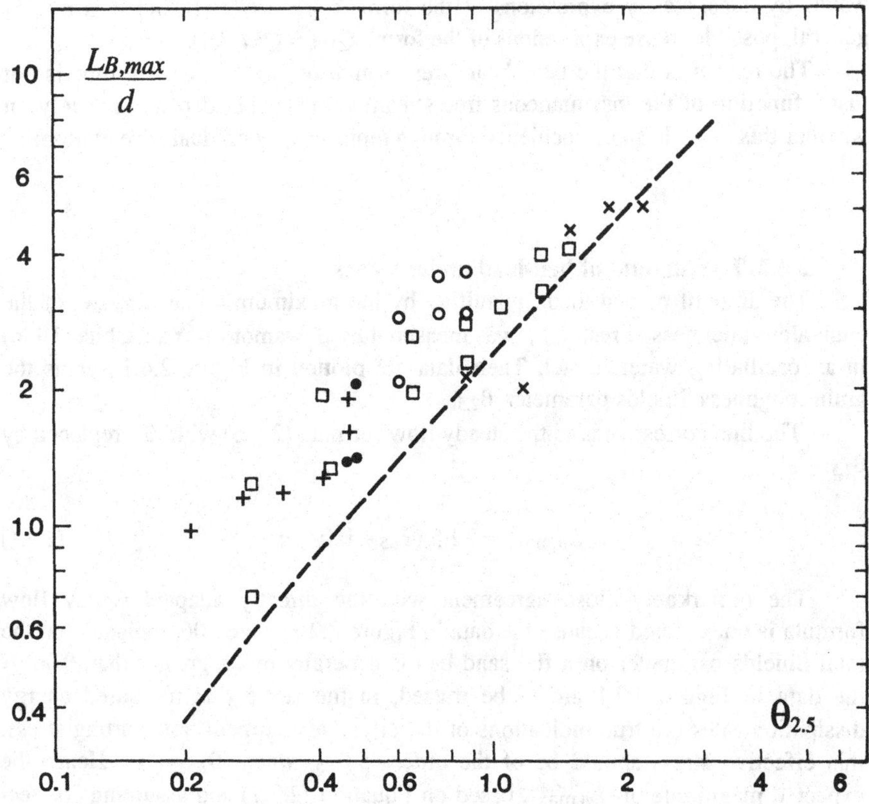

Figure 2.4.1: Measured values of the peak amount of bedload on a flat bed under "sine waves" in an oscillating water tunnel. The data are from Sawamoto & Yamashita (1986) and include relative sediment densities in the range $1.58<s<2.65$, and grain sizes in the range $0.2mm<d<1.6mm$. The curve corresponds to Equation (2.4.1).

with slightly adjusted numerical coefficients. Similarly, we shall see in the following, that the average sediment transport rate under half a sine wave can be estimated by a directly adapted version of Equation (2.3.11) or Equation (2.3.12).

Due to its symmetry, a complete sine wave must of course produce zero net sediment transport, but the average transport rate $Q_{T/2}$ during half a sine wave is of some theoretical interest. The corresponding, dimensionless $\Phi_{T/2}$ is plotted against $\theta_{2.5}$ in Figure 2.4.2.

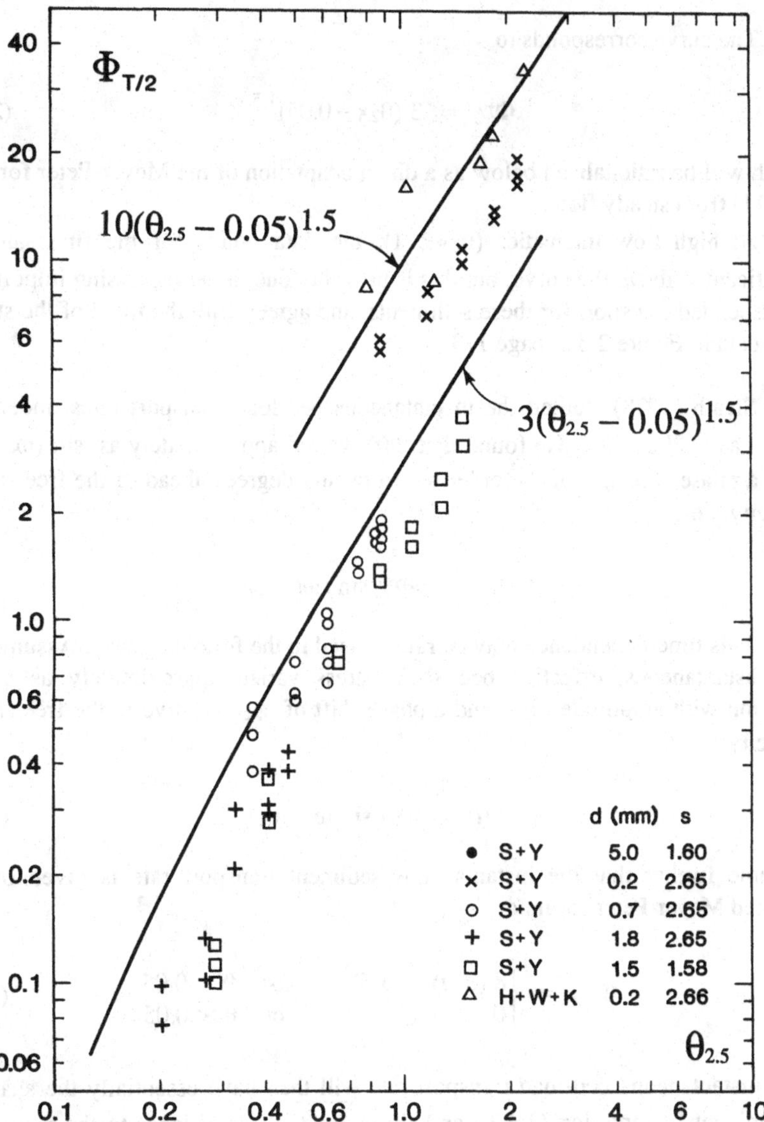

Figure 2.4.2: Averaged, dimensionless sediment transport rates under half sine waves as function of $\theta_{2.5}$. Data from Sleath (1978), Horikawa et al (1982), Sawamoto & Yamashita (1986) and King (1991). Sediment densities were in the range $1.14<s<2.66$, grain sizes in the range $0.135mm<d<4.24mm$ and periods in the range $0.5s<T<12s$. The crosses correspond to $d<0.25mm$, the squares to $d>0.25mm$.

The curve corresponds to

$$\Phi_{T/2} = 3\,(\theta_{2.5} - 0.05)^{1.5} \tag{2.4.2}$$

which will be rationalised below as a direct adaptation of the Meyer-Peter formula (2.3.11) from steady flow.

At high flow intensities ($\theta_{2.5} \gtrsim 1$), the data points for the fine sand lie significantly above the curve, but this is probably due to the increasing importance of suspended transport for these sediments, and agrees with the trend of the steady flow data in Figure 2.3.2, page 113.

Sleath (1978) studied the instantaneous bed-load transport rates under sine waves as well as $\Phi_{T/2}$. He found that $\Phi(t)$ varied approximately as $\sin^4(\omega t + \varphi_B)$ with a phase shift φ_B of the order ten to twenty degrees ahead of the free stream velocity, i e

$$\Phi(t) = \frac{8}{3}\Phi_{T/2}\sin^4(\omega t + \varphi_B) \tag{2.4.3}$$

This time-dependence may be rationalised in the following way. Assume that the instantaneous, effective bed shear stress varied approximately as a sine function with amplitude $\theta_{2.5}$ and a phase shift of φ_τ relative to the free stream velocity:

$$\theta'(t) = \theta_{2.5}\sin(\omega t + \varphi_\tau) \tag{2.4.4}$$

Assume further that the instantaneous sediment transport rate is given by the adapted Meyer-Peter formula

$$\Phi(t) = \begin{cases} 8\,(\theta'(t) - 0.05)^{1.5} & \text{for } \theta' > 0.05 \\ 0 & \text{for } \theta' < 0.05 \end{cases} \tag{2.4.5}$$

The instantaneous bed-load transport rate will then have essentially the shape of Sleath's $\sin^4$-expression (2.4.3), and since $\overline{\sin^4 t} = 3/8$ it leads to the expression (2.4.2) for the averaged transport rate.

We note that since Equation (2.4.5) includes no phase lag between $\theta'(t)$ and $\Phi(t)$, we get $\varphi_B = \varphi_\tau$. In reality however, it would be reasonable to expect some time lag of $\Phi(t)$ behind $\theta'(t)$ and hence, in general $\varphi_B < \varphi_\tau$.

2.4.4 Bed-load and sheet-flow under skew waves.

In order to model bed-load transport under irregular waves of arbitrary shapes (arbitrary $u_\infty(t)$) it seems reasonable, at this stage, to make use of the assumption of quasi steadiness in some form.

The simplest quasi-steady transport models may be written in the form

$$Q_B(t) \;=\; F(u_\infty[t]) \tag{2.4.6}$$

expressing that the bed-load/sheet-flow transport rate is at all times determined by the instantaneous velocity above the boundary layer.

Several models of this type, have been suggested, e g by Bailard (1981) and by Ribberink & al Salem (1990), and they lead to transport formulae of the type

$$\overline{Q_B} \;=\; Const \cdot \overline{|u_\infty(t)|^n \, u_\infty(t)} \tag{2.4.7}$$

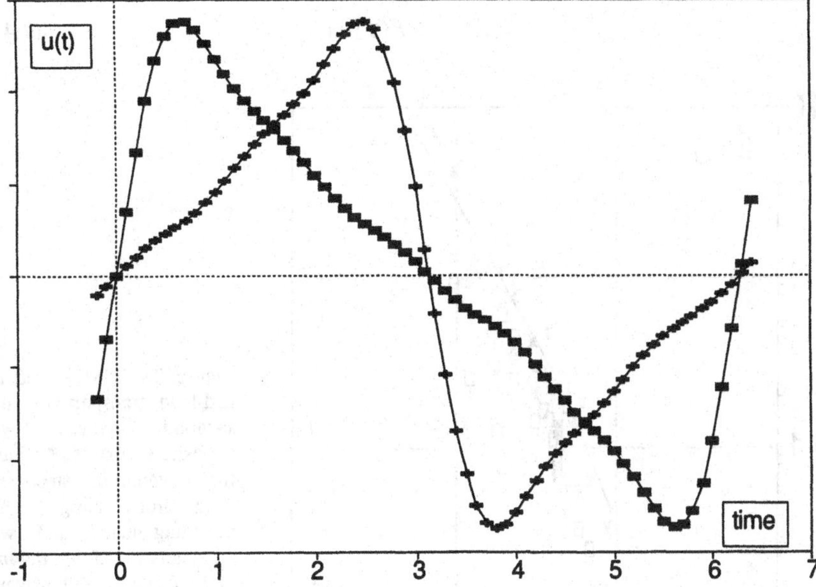

Figure 2.4.3: Sawtooth waves with zero mean flow and opposite asymmetry. These waves would both generate zero net bed-load transport according to formulae of the type (2.4.7).

While such formulae have the advantage of simplicity, they have the disadvantage of not being able to model the net sediment transport rate which results from certain types of wave asymmetry.

That is, asymmetry which are described essentially by the velocity moments such as that of a 2nd order Stokes wave can be represented, while "sawtooth asymmetry" is overlooked. Thus, the two "opposite" sawtooth waves in Figure 2.4.3 would lead to exactly the same transport rate according to formulae of the type (2.4.7).

This is important to note, because several experiments, e g those of King (1991), clearly show that the sawtooth wave with the steep front tends to generate shoreward bed-load transport while the one with the steep rear generates seaward bed-load transport. See Figure 2.4.4.

King's observations show that sawtooth skewness matters, and hence that the approach which leads to bed-load transport formulae of the type (2.4.7) is too simplistic. The "next one up" from Equation (2.4.6) as a starting point is to assume that the instantaneous transport rate is determined by the instantaneous effective bed shear stress, i e

$$Q_B(t) = F(\tau'[o,t]) \tag{2.4.8}$$

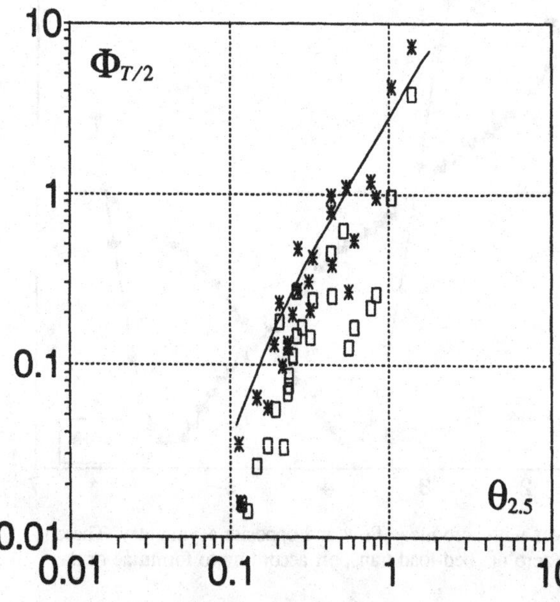

Figure 2.4.4: Measured net bed-load transport for half sawtooth waves with opposite steepness. * : Steep front, *rectangle* : steep rear. Data from King (1991) including quartz sand with diameters *0.135mm, 0.44mm and 1.1mm*, and "wave" periods in the range *3s<T<11s*.

where the findings in Sections 2.4.2 and 2.4.3, page 117 ff, seem to indicate that the effective stress τ' can be calculated as the grain roughness stress $\tau_{2.5}$.

Basing the transport rate on $\tau'(o,t)$ instead of $u_\infty(t)$ gives the potential of capturing the effect of sawtooth asymmetry because the bed shear stress under a sawtooth wave is likely to have the shape shown in Figure 2.4.5, with higher (absolute) stress values under the rapidly accelerated half cycle.

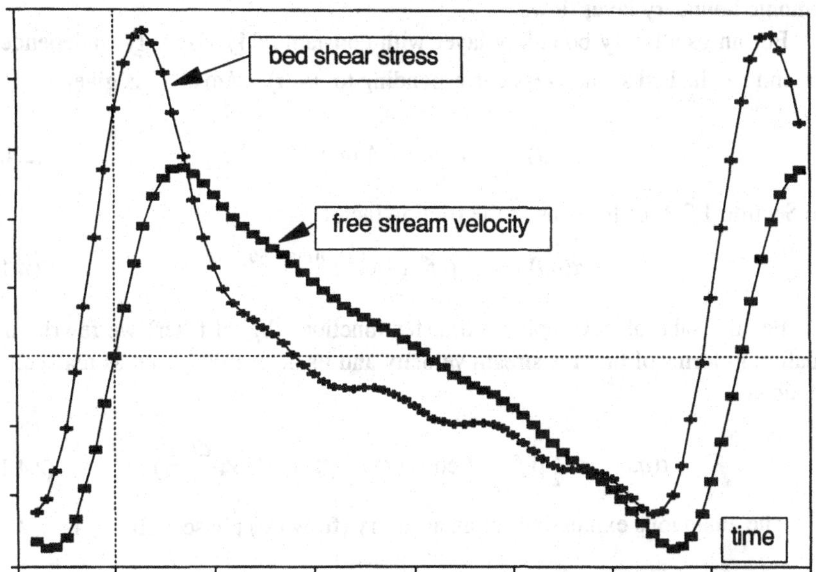

Figure 2.4.5: Bed shear stress under a sawtooth wave for a boundary layer which is laminar or has a constant (independent of z and t) eddy viscosity. Although the shoreward and seaward peak velocities are of equal magnitude, the shoreward peak shear stress is considerably larger than its seaward counterpart. Qualitatively, the reason is that the boundary layer has had less time to grow during the rapid shoreward acceleration.

The important question is then: How can we estimate instantaneous, effective bed shear stresses under waves of arbitrary shape? - A hard question since we know very little about instantaneous stresses on movable beds at all. What has been measured are total stresses under sine waves, and the discussion of sediment transport rates in Section 2.4.3 indicates that the effective stresses may be considerably smaller than the total, even on flat beds.

The influence of the steady flow component on the bed shear stress is difficult to estimate from a time series of $u_\infty(t)$ at a single level, and if $\overline{u}$ is relatively large as in some rip currents or in a strong undertow, this calls for special consideration, see e g Section 1.4.5 and Example 1.5.2.

If the steady flow component is weak however, in the sense that its influence on the bed shear stress is small, it seems reasonable to try and derive $\tau(o,t)$ from $u_\infty(t)$ by means of a simple transfer function based on our knowledge from simple harmonic boundary layer flows.

For an oscillatory boundary layer with constant eddy viscosity (independent of z and t), the bed shear stress corresponding to $u_\infty(t) = A\omega\, e^{i\omega t}$ is given by

$$\tau(o,t) \;=\; \rho\sqrt{\omega\,\nu_t}\;A\,\omega\, e^{i(\omega t + \pi/4)} \tag{2.4.9}$$

see Section 1.2.4, or in terms of the friction factor

$$\tau(o,t) \;=\; \tfrac{1}{2}\,\rho\, f_w\,(A\omega)^2\, e^{i(\omega t + \pi/4)} \tag{2.4.10}$$

For the sake of developing a transfer function (digital filter) we rewrite this equation in terms of the free stream velocity and its derivative given as real-valued functions

$$\tau(o,t) \;=\; \tfrac{1}{2}\,\rho\, f_w\, A\,(\,\cos\pi/4\;\omega\, u_\infty(t) \;+\; \sin\pi/4\,\frac{du_\infty}{dt}\,) \tag{2.4.11}$$

The analogous expression for an arbitrary (forward) phase shift φ_τ is

$$\tau(o,t) \;=\; \tfrac{1}{2}\,\rho\, f_w\, A\,(\,\cos\varphi_\tau\;\omega\, u_\infty(t) \;+\; \sin\varphi_\tau\,\frac{du_\infty}{dt}\,) \tag{2.4.12}$$

The phase shift φ_τ is likely to depend on the Reynolds number $A^2\omega/\nu$ and on the relative bed roughness r/A.

At present, it seems reasonable to approximate the instantaneous, effective bed shear stress under waves of arbitrary shape (*arbitrary* $u_\infty[t]$) by an adaptation of Equation (2.4.12), for example

$$\tau'(o,t) \;=\; \tfrac{1}{2}\,\rho\, f_{2.5}\, A_{rms}\,(\,\cos\varphi_\tau\;\omega_p\, u_\infty(t) \;+\; \sin\varphi_\tau\,\frac{du_\infty}{dt}\,) \tag{2.4.13}$$

Here, the friction factor $f_{2.5}$ should be based on the representative semi-excursion $A_{rms} = \sqrt{2}\,u_{rms}/\omega_p$, where ω_p is the spectral peak angular frequency $(2\,\pi f_p)$.

The bed shear stress time series corresponding to a given u_∞-record may then be constructed, either by using forward and inverse Fourier transformation, or alternatively, by using a simple digital filter in the time domain.

For the spectrum transformation approach, one might suggest the simple frequency response function

$$F(\omega) = \frac{1}{2}\rho f_{2.5} A_{rms}\, \omega_p\, e^{i\varphi_\tau}$$

based on the expression (2.4.13), in analogy with Equation (1.2.17), page 23.

As a simple digital filter to generate the time series $\tau'(o,t_n)$, on a flat bed, from $u_\infty(t_n)$ one might correspondingly suggest

$$\tau'(o,t_n) = \frac{1}{2}\rho f_{2.5} A_{rms} \left(\cos\varphi_\tau\, \omega_p\, u_\infty(t_n) + \sin\varphi_\tau \frac{u_\infty(t_{n+1}) - u_\infty(t_{n-1})}{2\,\delta_t} \right) \tag{2.4.14}$$

The corresponding time series for the instantaneous effective Shields parameter is then given by

$$\theta'(t_n) = \frac{\frac{1}{2} f_{2.5} A_{rms}}{(s-1)\,g\,d} \left(\cos\varphi_\tau\, \omega_p\, u_\infty(t_n) + \sin\varphi_\tau \frac{u_\infty(t_{n+1}) - u_\infty(t_{n-1})}{2\,\delta_t} \right) \tag{2.4.15}$$

As indicated by the review of bed-load transport under sine waves in Section 2.4.3, page 117 ff, a respectable estimate of the instantaneous bed-load transport rate is given by the Meyer-Peter formula with the effective Shields parameter based on $\theta_{2.5}$, see Equations (2.4.4) and (2.4.5), page 120.

In order to incorporate information about the changing direction of the bed shear stress, Equation (2.4.5) must now be augmented to

$$\Phi(t) = \begin{cases} 8\,(\,\theta'(t) - 0.05\,)^{1.5} \dfrac{\theta'(t)}{|\,\theta'(t)|} & \text{for} \quad |\theta'(t)| > 0.05 \\ 0 & \text{for} \quad |\theta'(t)| < 0.05 \end{cases} \tag{2.4.16}$$

The factor *8*, which corresponds to the Meyer-Peter formula for bed-load transport, may not be appropriate for fine sand at high flow intensities as indicated by the data in Figures 2.3.2 and 2.4.2. For sand with $d_{50} < 0.25mm$ at $\theta_{2.5} \gtrsim 1.0$ a factor *12* might be more appropriate.

For half sine waves, Equation (2.4.16) is identical to (2.4.5) which is in close agreement with instantaneous bed-load transport rates observed by Sleath (1978).

For the half cycle average $\Phi_{T/2}$, Equation (2.4.16) leads to results very similar to Equation (2.4.2) which corresponds to the curve in Figure 2.4.2 with minor deviations depending on the relative magnitude $\theta_{2.5}/\theta_c$ of the peak effective stress.

The expressions (2.4.15) and (2.4.16) are easily applied to any u_∞-record. However, since the derivation of the model involves several assumptions and simplifications, which cannot be checked completely at present, the model should be calibrated as more data become available.

Application of the sediment transport model consisting of Equations (2.4.15) and (2.4.16) to skew waves over flat beds is illustrated in the following example.

Example 2.4.1: **Application to King's data.**

King (1991) performed a comprehensive set of experiments on bed-load transport under half sine waves and under half sawtooth waves with some pairs mirror imaged. King's sawtooth waves had the shape of the first halves of the two curves in Figure 2.4.3.

For the half sawtooth waves he found typically that of a pair of mirror images, the one with steep front would transport of the order *1.7* times more sediment as the one with the steep rear.

The reason for this was given qualitatively in connection with Figure 2.4.5. A quantitative example is given in the following by applying the model outlined above to a pair of King's sawtooth waves.

The example corresponds to the average of runs 473-485 and 478-520, which had the following characteristics: $(T, u_{rms}, d_{50}, \Phi_{T/2}) = (5s, 0.89m/s, 0.44mm, 1.09)$ for the steep-front-case, and $(5s, 0.88m/s, 0.44mm, 0.60)$ for the steep-rear-case.

The two cases were modelled as one complete (periodic) wave with the shape shown in Figure 2.4.6 corresponding to both halves of the wave having the rms-velocity $0.89m/s$.

The calculation of the ingredients for the formula (2.4.15) and hence (2.4.16) goes as follows.

The peak period is $5s$ corresponding to the main harmonic component, and hence $\omega_p = 2\pi/T_p = 1.257s^{-1}$.

From the rms-velocity and ω_p one finds $A_{rms} = \sqrt{2}\, u_{rms}/\omega_p = 0.706m$, and with this inserted for A in Equation (2.2.6), page 105, we find $f_{2.5} = 0.0112$.

The instantaneous dimensionless transport rates were then calculated using Equations (2.4.15) and (2.4.16) with four different values of the phase shift φ_τ : $20°$, $30°$, $40°$, and $45°$. The results are shown in Figure 2.4.6.

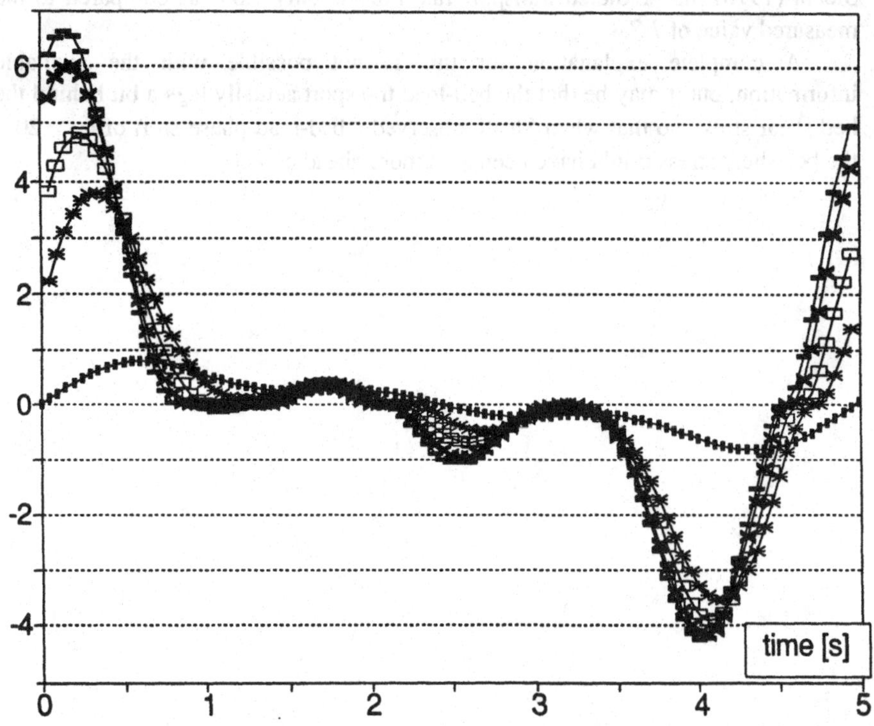

Figure 2.4.6: Free stream velocity and calculated, dimensionless sediment transport rates corresponding to the conditions in runs 473-485 and 478-520 of King (1991). Legend: ▮: free stream velocity, * : $\Phi(t)$ with $\varphi_\tau = 20^o$, *rectangle*: $\Phi(t)$ with $\varphi_\tau = 30^o$, **x** : $\Phi(t)$ with $\varphi_\tau = 40^o$, - : $\Phi(t)$ with $\varphi_\tau = 45^o$,

The calculations give the correct general magnitudes for $\Phi_{T/2}$. Thus, for the steep-front-halves ($0<t<2.5s$) the calculations give *1.09, 1.13, 1.18,* and *1.19* in order of increasing φ_τ. The corresponding numbers for the last half of the waves ($2.5s<t<5s$) are *-1.04, -0.985, -0.855, -0.764*. King's measured values are *1.09* and *-0.60* respectively.

Inspecting the ratios between corresponding values reveals that the right order of transport skewness is achieved by the model only when the phase shift φ_τ is set the maximum value of 45°.

For a phase shift of 20° which corresponds roughly to the observations of

Sleath (1978) the predicted transport rate ratio is only *1.05* as compared to the measured value of *1.7*.

A complete explanation for this is not possible with the available information, but it may be that the bed-load transport actually lags a bit behind the bed shear stress, so that when Sleath observed a bed-load phase shift of $\varphi_B = 20^o$, the bed shear stress might have been a bit more ahead of $u_\infty(t)$.

CHAPTER 3

BEDFORMS AND HYDRAULIC ROUGHNESS

3.1 INTRODUCTION

The sea bed is very rarely flat. On the contrary, it tends to be covered by sedimentary structures with a large range of sizes and of many different shapes, and these structures: bars; dunes; ripples and animal mounds interact with the flow in different ways.

The larger sedimentary structures such as bars modify the main flow pattern, i e they make the incoming waves refract, diffract or break, and they reflect part of the wave energy back into deep water. They may also determine the positions of rip currents.

The small scale structures, such as ripples, have no immediate impact on the main flow patterns, but they strongly influence the boundary layer structure and the turbulence intensity near the bed. Hence, they have great influence on the sediment transport.

The interaction between the large scale topography and the main flow will not be considered here, but the dynamics and geometry of the bedforms (mainly vortex ripples) will be discussed in detail together with their influence on the boundary layer structure. Finally, an attempt is made to quantify the influence of a given bed topography on the flow by a single linear measure, namely the *equivalent Nikuradse roughness*.

3.2 COASTAL BEDFORM REGIMES

The type of bedforms which prevail in a certain area at a certain time depends mainly on the flow strength at the time.

If the flow is too weak to cause appreciable sediment motion ($\theta_{2.5} \lesssim 0.05$), the bed topography will be dominated by relict bedforms from previous more vigorous events and if no such events have occurred recently, the topography will be dominated by bioturbation.

Under flows of intermediate strength ($0.05 \lesssim \theta_{2.5} \lesssim 1.0$) the bed will be active and will be covered with bedforms which are more or less in equilibrium with the flow conditions. However, the shape of these bedforms will depend on the detailed nature of the flow. If the flow is purely oscillatory and almost symmetrical, the bed will be covered by regular, long crested vortex ripples, which are so named because a vortex is formed twice every wave period in the lee of their crests.

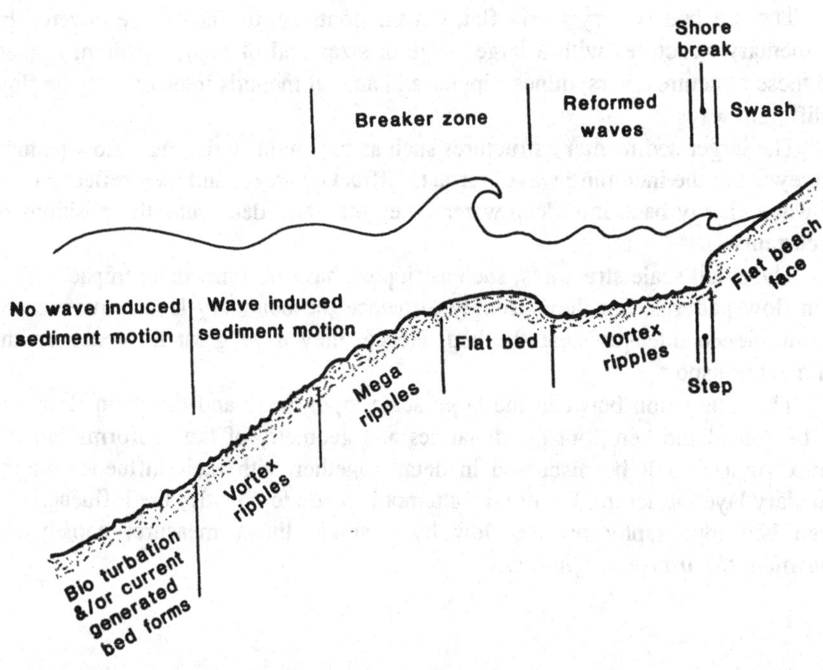

Figure 3.2.1: Example of bedform distribution on an accreting, barred beach. Note that the situation may be very different under storm conditions and that special types of bedforms exist in rip channels.

The regular, symmetrical flows required for the formation of vortex ripples occur in nature seaward of the breaker zone and in bar troughs where the waves have reformed after breaking on the bar, see Figure 3.2.1.

Vortex ripples are rarely seen in the breaker zone. Here, the bed tends to be either flat or covered by megaripples. Megaripples or lunate megaripples (Clifton, 1976), are irregular features with typical lengths of the order one to two metres and heights of the order ten to twenty centimetres, and with rounded crests. There is no regular, rhythmic vortex shedding associated with megaripples, but occasionally a large plume of turbulent, sediment laden water rises from a certain spot due to complicated instabilities. These sand plumes can be of the order one metre high.

Vortex ripples are sometimes found superimposed on megaripples (Southard et al 1990).

Under very vigorous flows ($\theta_{2.5} \gtrsim 1.0$) vortex ripples cannot exist even under perfectly regular, oscillatory flows. However, megaripples have been found to exist under such flow conditions ($\theta_{2.5}$ up to about 2.4) both in large wave flumes and in oscillating water tunnels, see e g Ribberink & Al-Salem (1989).

3.3 BEDFORM DYNAMICS

3.3.1 The continuity principle

The relationship between gradients of the sediment transport rate Q and changes to the bed level z_b is derived by expressing the conservation of sediment volume, see Figure 3.3.1.

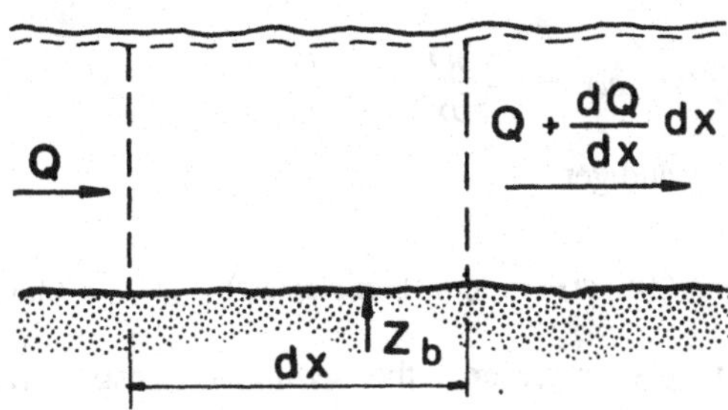

Figure 3.3.1: If there is more sediment entering than there is leaving a certain control volume, the bed level must go up and vice versa.

We quantify the sediment transport rate Q as the volume flux of sediment per unit width of the channel and the dimension of Q is therefore L^2/T, and the units usually m^2/s.

The volume of sediment which enters the control volume in Figure 3.3.1 per unit time is therefore $Q(x)$ per unit width and the volume that leaves is $Q(x) + \frac{dQ}{dx}dx$, and hence, the accumulated volume of solid sediment is $-\frac{dQ}{dx}dx$ per unit time.

A bed level change δz_b corresponds to a sediment volume of $n\,\delta z_b$ per unit area, where n is the volume of solid grain material in a unit volume of the bed; n is usually about 0.7.

Hence in order to conserve the volume of sediment we must have

$$n\frac{dz_b}{dt} = -\frac{dQ}{dx} \qquad (3.3.1)$$

Note that this equation is only strictly valid when sediment storage in the water column is insignificant. See Equation 5.3.1, page 223.

3.3.2 Bedforms migrating with constant shape

The continuity principle can be used very easily to derive the sediment transport rate $Q(x, t)$ from the shape and speed of bedforms if these are known to be moving without change of shape. The argument is as follows.

Consider bedforms which move with constant shape at speed c so that the local sand level can be described by

$$z_b(x,t) = f(x-ct)$$

Inserting this on the left hand side of the continuity equation, we get

$$n\,(-c)\,f'(x-ct) = -\frac{\partial Q}{\partial x}$$

which we integrate with respect to x and get

$$Q(x,t) = Q_o + n\,c\,f(x-ct) = Q_o + n\,c\,z_b(x,t) \qquad (3.3.2)$$

where the constant of integration Q_o represents the sediment transport rate through cross sections where $z_b(x,t) = 0$. The result (3.3.2) shows that if the bedforms move downstream $(c > 0)$ with constant shape, the sediment transport rate must be maximum over the crest and minimum over the trough, i e Q varies in step with z_b, and the difference in transport rates between the crest and trough sections is

$$Q_{max} - Q_{min} = nc\,(z_{crest} - z_{trough}). \qquad (3.3.3)$$

For bedforms which move upstream (e g antidunes) with constant shape, Equation (3.3.2) says correspondingly that the maximum transport rate occurs over the trough, but Equation (3.3.3) is still valid $(c<0)$.

3.3.3 Migration and growth of sinusoidal bedforms

Consider sinusoidal bedforms with length λ and height $2A$, which are migrating with speed c, so that the local, instantaneous sand level can be given by

$$z_b(x,t) = A\cos\left(\frac{2\pi}{\lambda}(x-ct)\right) = A\cos k(x-ct) \qquad (3.3.4)$$

or by introducing the complex exponential $e^{iz} = \cos z + i\sin z$ and only attaching physical meaning to the real part

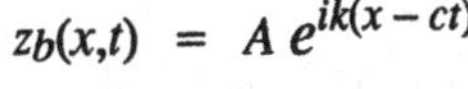

$$z_b(x,t) = A\,e^{ik(x-ct)}$$

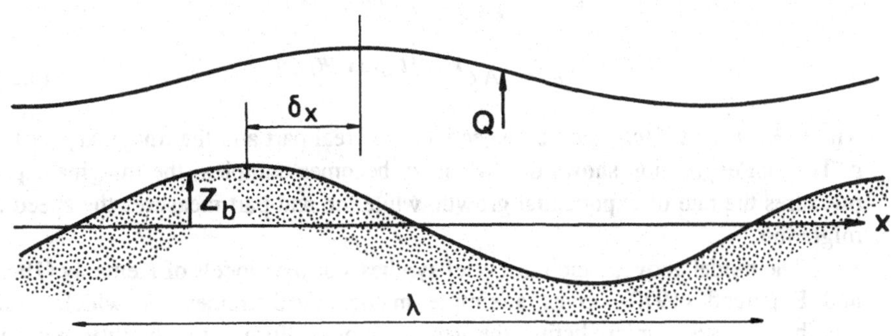

Figure 3.3.2: If the peak in the sediment transport rate is shifted δ_x away from the crest, the bedforms will be growing or diminishing at a rate determined by δ_x/λ.

Assuming then, that the sediment flux Q varies in a similar way to z_b i e

$$Q(x,t) \;=\; Q_o + Q_1\, e^{ik(x-ct-\delta_x)}$$

where the shift δ_x is the downstream distance between the bedform crest and the point of maximum sediment flux, see Figure 3.3.2.

Under these conditions, the continuity Equation (3.3.1) gives

$$n\,A\,ik\,(-c)\,e^{ik(x-ct)} \;=\; -\,Q_1\,ik\,e^{ik(x-ct-\delta_x)}$$

where we can cancel the common factor $-\,ik\,e^{\,ik\,(x-ct)}$ and solve for the speed of migration to get

$$c \;=\; \frac{Q_1}{n\,A}e^{-ik\delta_x} \;=\; \frac{Q_1}{n\,A}\,[\cos k\delta_x \;-\; i\sin k\delta_x)] \qquad (3.3.5)$$

This expression is most easily interpreted when $\delta_x = 0$ *or* $\delta_x = \lambda/2$ so that c is real-valued and equal to Q_1/nA and to $-Q_1/nA$ respectively. That is: the bedforms move downstream with speed Q_1/nA when Q peaks over the crest and, they move <u>upstream</u> with the same speed if Q peaks over the trough.

We can however also give a physical interpretation of the complex c-values which result when $\delta_x \neq (0, \lambda/2)$. If we insert the expression (3.3.5) for the speed of migration c into $z_b(x,t) \;=\; A\,e^{ik(x-ct)}$, we get

$$z_b \;=\; A\;\exp\,[-k\frac{Q_1\,\sin k\delta_x}{nA}\,t\,]\;\exp\,[ik\,(x - \frac{Q_1\,\cos k\delta_x}{nA}\,t\,)]$$

$$z_b \;=\; A\,e^{kI\{c\}t}\,e^{ik(x-R\{c\}t)} \qquad (3.3.6)$$

where $R\{c\}$ and $I\{c\}$ denote respectively the real part and the imaginary part of c. The last expression shows that when c becomes complex, the imaginary part measures the rate of exponential growth, while the real part measures the speed of migration.

The model above, which is a brief synthesis of the models of Kennedy (1963) and Engelund (1970), gives a simple mathematical framework which could possibly be used for predicting the geometry of bedforms via stability analysis. Unfortunately however, the model does not apply to vortex ripples because their length is known to grow with their height while the model above assumes a fixed wave length during the growth process.

Secondly, it is hard to predict the shift δ_x between the bedform crest and the point of maximum transport.

For a comprehensive discussion of the formation and growth of vortex ripples see Lofquist (1978) and Sleath (1984).

3.4 VORTEX RIPPLES

3.4.1 Introduction

Vortex ripples are of special interest for coastal sediment transport studies because their influence on the boundary layer structure and the sediment transport mechanisms is very strong. That is, over vortex ripples, the suspended sediment distribution will scale on the ripple height, while for other bedforms like megaripples and bars, the suspension distribution will scale on the flat bed boundary layer thickness which is much smaller than the height of those bedforms.

The shape and size of vortex ripples was first studied in detail by Bagnold (1946), who also described the flow and the sediment transport mode above them. Inman (1957) investigated their natural occurrence in a comprehensive field study, and their development from a flat bed and their adaptation to new flow conditions have been studied comprehensively by Sleath (1984) and Lofquist (1978) respectively. All of these sources provide excellent illustrations of ripple shapes and associated flow patterns. The occurrence of different overall ripple patterns, i e neatly two-dimensional versus confused three-dimensional has been discussed by Carstens et al (1969), Nielsen (1979) and Sleath (1984).

3.4.2 Ripple length

Vortex ripples are unique to the wave environment, and their scaling is closely tied to the wave motion. With respect to sedimentary structures the most important difference between waves and unidirectional flows is that wave flows have a well defined horizontal scale namely the wave-induced water particle excursion $2A$, see Figure 3.4.1.

It turns out that, under an important range of flow conditions, the ripple length λ is a constant fraction of $2A$

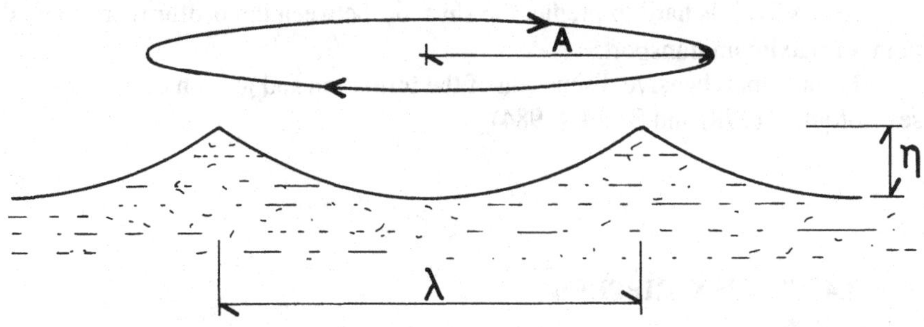

Figure 3.4.1: The size of vortex ripples is closely linked to the orbit length $2A$ of the wave-induced fluid motion near the bed.

$$\lambda \approx 1.33\,A \qquad \text{for } \psi < 20 \tag{3.4.1}$$

see Figure 3.4.2.

Under more vigorous flow conditions, the relative ripple length λ/A tends to be smaller than *1.33*, but the details of the mechanisms which determine the ripple length in this regime are not well understood. Nielsen (1981) suggested the following simple formula

$$\frac{\lambda}{A} = 2.2 - 0.345\,\psi^{0.34} \qquad \text{for } 2 < \psi < 230 \tag{3.4.2}$$

which describes the behaviour of λ/A reasonably well for a large range of wave periods, grain sizes and sediment densities as long as the waves are regular.

The validity of Equation (3.4.2) has recently been questioned (Ribberink & Al-Salem, 1989) because bedforms with lengths of the order $2A$ have been observed in tunnels and large wave flumes at very high flow intensities ($\psi \approx 1000$).

However, those bedforms are not vortex ripples, they tend to have rounded crests like the megaripples which are sometimes found in the breaker zone in the field (Clifton 1976 and Figure 3.2.1, page 130). They do not shed vortices regularly like vortex ripples, and they are in fact some times superimposed with vortex ripples with lengths of the order predicted by Equation (3.4.2).

Wave irregularity tends to result in smaller ripples in the sense that λ/A is smaller than estimated by Equation (3.4.2) when A is based on for example, *significant wave height* or the *rms wave height.*

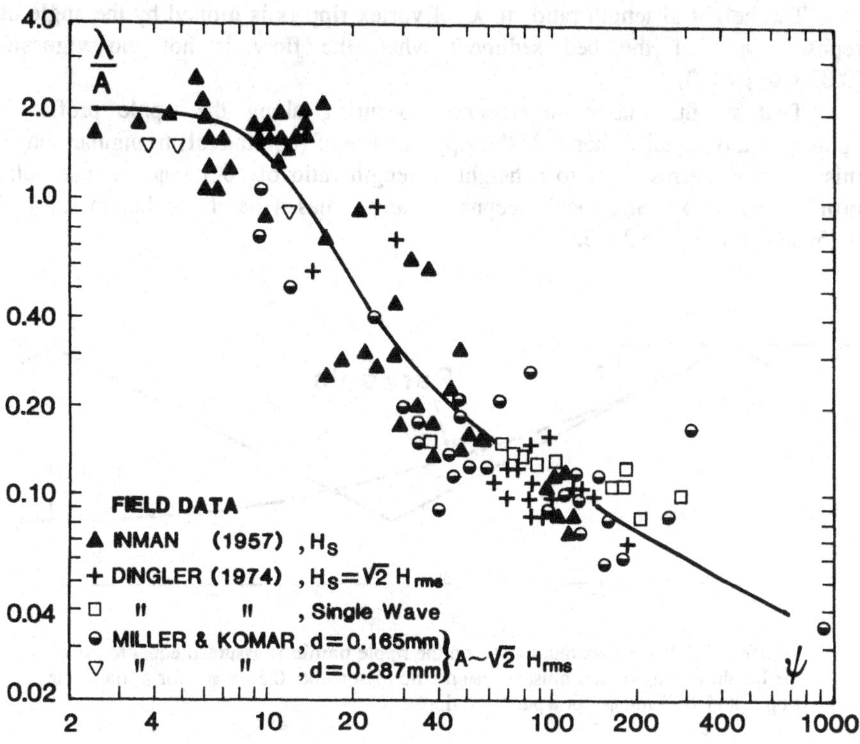

Figure 3.4.2: Field data of relative ripple length versus mobility number based on significant wave heights for field data. The curve corresponds to Equation (3.4.3).

Based on the field data of Inman (1957), Dingler (1974) and Miller & Komar (1980), Nielsen (1981) suggested

$$\frac{\lambda}{A} = \exp\left(\frac{693 - 0.37\ln^8\psi}{1000 + 0.75\ln^7\psi}\right) \tag{3.4.3}$$

for ripples under field conditions, i e, under irregular waves.

3.4.3 Ripple steepness

The height to length ratio η/λ of vortex ripples is limited by the angle of repose φ of the bed sediment when the flow is not too vigorous $(0.05 < \theta_{2.5} < 0.2)$.

That is, the maximum steepness occuring along the ripple profile is approximately equal to $\tan\varphi$. If the ripples were of (equilateral) triangular shape, this criterion corresponds to a height to length ratio of $0.5\tan\varphi$. A parabolic profile with maximum, local steepness $\tan\varphi$ has a height to length ratio of $0.25\tan\varphi$, see Figure 3.4.3.

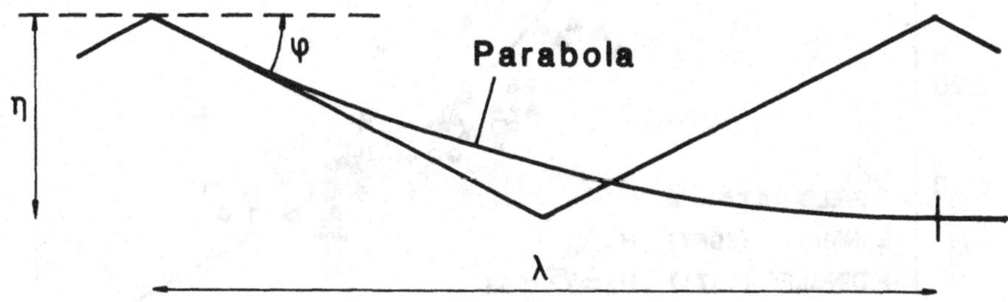

Figure 3.4.3: If the maximum slope on the ripple profile is assumed equal to $\tan\varphi$, the height to length ratio must for geometric reasons be $0.5\tan\varphi$ for a triangular ripple, and $0.25\tan\varphi$ for a parabolic ripple.

It turns out that vortex ripples have a maximum height to length ratio (η/λ) which falls within the range corresponding to these two idealised geometries, see Figure 3.4.4. The data indicate a maximum steepness, at vanishing flow speed, of

$$(\eta/\lambda)_{max} \approx 0.32 \tan\varphi \tag{3.4.4}$$

When the grain roughness Shields Parameter $\theta_{2.5}$ exceeds 0.2, the flattening effect of the contracted flow over the crest tends to increasingly overpower the constructive effect of the lee vortices which build up the crests by scooping sand towards them.

As a result, η/λ becomes a decreasing function of $\theta_{2.5}$, see Figure 3.4.5, and eventually the ripples are flattened completely. This happens at a $\theta_{2.5}$-value of approximately *1.0*.

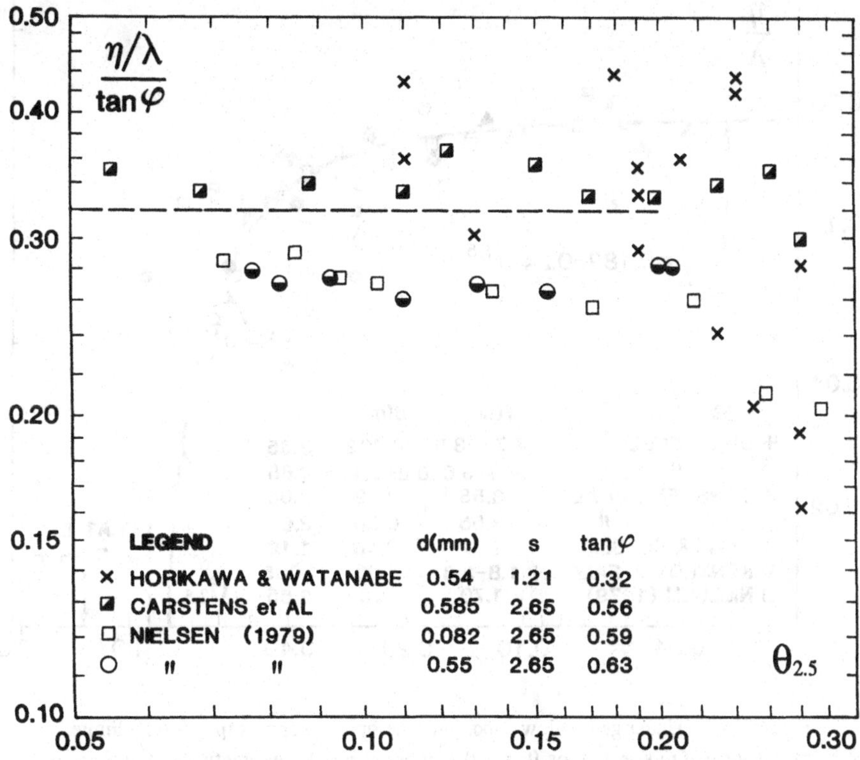

Figure 3.4.4: The maximum value of η/λ for a given bed material lies in the range $0.25 \tan\varphi < (\eta/\lambda)_{max} < 0.5 \tan\varphi$, where the two limiting values correspond to an equilateral triangle and to a parabola respectively, both with a maximum local steepness of $\tan\varphi$, see Figure 3.4.3.

The trend of the laboratory data (regular waves) shown in Figure 3.4.5 is reasonably described by

$$\frac{\eta}{\lambda} = 0.182 - 0.24\,\theta_{2.5}^{1.5} \tag{3.4.5}$$

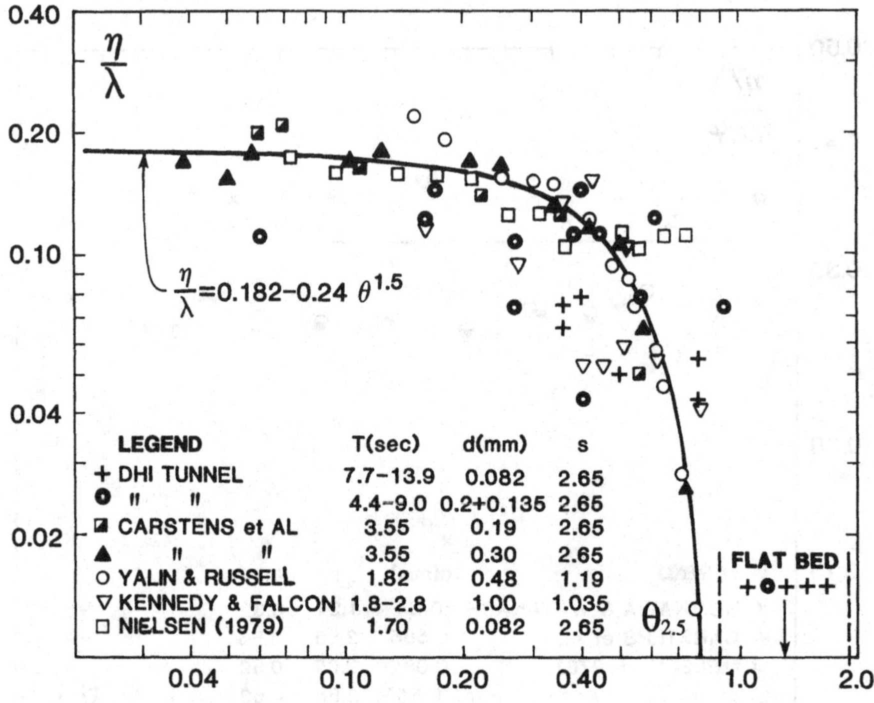

Figure 3.4.5: Under gentle flow conditions, the ripple steepness η/λ is only limited by the angle of repose, but for $\theta_{2.5} \gtrsim 0.2$ it becomes a decreasing function of $\theta_{2.5}$.

For field conditions (irregular waves), the ripple steepness tends to be smaller than for regular laboratory waves. On the basis of the field observations of Inman (1957) and of Dingler (1974), Nielsen (1981) suggested the formula

$$\frac{\eta}{\lambda} = 0.342 - 0.34\sqrt[4]{\theta_{2.5}} \qquad (3.4.6)$$

for irregular waves, with $\theta_{2.5}$ based on the significant wave height.

It would be convenient if the concept of the nominal grain roughness Shields parameter $\theta_{2.5}$ and the corresponding friction factor $f_{2.5}$ could be avoided. In that case, the most likely replacement for $\theta_{2.5}$ would be the mobility number ψ, and this was tried by Dingler (1974).

Dingler suggested, on the basis of his field observations, that the ripples disappear when ψ reaches a value of approximately *240*, and that η/λ could, in general, be described as a function of the mobility number ψ.

However, while this may be true, when only quartz sand is considered, it does not hold in general. This can be seen by comparing Figures 3.4.6 and 3.4.5.

In Figure 3.4.6 η/λ has been plotted as a function of ψ for the same data as in Figure 3.4.5 and it is quite obvious, that ripples formed by sediments of different densities form different trends in this plot.

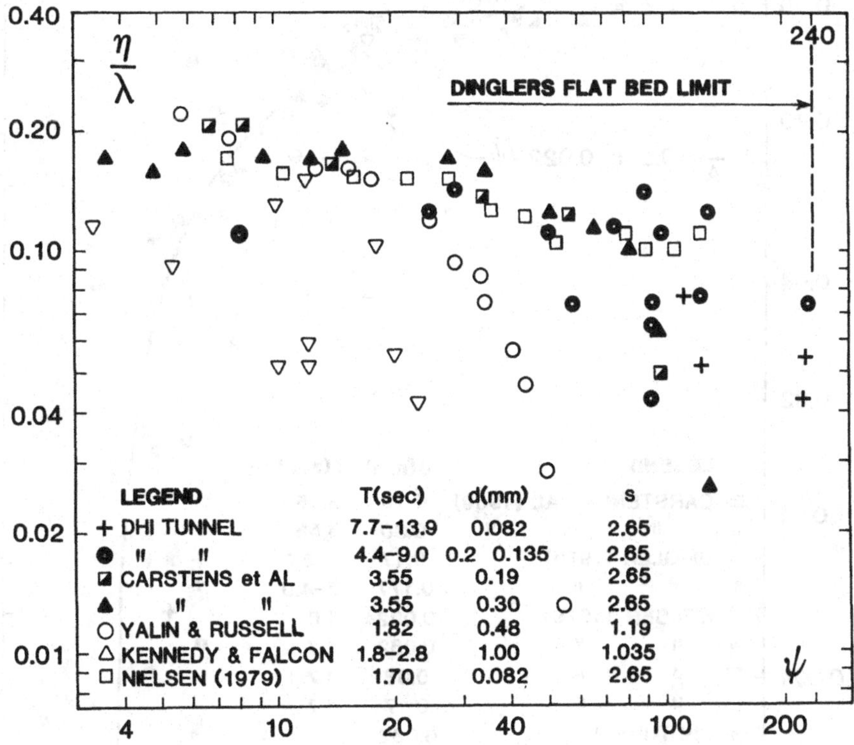

Figure 3.4.6: Ripple steepness as function of the mobility number for sediments of different densities.

3.4.4 Ripple height

Figure 3.4.6 shows that the mobility number cannot be used as a predictor for ripple steepness when widely different sediment densities are considered. For field

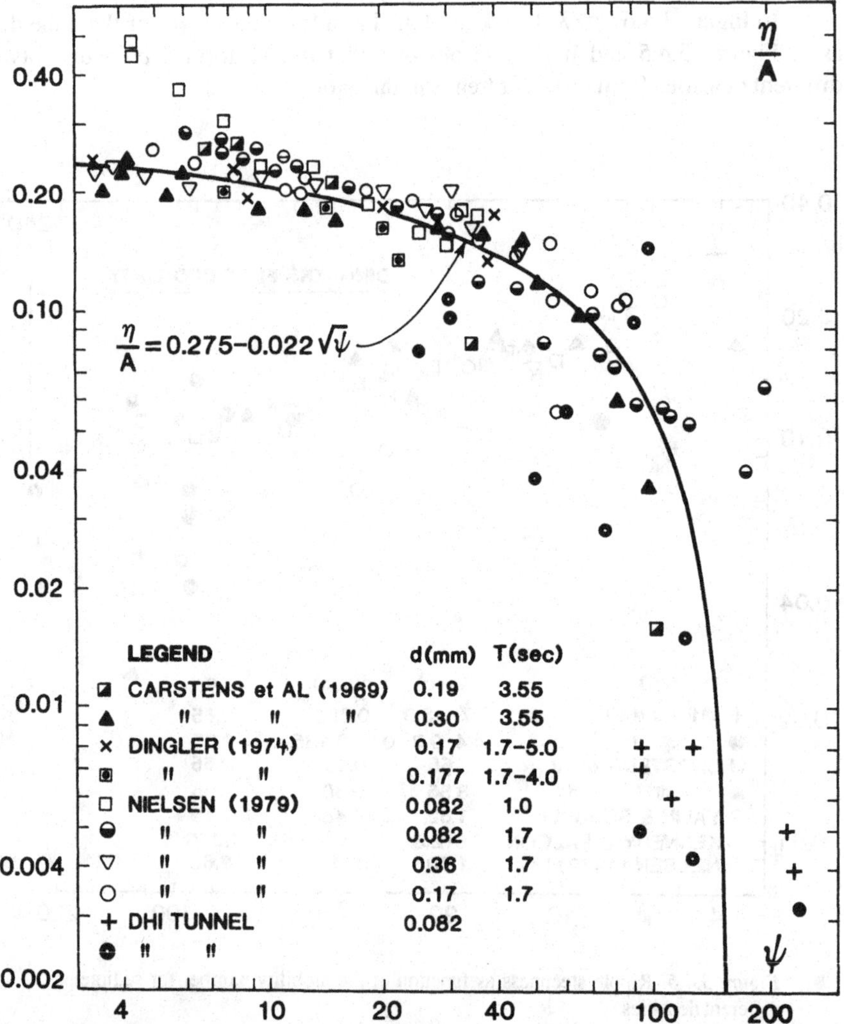

Figure 3.4.7: Relative ripple height as function of the mobility number for laboratory data (regular waves) with quartz-density sediments.

conditions however, where most sediments will have densities close to that of quartz ($s \approx 2.65$), it is quite reasonable to describe η/λ as a function of the mobility number.

Consequently, since the relative ripple length λ/A, is also reasonably well described as a function of ψ (cf Figure 3.4.2, page 137) one might expect that the relative ripple height η/A can be described as a function of ψ. Figure 3.4.7 shows that this is indeed the case.

Nielsen (1981) suggested the formula

$$\frac{\eta}{A} = \begin{cases} 0.275 - 0.022\,\psi^{0.5} & \text{for } \psi < 156 \\ 0 & \text{for } \psi > 156 \end{cases} \qquad (3.4.7)$$

to model the trend of the regular wave data in Figure 3.4.7.

For irregular waves, the formula

$$\frac{\eta}{A} = 21\,\psi^{-1.85} \qquad (\psi > 10) \qquad (3.4.8)$$

was suggested, where ψ and A should be based on the significant wave height.

3.5 BEDFORMS IN COMBINED WAVE-CURRENT FLOWS

Bedforms generated by combined wave-current flows occur in a number of different coastal sub-environments: inner shelf subjected to waves and tides, rip channels; rip feeder channels; surf zones with longshore currents and finally in the mouths of estuaries.

They have been studied in the laboratory by Bijker (1967), Natarajan (1969) and Arnott & Southard (1990), and in the field by Amos & Collins (1978), Coffey (1987) and Amos et al (1988).

Qualitatively, all observers agree that when co-directional currents are superimposed, the ripples tend to migrate in the direction of the current (while the net sediment transport may be either with or against the current) and they become asymmetrical with steeper downstream faces.

When perpendicular or near-perpendicular currents are superimposed, two ripple systems may coexist with one or the other being better developed according to the relative current strength, see Figure 3.5.1 and Amos et al (1988).

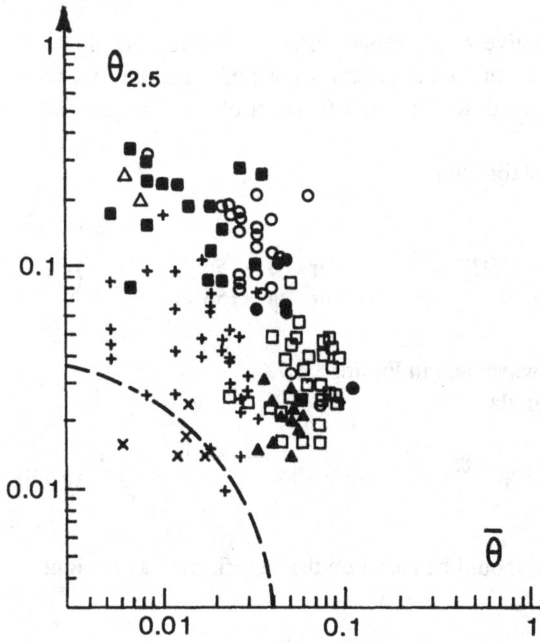

Figure 3.5.1: Ripple types mapped in terms of wave and current Shields parameters. *Open triangle*: wave ripples, *filled square*: wave ripples with subordinate current ripples superimposed, o : wave ripples and current ripples coexisting, *open square*: linguoid current ripples, • : current ripples with subordinate wave ripples superimposed, + : poorly developed ripples, x : flat bed. The curve corresponds to the initiation of motion criterion $\theta_{2.5} + \bar{\theta} = 0.04$. After Amos et al (1988).

From their field study (weak to moderate flows in *22m* depth on the shelf off Sable Island), Amos et al identified eight different bedform regimes and mapped them in terms of the grain roughness Shields parameter $\theta_{2.5}$ versus a current Shields parameter given by

$$\bar{\theta} = \frac{\frac{1}{2}\rho\, 0.003\, \bar{u}_{100}^2}{\rho\,(s-1)\, g\, d}$$

where $\bar{u}_{100}$ is the mean current velocity measured one metre above the bed. See Figure 3.5.1.

The bedforms in strong rip currents are not well researched for the good reason that such rips are extremely dangerous work environments. However, qualitative observations indicate that the bedforms tend to be current-dominated, i e , they have steeper offshore faces and migrate offshore.

In fairly deep bar troughs over which waves have reformed after breaking on the bar, very regular wave ripples are often found. One such case was reported by Nielsen (1983) and Wright et al (1986). The depth was *1.2m* to *1.5m*, the significant height and period of the reformed waves were *0.5m* and *7s* respectively, and the depth averaged longshore current was approximately *0.5m/s*.

In shallower bar troughs or rip feeder channels however, the conditions are often quite obviously current-dominated, with the bedforms resembling current dunes. This may be due to relatively weaker wave activity, but it may also be due to the shallower depth itself via the current Froude number V^2/gD. The stability analysis of Engelund (1970) indicates that current dunes are generally formed at Froude numbers between *0.5* and *2.0*, but are unlikely to form at Froude numbers below *0.25*.

With respect to the megaripples which are sometimes found in oscillatory flows of high intensity ($\theta \gtrsim 1$, see Section 3.4.2) , Arnott and Southard (1990) found that they generally disappeared if a moderate current ($<\bar{u}> > 0.05m/s$) was superimposed. Their experiments were carried out in an oscillating tunnel with *0.09mm* sand and a fixed wave period of *8.5s*.

Coffey (1987) observed bedforms, current profiles and sediment suspension profiles in depths between one and two metres in the mouth of the Port Hacking estuary south of Sydney and found ripples superimposed on the shoals. These ripples were in the relative height and length ranges : $50 < \eta/d < 200$ and $300 < \lambda/d < 1540$, which corresponds to the expected ranges for current ripples. The relative current strengths $\overline{u_*}/\hat{u}_*$ for these observations were in the range [*0.15*; *0.72*].

3.6 HYDRAULIC ROUGHNESS OF NATURAL SAND BEDS

3.6.1 Introduction

In order to formulate simple models of natural flows, it is generally necessary to apply a simplified description of the bed geometry, and in the extreme, we often try to summarise the bed geometry in terms of a single length. Most commonly, the chosen length is the *equivalent Nikuradse roughness* or briefly the *hydraulic roughness, r*.

The only bed geometry for which the definition of the roughness is obvious is a layer of densely packed spheres for which the roughness equals the grain diameter, $r=d$. For all other geometries, the definition is indirect.

That is, the roughness is determined from the structure of a purely steady flow above the bed, see Equation (1.5.9), page 66.

The following section deals with the concept of the equivalent Nikuradse roughness of natural sand beds, particularly those exposed to oscillatory flows. It is found from the available friction and energy dissipation data that the roughness of such beds are generally one to two orders of magnitude larger than that of sand paper with the same sand size.

This great roughness is in many cases obviously due to bedforms which generate roughness of the order of their height ($r \approx \eta$).

However, also flat, mobile sand beds dissipate wave energy at a high rate and thus, in this sense, appear very rough, particularly at high flow intensities where a substantial layer of sediment is in motion. Thus, flat beds in oscillatory sheet-flow ($\theta_{2.5} \gtrsim 1.0$) generally exhibit roughness values (based on total friction or on total energy dissipation rates) of the order 100 to 200 grain diameters.

This is somewhat surprising because the corresponding roughness in steady sheet-flows are generally one order of magnitude smaller. It therefore prompts the question whether a substantial part (more than half) of the energy dissipation measured by Carstens et al (1969) and Lofquist (1986) on flat sand beds, could have been due to other mechanisms than skin friction. If that is the case, the obvious candidate for an additional dissipation mechanism under waves is percolation, see e g Sleath (1984) or Dean & Dalrymple (1991).

The suspicion above is reinforced by the observations of sediment transport rates over flat beds under waves, which indicate that the effective stress, with respect to sediment transport, is considerably smaller than the bed shear stress measured either directly or via the rate of energy dissipation.

Thus, the bed-load data analysed in Sections 2.4.2 and 2.4.3, page 117 ff, indicated that the effective sediment transporting stress corresponds to a roughness of only about $2.5d_{50}$, while energy dissipation measurements under similar conditions indicate roughness values of the order $100d_{50}$.

This difference between steady and oscillatory flows remains unexplained, except for the possibility that it may be due to percolation.

3.6.2 Sand bed friction factors

The hydraulic roughness is closely related to the rate of momentum transfer at the bed and hence to the friction factor f.

Measured bed shear stresses and energy dissipation rates for sand beds exposed to oscillatory flows are presently available from two main sources:

Carstens et al (1969) and Lofquist (1986).

Carstens et al measured the rate of energy dissipation, while Lofquist measured the total drag in terms of the total pressure gradient minus the inertia of the fluid.

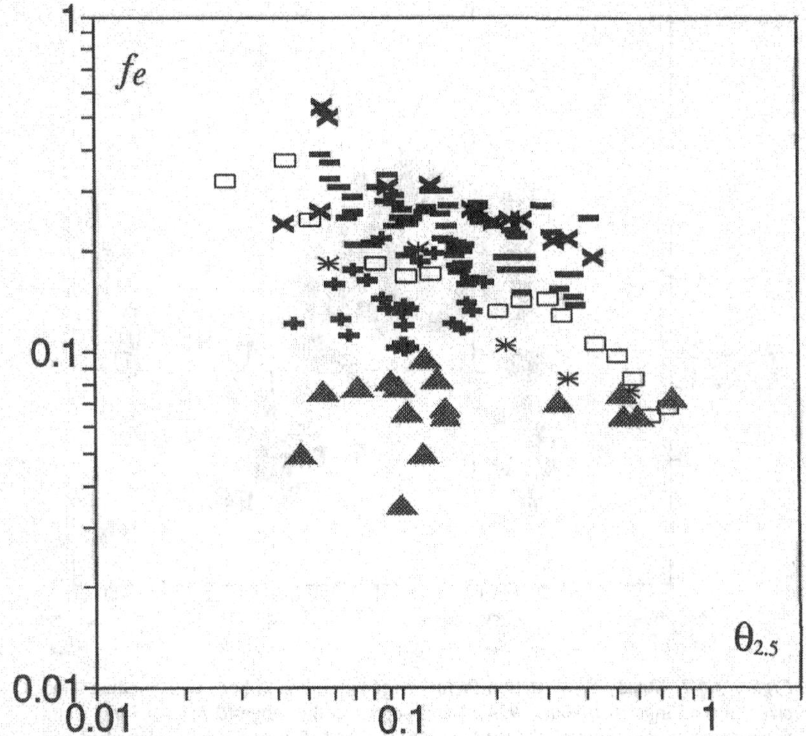

Figure 3.6.1: Measured wave energy dissipation factors from Carstens et al (1969) and from Lofquist (1986). Legend: bar: Lofquist *0.55mm* sand, +: Lofquist *0.18mm* sand, *: Carstens et al *0.19mm*, *rectangle*: Carstens et al *0.30mm*, x: Carstens et al *0.59mm*. All of the above correspond to equilibrium ripples while the *triangles* correspond to artificial flat beds where measurements were taken before ripples had time to form (Carstens et al).

By assuming that alternate dissipating mechanisms than surface friction were negligible in the experiments, the results of Carstens et al (1969) and of Lofquist (1986) have been translated (Equation 1.2.26) into the energy dissipation factors

(f_e) which are plotted in Figure 3.6.1. f_e is expected to be smaller than f_w, but the differences are generally small compared to the scatter see Figure (1.2.9).

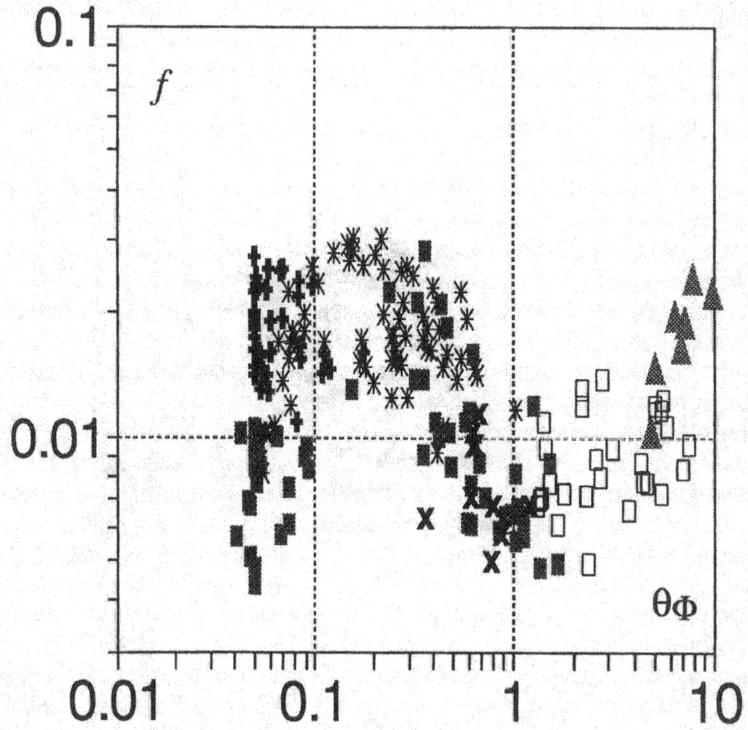

Figure 3.6.2: Steady flow friction factors over natural sand beds with median sand sizes in the range [*0.19mm; 0.93mm*] and depths in the range [*0.14m; 0.33m*]. Data from flume experiments by Guy et al (1966). Legend: *filled rectangle*: plane bed and transition., + : rippled beds, *: dunes, *open rectangle*: antidunes, **x** : standing waves, *triangle*: chute and pool flow.

Roughly speaking, these friction factors for sand beds in oscillatory flows are an order of magnitude larger than those found in steady flows; namely between *0.04* and *0.4*, compared to a range of about [*0.005; 0.04*] for steady flows, see Figures 3.6.1. and 3.6.2.

When data from artificially flattened beds at low flow intensities ($\theta_{2.5} < 0.25$) are excluded, the range of observed energy dissipation factors for sand beds under oscillatory flows is

$$0.06 < f_e < 0.5 \tag{3.6.1}$$

In order to provide a steady flow comparison, the steady flow friction factors in Figure 3.6.2 are derived from the flume data of Guy et al (1966) through the definition

$$\tau = \tfrac{1}{2}\rho f \langle\bar{u}\rangle^2 \quad , \qquad \tau = \rho g D I$$

The abscissa θ_Φ is an effective Shields parameter derived from measured transport rates via Equation (2.3.12), page 114, which is a reasonably reliable relationship between Shields parameter and sediment transport rates over flat beds see Figure 2.3.2, page 113.

At high rates of sediment transport, the momentum transfer between the flow and the bed is partly due to sediment particles receiving momentum from the flow, when they are picked up and accelerated, and transferring it to the bed, when they are deposited.

In this case the friction coefficient contains a contribution f_s from the moving sediment as well as the usual f_f from the form drag and the surface drag on a fixed bed.

Correspondingly, the roughness of a mobile sand bed may be seen as including a "fixed bed component" r_f and a component r_s, which accounts for the momentum transfer by moving sediment.

We shall see in the following, that the fixed bed contributions are similar in steady and oscillatory flows over rippled beds; while the moving sediment contributions, which become dominant over flat beds may be much larger in oscillatory flows with similar grain sizes and Shields parameters.

3.6.3 Sand bed roughness in steady flows

The following contains a brief discussion of the hydraulic roughness of sand beds exposed to steady flows. Since the main aim of the section is to provide a basis for comparison with subsequent results for sand beds under waves, the approach is somewhat different from previous treatments such as those of Engelund & Hansen (1972), van Rijn (1984) and Wilson (1989).

Nikuradse (1933) noted that if the formula

$$u(z) = \frac{u_*}{\kappa} \ln\frac{z}{z_0} \tag{3.6.2}$$

was fitted to steady flows over beds of densely packed sand grains, at sufficiently large grain Reynolds numbers, then the zero intercept level z_o was approximately one thirtieth of the sand size (cf Schlichting 1979)

$$z_o = d/30 \qquad \text{for} \quad u_* d/\nu > 70$$

Correspondingly, if a value z_o has been found by fitting a distribution of the form (3.6.2) to a measured velocity profile, the bed geometry is said to have the *equivalent Nikuradse roughness*

$$r = 30\, z_o \tag{3.6.3}$$

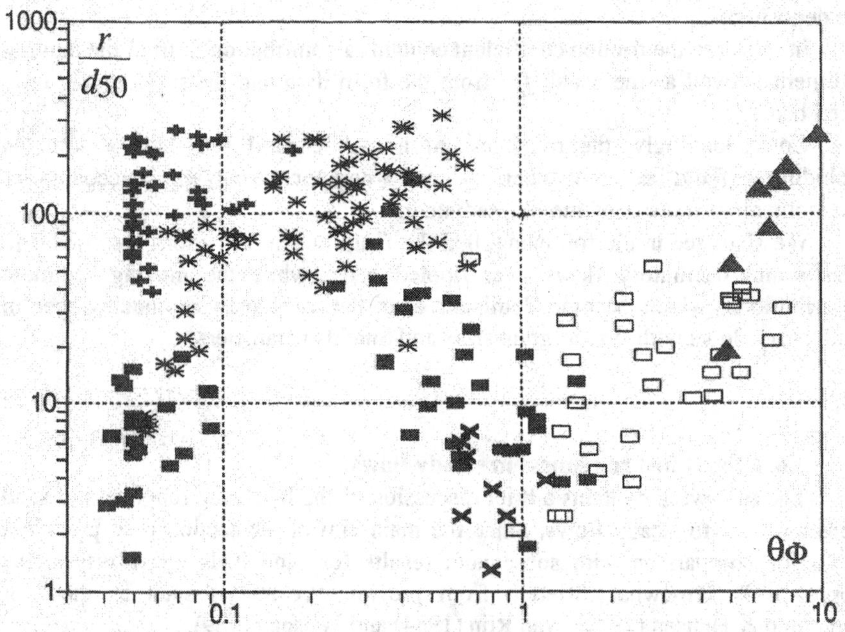

Figure 3.6.3: Hydraulic roughness derived through Equation (3.6.5) for sand beds with different grain sizes in a variety of steady flows. Same legend as for Figure 3.6.2.

If not the whole velocity distribution, but only the depth-averaged velocity $<\bar{u}>$ together with the depth D and the hydraulic gradient I has been measured, a crude estimate of the roughness can be obtained based on a total-flow relationship of the form

$$<\bar{u}> = \frac{1}{D}\int_{z_0}^{D} \frac{u_*}{\kappa} \ln\frac{z}{z_0}\, dz \ , \qquad u_* = \sqrt{g\,D\,I} \tag{3.6.4}$$

which leads to

$$r = 30 z_0 \approx 30\,D \exp\left[\frac{-\kappa <\bar{u}>}{\sqrt{g\,D\,I}} - 1\right] \tag{3.6.5}$$

A sample of roughness values derived via Equation (3.6.5) from a variety of flow regimes over beds with different sand sizes are shown in Figure 3.6.3.

These data show that for a *plane bed* and for *standing waves*, the roughness is of the order one to ten grain diameters, while for *rippled beds*, it is typically of the order $100d_{50}$. The latter corresponds to the typical height of current ripples.

Note that the depth range of the shown data was only about *[0.14m; 0.33m]* and that deeper flows may form dunes and antidunes of greater heights, corresponding to greater values of r/d_{50}. For a general discussion of the size and shape of steady flow bedforms, see e g Allen (1982) or Sleath (1984).

Roughness values extracted with Equation (3.6.5) should not be interpreted too closely for at least two reasons. Firstly, most of the flows are highly non-uniform with depth and boundary layer structure varying strongly from bedform crest to bedform trough. Secondly, the logarithmic velocity distribution on which Equation (3.6.5) is based, is not an accurate description over the full depth - only the bottom twenty to thirty percent of the velocity distribution is well described by the logarithmic law of the wall.

Nevertheless, these crude roughness values are of interest in relation to the roughness values for sand beds under waves, which will be discussed in the following.

For steady-sheet flow in closed conduits, Wilson (1989) found remarkable agreement between his own (1966) data and the simple formula

$$r = 5\,\theta\, d \tag{3.6.6}$$

corresponding to $r = 2L_B$, where L_B is the equivalent bed-load thickness which is defined in Section 2.3.3, page 111.

3.6.4 Hydraulic roughness from oscillatory flows

For oscillatory flows, similar procedures can be applied in order to extract the hydraulic roughness of a sand bed from velocity or friction measurements.

If the velocity distribution has been measured so that the velocity defect D_1 can be plotted against z, then a value for z_1 (see Figure 1.3.2, page 44) can be found. Subsequently, the roughness can be found from Equation (1.3.10), page 47.

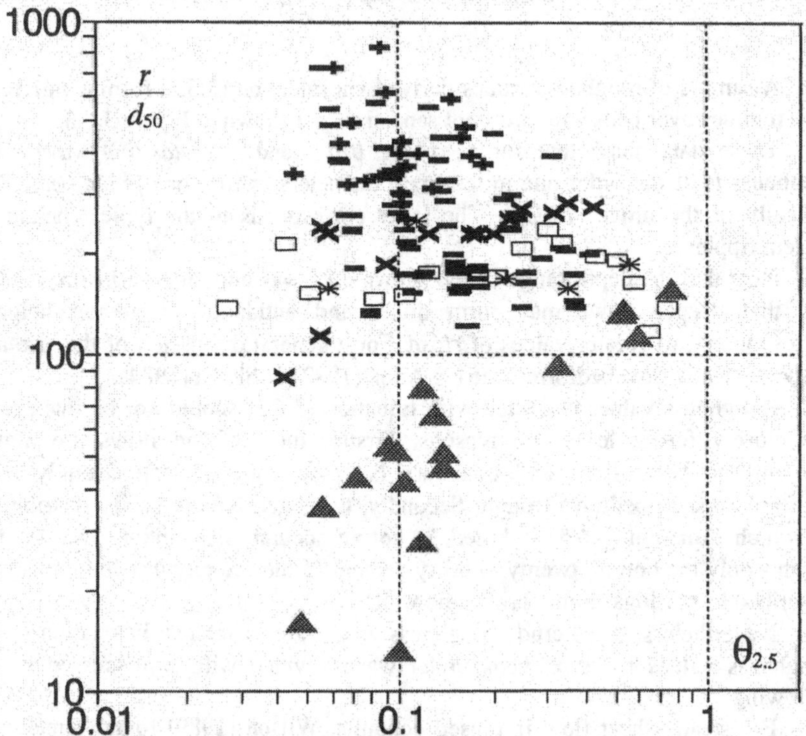

Figure 3.6.4: Hydraulic roughness corresponding to the f_e-data in Figure 3.6.1. For rippled beds the roughness is generally in the range [$100d_{50}$; $1000d_{50}$], and if the trend of the limited flat bed data can be extrapolated beyond $\theta_{2.5} \approx 1$, the bed roughness under oscillatory sheet-flow is also of the order $100d_{50}$ or more. Same legend as in Figure 3.6.1, page 147.

$$r = \frac{z_1^2}{0.0081\,A} \tag{3.6.7}$$

If only the peak bed shear stress $\hat{\tau}$ or the time-averaged wave energy dissipation D_E has been measured, the hydraulic roughness may be inferred via the wave friction factor f_w or the energy dissipation factor f_e. That is, f_w or f_e is found from Equation (1.2.18) or Equation (1.2.26) respectively, and the roughness is then found by applying a wave friction factor formula like Equation (1.2.22) in reverse i e

$$r = A\left(\frac{\ln f + 5.977}{5.213}\right)^{5.15} \tag{3.6.8}$$

The roughness values derived with this formula from the energy dissipation factor data in Figure 3.6.1 are shown in Figure 3.6.4.

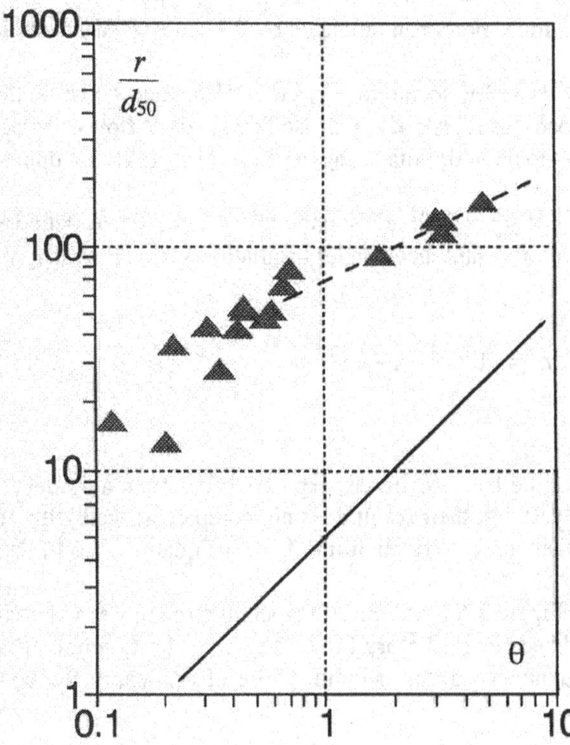

Figure 3.6.5: Roughness of flat sand beds in oscillatory flows (flat bed data from Carstens et al). The full line corresponds to Wilson's (1989) formula (3.6.6) for steady sheet-flow, and the dotted line corresponds to Equation (3.6.9)

We see that, except for (artificially) flat beds at low flow intensities ($\theta_{2.5} \lesssim 0.20$), this roughness is of the order $100d_{50}$ to $1000d_{50}$.

This is interesting because, the steady flow data in Figure 3.6.3 indicate roughness values as low as $2d_{50}$ occurring in the upper (steady) flow regimes, and Wilson's (1989) formula (3.6.6) predicts only $r \approx 5\,d$ for Shields parameters of the order unity, see Figure 3.6.5.

3.6.5 The roughness of flat sand beds under waves

The roughness values in Figure 3.6.5 indicate that, in terms of the Shields parameter θ (= θ'), the energy dissipation over flat, mobile sand beds in oscillatory flows corresponds to a roughness of the order

$$r \approx 70\sqrt{\theta}\,d \qquad \text{for } \theta \gtrsim 0.5 \tag{3.6.9}$$

This is of the order ten times more than predicted by Wilson's (1989) formula for steady flows.

No explanation has been given so far to this large difference between the roughness values of sand beds under steady and oscillatory sheet-flows, except that part of the energy dissipation or total drag under waves may be due to percolation.

In terms of the grain roughness Shields parameter $\theta_{2.5}$ the roughness, corresponding to total drag on flat sand beds under oscillatory sheet-flow, may be expressed as

$$r = 170\sqrt{\theta_{2.5} - 0.05}\;d \tag{3.6.10}$$

see Figure 3.6.6.

Grant & Madsen (1982) derived a different expression based on a model for saltation in air by Owen (1964), but their result does not compare as well with the presently available data when r is derived from f via Equation (3.6.8). For details, see Nielsen (1983).

The friction factors in Figure 3.6.1 and the corresponding roughness values in Figure 3.6.4 show that sand beds in oscillatory flows are always fairly rough. That is, the range of relative roughness values, including those of artificially flat beds with $\theta_{2.5} > 0.3$ is

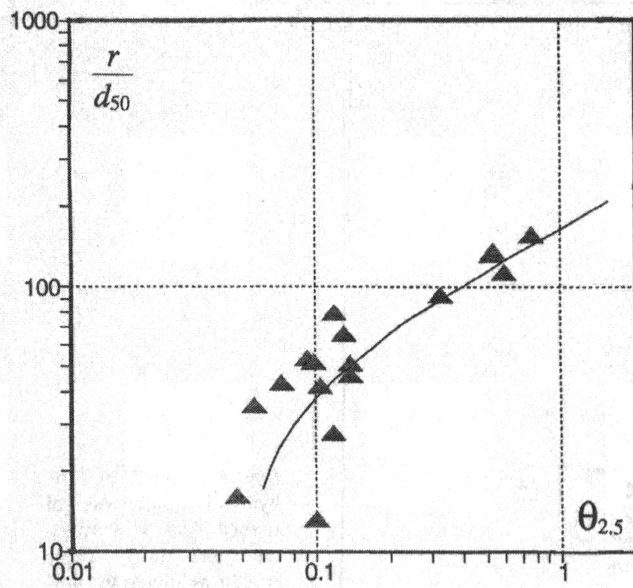

Figure 3.6.6: Relative roughness of flat sand beds in oscillatory flow as function of $\theta_{2.5}$. The curve corresponds to Equation (3.6.10). The roughness values are obtained from Equation (3.6.8) applied to the energy dissipation factors of Carstens et al (1969).

$$0.085 < r/A < 1.15 \tag{3.6.11}$$

If this is a valid indication of the boundary layer structure in general over flat beds of loose sand, it has important implications for the relevance to flow over natural sand beds of many of the existing models of oscillatory boundary layers.

The reason is that many of these have been developed to match experimental evidence from flows with a relative bed roughness of less than *0.03*, and are not suited for flows with $r/A > 0.1$, see Section 1.3, pp 40-52.

3.6.6 The roughness of rippled beds under waves

For sand beds covered by vortex ripples, the shape of which varies little for $\theta_{2.5} < 0.25$ (see Figures 3.4.4 and 3.4.5, pp 139 and 140), it is natural to expect the hydraulic roughness to be closely related to the ripple height. This is confirmed by Figure 3.6.7.

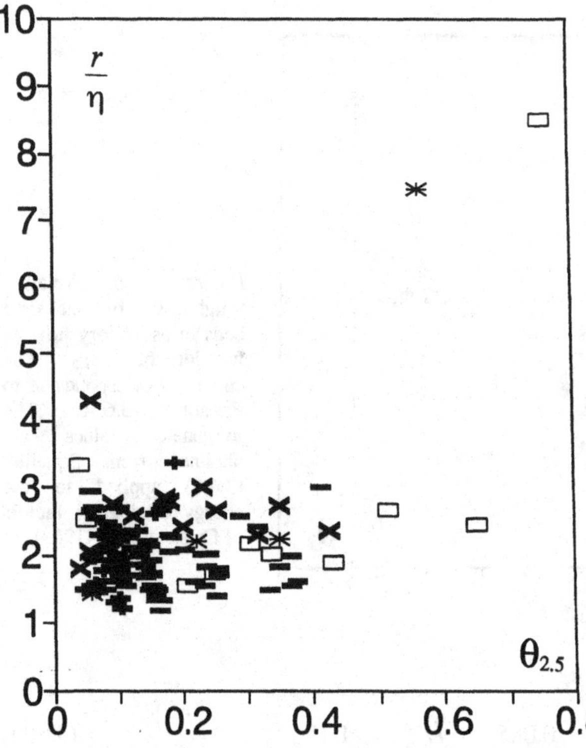

Figure 3.6.7: The hydraulic roughness of rippled beds is between one and three ripple heights as long as the flow is not too strong ($\theta_{2.5} \lesssim 0.5$). Same legend as in Figure 3.6.1.

Indeed , r/η is seen to vary very little up to $\theta_{2.5} \approx 0.5$ even though the ripple steepness generally starts to decrease from about $\theta_{2.5} = 0.25$. See Figures 3.4.5 and 3.4.7.

The momentum transfer between a sediment laden flow and the bed can be seen as consisting of two components: The usual drag contribution, which would also be there for a fixed bed with the same geometry, and a component due to the transfer of momentum by accelerating and subsequently crashing sediment particles.

Correspondingly the roughness may be seen as consisting of a fixed bed part r_f and of a moving sediment part r_s:

$$r = r_f + r_s \tag{3.6.12}$$

The magnitude of the fixed bed contribution r_f for rippled beds may be

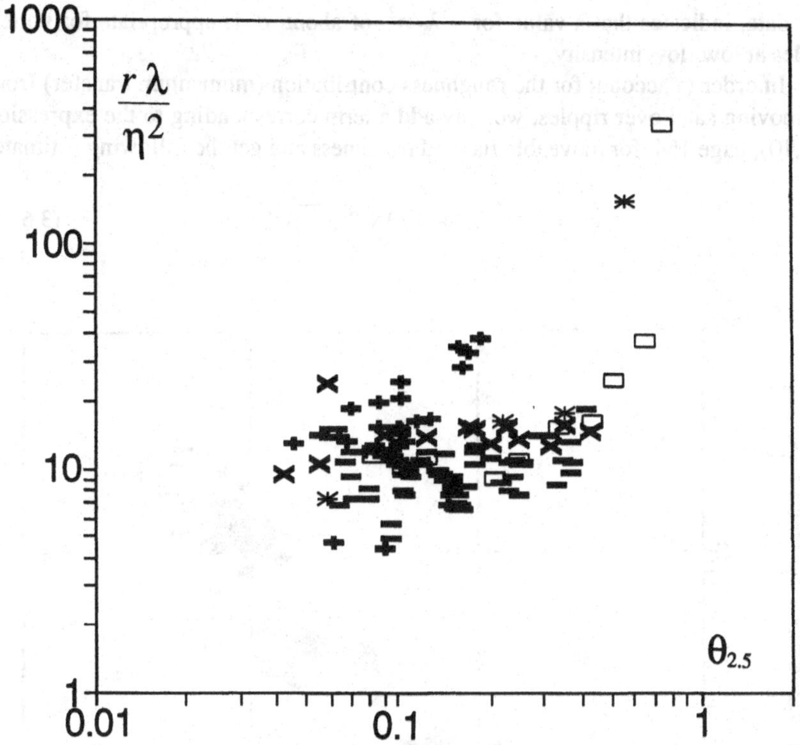

Figure 3.6.8: Relative ripple roughness as function of the grain roughness Shields parameter for oscillatory flow data. Same legend as in Figure 3.6.1.

inferred from some of the experiments which have been carried out with ripple-like fixed beds.

It has generally been assumed that the roughness of a rigid ripple profile would be proportional not only to the ripple height, but also to the steepness and hence relations of the form $r = \text{const} \cdot \eta \; \eta/\lambda$ or $r\lambda/\eta^2 = \text{const}$ have been sought.

Thus, Bagnold's (1946) sharp crested, parabolic ripples correspond to $r\lambda/\eta^2 = 20.3$, while the triangular concrete ripples of Jonsson & Carlsen (1976) correspond to $r\lambda/\eta^2 = 10.9$ and 7.4 for Test I and Test II respectively.

Figure 3.6.8, where $r\lambda/\eta^2$ is plotted versus $\theta_{2.5}$ for the available movable bed data, indicates that a value for $r\lambda/\eta^2$ of about *8* is appropriate for vortex ripples at low flow intensity.

In order to account for the roughness contribution (momentum transfer) from the moving sand over ripples, we may add a term corresponding to the expression (3.6.10), page 154, for moveable flat bed roughness and get the following estimate

$$r = 8\,\eta^2/\lambda + 170\sqrt{\theta_{2.5} - 0.05}\;d \tag{3.6.13}$$

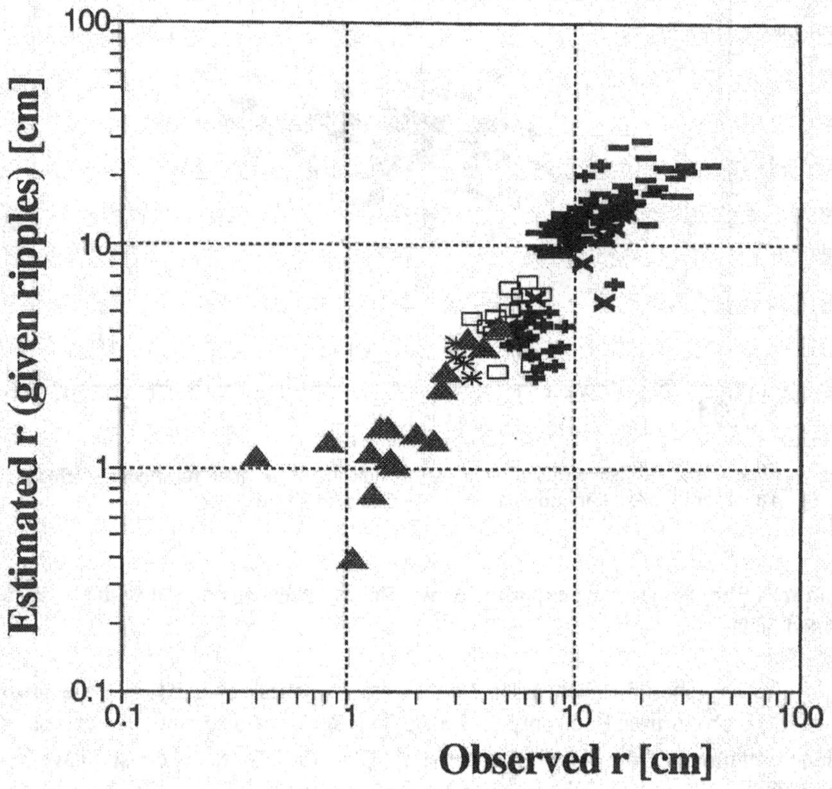

Figure 3.6.9: Predicted versus measured roughness values corresponding to the f_e-data of Carstens et al (1969) and of Lofquist (1986). The predictions are based on Equation (3.6.13) with measured ripple data inserted.

where the ripple geometry may be measured or estimated from Equations (3.4.5, 3.4.7) or (3.4.6, 3.4.8).

Figure 3.6.9 shows a comparison between this formula and the roughness values derived via Equation (3.6.8), page 153, from the f_e-data of Carstens et al (1969) and Lofquist (1986). The figure shows reasonable agreement.

We note that the nature of the first term makes it impossible to estimate the roughness by a simple formula of the form $r/d = F(\theta_{2.5})$ or $r/d = F(\psi)$ as suggested by Raudkivi (1988). The reason is that the ratio $\eta^2/\lambda d$ varies strongly with the wave period for a fixed value of $\theta_{2.5}$ or of ψ, see the discussion of Nielsen et al (1990).

While, according to Figure 3.6.9, energy dissipation rates based on Equation (3.6.13) are in good agreement with experiments, there is reason to suspect that the roughness determined from (3.6.13) and (3.6.10) for flat beds are too large for the estimation of the vertical scale of suspended sediment profiles, see Section 5.4.9, page 255.

Also, the discussion of bed-load transport rates and corresponding effective bed shear stresses for flat beds under waves in Section 2.3 indicates that bed shear stresses determined from a roughness given by (3.6.13) or (3.6.10) for flat beds are too large.

The optimal choice of roughness for both bed-load transport rates and suspension distributions over flat sand beds under waves lies closer to Wilson's steady flow formula (3.6.6), page 151, with θ replaced by $\theta_{2.5}$.

Thus, for purposes other than estimation of frictional dissipation, it might be appropriate to replace Equation (3.6.13) by an equivalent formula with the "moving grain contribution" given by Wilson's expression for example

$$r = 8\,\eta^2/\lambda + 5\,\theta_{2.5}\,d \qquad (3.6.14)$$

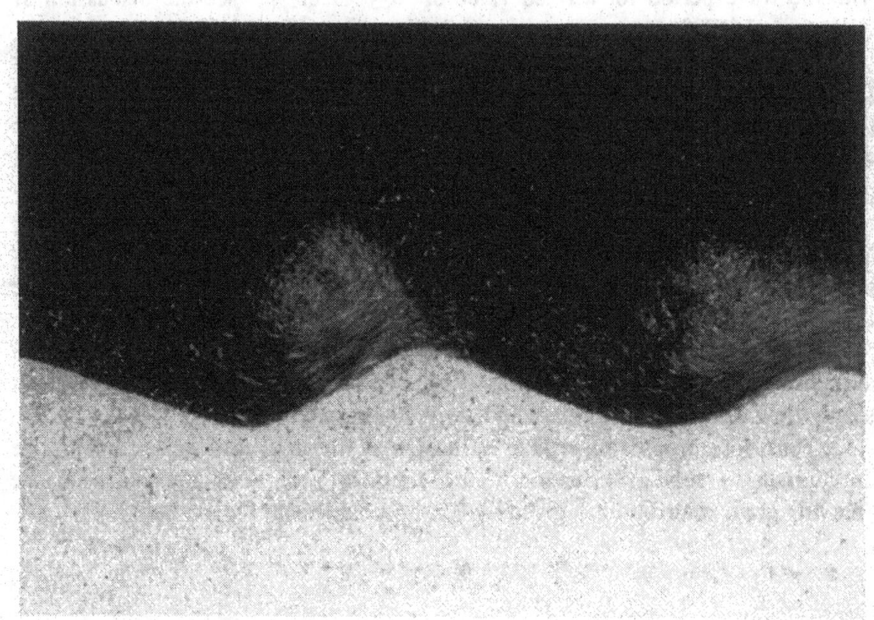

CHAPTER 4

THE MOTION OF SUSPENDED PARTICLES

4.1 INTRODUCTION

Transport of suspended sand is of great practical importance in coastal and fluvial engineering. Therefore, the mechanics of suspended sediment transport have been the object of substantial research efforts through several decades. However, there is still much to be learned about the concentration magnitude and about the distribution of suspended sediment in waves, in currents, and in combined wave-current flows.

In order to describe suspended sediment distributions we must first seek to understand the behaviour of suspended sand grains in different types of flows. That is the object of the following sections.

The most important flow structure in connection with suspended sediment is that of a vortex with horizontal axis because such vortices are able to trap sand grains and carry them along over considerable distances, see Figure 4.1.1 (right).

In contrast, a wave motion where the fluid orbits are also circles or ellipses, has no trapping capability, see Figure 4.1.1 (left). The mechanism of sand trapping in vortices was first pointed out by Tooby et al (1977), but we shall discuss it in detail in Section 4.6, p 181, and refer to some empirical evidence of its importance.

In coastal sediment transport the nearly homogeneous oscillating flow field induced by surface waves is, of course, important, and there have been many attempts to derive a suspension-maintaining mechanism from this flow structure. For example, the delay distance idea of Hattori (1969) and the time lag hypothesis of Bhattacharya (1971).

The former of these has little physical substance but the latter will be proven and quantified in Section 4.5.

A third mechanism, which might influence the sediment settling rate under waves, is the possible settling velocity reduction due to nonlinearity of the drag force. We shall see in Section 4.5.5, through an analytical approach to this mechanism, that the changes in settling rate due to the wave motion itself are either purely oscillatory, and therefore without net effect, or they are so small that they have no practical importance.

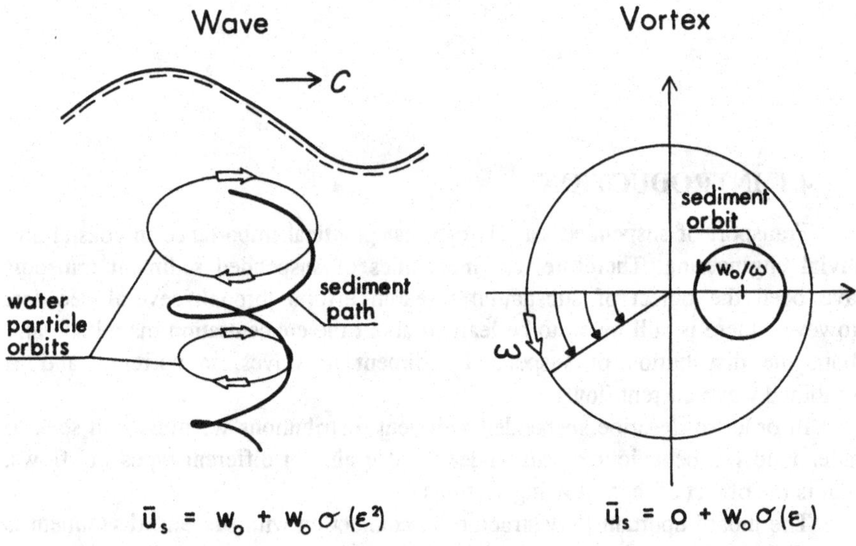

Figure 4.1.1: While vortices and wave motions are similar in so far as both have circular or elliptical fluid orbits, they differ greatly with respect to their influence on suspended sediment particles. Sand will settle almost unhindered through the wave motion, but it will tend to get trapped in vortices.

The analysis of the motion of suspended particles in the following sections takes a stepwise, analytical approach.

We start with the simplest case, where the fluid accelerations are negligible compared to the acceleration of gravity. In this case, the relative velocity between sand and water is everywhere equal to the still water settling velocity w_o.

The second step takes terms of magnitude εw_o into account, where ε is the relative magnitude of the fluid accelerations

$$\varepsilon = \frac{1}{g}\left|\frac{du}{dt}\right| \tag{4.1.1}$$

Finally, as a third step, effects of the order $\varepsilon^2 w_o$ are considered. Hence the sediment particle velocity vector $\boldsymbol{u}_s$ is built up as a series of the form

$$\begin{aligned} \boldsymbol{u}_s &= \boldsymbol{u} + \boldsymbol{w}_o + \boldsymbol{u}_{r1} + \boldsymbol{u}_{r2} + \ldots \\ &= \boldsymbol{u} + w_o\left[\begin{pmatrix} 0 \\ -1 \end{pmatrix} + \varepsilon U_{r1} + \varepsilon^2 U_{r1} + \ldots\right] \end{aligned} \tag{4.1.2}$$

where $\boldsymbol{u}$ is the fluid velocity vector, see Figure 4.1.2.

The justification for this approach lies very much in its usefulness. That is, in the fact that the most important suspension maintaining mechanism, namely trapping in vortices, can be derived from the zero order approximation

$$\boldsymbol{u}_s = \boldsymbol{u} + \boldsymbol{w}_o \tag{4.1.3}$$

Secondly, the fluid accelerations induced by waves are generally much smaller than the acceleration of gravity so that the perturbation parameter ε is conveniently small.

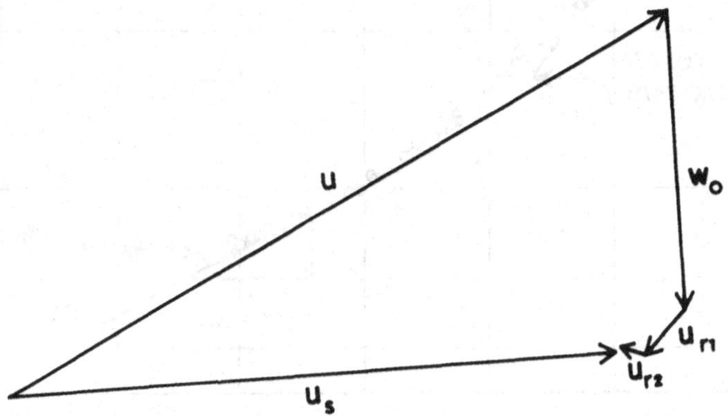

Figure 4.1.2: The sediment particle velocity vector, $\boldsymbol{u}_s$ is described in terms of a perturbation series.

4.2 THE SETTLING VELOCITY

By the sediment settling velocity w_o we understand the terminal velocity of a single sediment particle which settles through an extended, resting fluid.

In that situation the fluid drag on the particle balances exactly the force of gravity. For a spherical particle with relative density s and diameter d this can be expressed by

$$\frac{1}{2}\rho\frac{\pi}{4}d^2 C_D w_o^2 = \rho(s-1)\frac{\pi}{6}d^3 g \tag{4.2.1}$$

which corresponds to

$$w_o = \sqrt{\frac{4(s-1)\,g\,d}{3C_D}} \tag{4.2.2}$$

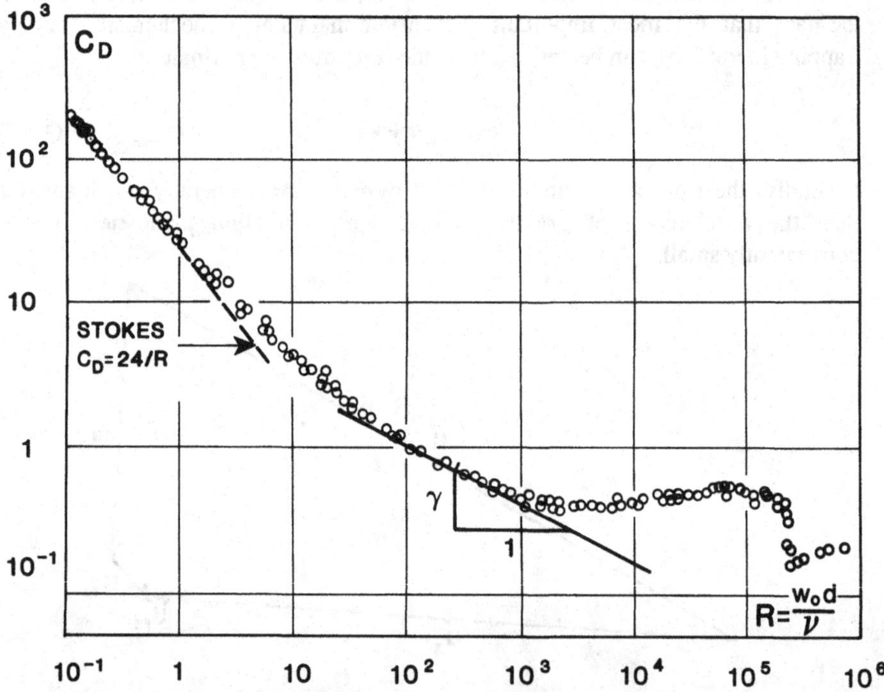

Figure 4.2.1: Drag coefficients for spherical particles.

The drag coefficient C_D is a function of the particle Reynolds number $R = w_o d/\nu$ and of the particle shape. The relationship for spherical particles is shown in Figure 4.2.1. The inclined asymptote

$$C_D = \frac{24}{w_o d/\nu} \tag{4.2.3}$$

corresponds to the linear drag law, Stokes law

$$F_D = 3\pi\rho\nu d w_o \tag{4.2.4}$$

which applies for very low Reynolds numbers.

Balancing the linear drag force given by equation (4.2.4) with gravity leads to the alternative expression

$$w_o = \frac{(s-1)\, g\, d^2}{18\,\nu} \tag{4.2.5}$$

which is valid for very small $(R = w_o d/\nu < 1)$, spherical particles.

Gibbs et al (1971) provided the following empirical formula for $w_o(d)$

$$w_o = \frac{-3\nu + \sqrt{9\nu^2 + g\, d^2 (s-1)(0.003869 + 0.02480\, d)}}{0.011607 + 0.07440\, d} \tag{4.2.6}$$

where w_o is measured in *cm/s*, d in *cm*, and the viscosity ν in cm^2/s. This formula is applicable for spheres in water for the diameter range [*0.0063cm; 1.0cm*].

Natural sand particles are of course not spherical, but more or less angular or even disc shaped in the case of shell hash. For such natural grain shapes the formula (4.2.6) should represent the upper limit for the settling velocity as function of size. An example of measured settling velocities as function of "sieve size" for a sample of natural beach sand is shown in Figure 4.2.2.

Many particles which settle together in a thick suspension will settle somewhat slower than single particles if the suspension is homogeneous. The reason is, that the downward motion of one sediment particle will generate a compensating upward flow elsewhere, which will delay the downward motion of the sediment particles.

On the other hand, a dense cloud of sediment in an otherwise clear fluid will settle faster than the typical settling velocity of the individual particles. This effect may be the reason for the unexpectedly large, measured settling velocities for the smaller (*d<0.04 cm.*) particles in Figure 4.2.2.

These small, angular sand grains may have been settling in clouds. Alternatively they may have obtained abnormally large averaged settling velocities by hitch-hiking during parts of their trip in the wakes of larger particles.

These effects must be considered when settling velocity measurements from settling tubes, where many sand sizes are dropped together, are interpreted.

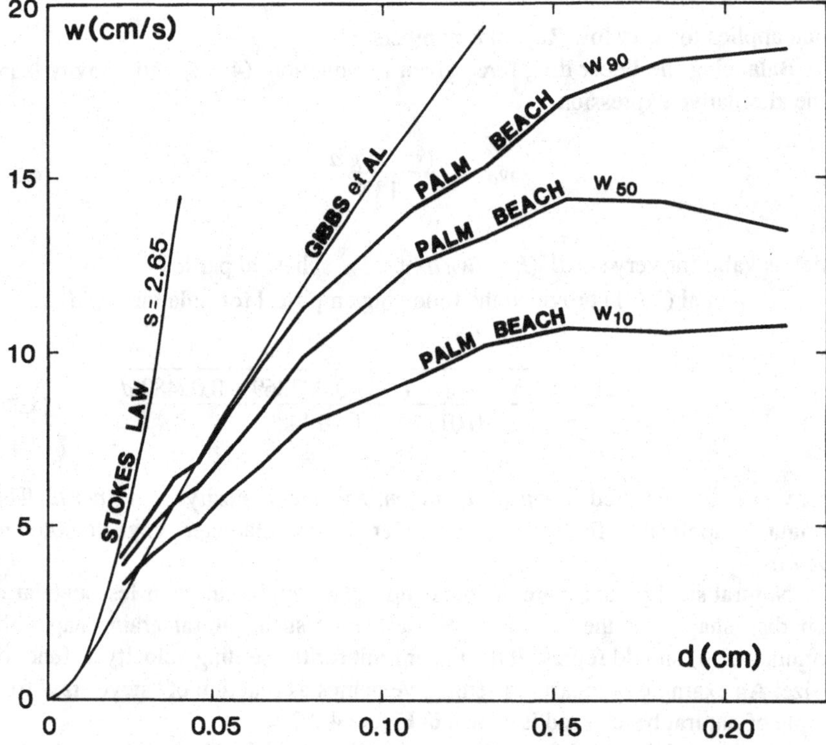

Figure 4.2.2: Settling velocities (w_{10}, w_{50}, w_{90}) determined with a settling tube for narrow sieve fractions of a sediment sample from the swash zone at Palm Beach, Sydney, Australia. This sample contained a large fraction of shell hash, so the measured settling velocities lie considerably below the curve given by Gibbs et al for quartz spheres, Equation (4.2.6). The relatively higher, measured values for the smaller grain sizes are probably due to the particles settling as a cloud rather than as individual particles.

4.3 EQUATION OF MOTION FOR SUSPENDED PARTICLES

The following section briefly discusses the various terms in the equation of motion for suspended sediment particles. Subsequently, the equation is brought into the most convenient form for deriving a perturbation solution in the form of Equation (4.1.2), page 163.

Consider a quasi-spherical particle with diameter d and relative density s which moves under the action of gravity and pressure gradients and drag from the surrounding fluid.

We neglect the history term of Basset (1888) which accounts for changes in the fluid drag due to changes in the flow structure around the particle, and neglect any effects of the particles rotation such as the Magnus effect (Magnus 1853). The reader is referred to, for example, Clift et al (1978) for a more detailed discussion.

With those simplifications applied, the equation of motion reads

$$\rho s \frac{\pi}{6} d^3 \frac{d\mathbf{u}_S}{dt} = \rho s \frac{\pi}{6} d^3 \mathbf{g} - \frac{\pi}{6} d^3 \nabla p + \frac{1}{2} \rho \frac{\pi}{4} d^2 C_D' \, |\mathbf{u} - \mathbf{u}_S| (\mathbf{u} - \mathbf{u}_S) + \frac{\pi}{6} \rho d^3 C_M \frac{d}{dt} (\mathbf{u} - \mathbf{u}_S) \tag{4.3.1}$$

The variable drag coefficient C_D' is defined by

$$\frac{C_D'}{C_D} = \left(\frac{|\mathbf{u}_S - \mathbf{u}|}{w_o} \right)^{-\gamma} \tag{4.3.2}$$

in accordance with the variation of C_D with changing particle Reynolds number, see Figure 4.2.1.

From Figure 4.2.1, we see that $\gamma = 1$ for small particles which move in accordance with Stokes Law, and $\gamma = 0$ for large particles which settle under fully developed turbulent conditions.

The usefulness of C_D' has been proven by the numerical study of Ho (1964) but the introduction of C_D' and γ in the present context also enables us to develop general expressions which cover both of the important, special cases $\gamma = 1$ and $\gamma = 0$.

The last term in Equation (4.3.1) corresponds to fluid pressure on the added hydrodynamic mass, and the coefficient C_M may be assumed to have a value close to *0.5* which is the theoretical C_M-*value* for a sphere.

By introducing the pressure gradient vector in the form

$$\nabla p = \rho g - \rho \frac{du}{dt} \tag{4.3.3}$$

and rearranging, the equation of motion becomes

$$(s+C_M)\frac{du_s}{dt} = (s-1)\,g + (1+C_M)\frac{du}{dt} - \frac{3}{4}\frac{C_D'}{d}|u_s - u|(u_s - u) \tag{4.3.4}$$

We wish to write the equation of motion in terms of the relative sediment particle velocity $u_r = u_s - u$ and for this purpose we note the following about the total derivative

$$\begin{aligned} \frac{du_s}{dt} &= \frac{\partial u_s}{\partial t} + u_s \cdot \nabla u_s \\ &= \frac{\partial}{\partial t}(u+u_r) + (u+u_r)\cdot\nabla(u+u_r) \\ &= \frac{\partial u}{\partial t} + \frac{\partial u_r}{\partial t} + u\cdot\nabla u + (u+u_r)\cdot\nabla u_r + u_r\cdot\nabla u \\ &= \frac{du}{dt} + \frac{du_r}{dt} + u_r \cdot \nabla u \end{aligned} \tag{4.3.5}$$

where we have applied

$$\frac{du_r}{dt} = \frac{\partial u_r}{\partial t} + (u + u_r)\cdot\nabla u_r \tag{4.3.6}$$

With Equation (4.3.5) inserted, the equation of motion (4.3.4) becomes

$$(s+C_M)\left(\frac{du_r}{dt} + u_r \cdot \nabla u\right) = (s-1)g + (1-s)\frac{du}{dt} - \frac{3}{4}\frac{C_D'}{d}|u_s - u|(u_s - u) \tag{4.3.7}$$

and with

$$\alpha = \frac{s-1}{s+C_M} \tag{4.3.8}$$

we get

$$\frac{du_r}{dt} + \frac{3}{4}\frac{C_D'}{d(s+C_M)} u_r u_r = \alpha g - \alpha \frac{du}{dt} \tag{4.3.9}$$

It is convenient to eliminate the drag coefficient C_D' from this equation. To this end, we apply the definition of C_D, Equation (4.2.1), page 164, which gives

$$\frac{3}{4}\frac{C_D}{d} = \frac{(s-1)g}{w_O^2} \tag{4.3.10}$$

and the definition, Equation (4.3.2), of C_D'. By inserting these into Equation (4.3.9) we find

$$\frac{du_r}{dt} + \alpha g \left(\frac{u_r}{w_O}\right)^{1-\gamma} \frac{u_r}{w_O} + u_r \cdot \nabla u = \alpha g - \alpha \frac{du}{dt} \tag{4.3.11}$$

This general differential equation for the relative sediment particle velocity u_r is the basis for the analysis of the motion of suspended particles in the following sections.

The nature of the solutions is found to depend in a very important way on the third term $u_r \cdot \nabla u$, which, in essence, represents the effect of the small scale flow structure.

This term is negligible in a pure wave motion (for $w_O T/L << 1$), and this is the reason why a pure wave motion has very little effect on the sediment settling rate, see Figure 4.1.1, page 162, and Section 4.5.5, page 178.

On the other hand, for a sediment particle inside a vortex with diameter of the order of ten centimetres, the term $u_r \cdot \nabla u$ is very significant, and it provides the vortex trapping mechanism which is discussed in Section 4.6, pp 181-189.

Equation (4.3.11) is linear for $\gamma = 1$, and exact analytical solution is therefore possible, for small particles for which $\gamma = 1$, see Figure 4.2.1.

For larger sediment particles, which have γ-values in the interval $[0;1]$, perturbation solutions of the form

$$u_s = u + u_r = u + w_O + u_{r1} + u_{r2} + \ldots.$$

are provided in Section 4.5 for homogeneous flows and in Section 4.6 for particles in vortex flows.

4.4 ACCELERATING SEDIMENT IN A RESTING FLUID

Consider the simplified equation for the relative sediment velocity

$$\frac{d\mathbf{u}_r}{dt} + \alpha\, g \frac{\mathbf{u}_r}{w_o} = \alpha\, \mathbf{g} \tag{4.4.1}$$

which results from Equation (4.3.11) when the fluid is at rest (or moving with a steady, uniform velocity), and the particle is so small that the flow around it is laminar and Stokes law applies ($\gamma = 1$).

This equation is equivalent to

$$\frac{d}{dt}(\mathbf{u}_r - \mathbf{w}_o) = -\frac{\alpha\, g}{w_o}(\mathbf{u}_r - \mathbf{w}_o) \tag{4.4.2}$$

where the vector $\mathbf{g}$ has been written as $g\mathbf{w}_o/w_o$, and this equation has the solution

$$\mathbf{u}_r(t) = \mathbf{w}_o + [\mathbf{u}_r(t_o) - \mathbf{w}_o]\, e^{-\alpha g (t - t_o)/w_o} \tag{4.4.3}$$

From this we see that deviations from the terminal settling velocity $\mathbf{w}_o$ decay exponentially with a time scale of $w_o/\alpha g$ which, for beach sand with relative density 2.65 and settling velocity $0.02 m/s$, is approximately $4 \cdot 10^{-3} s$.

This means that, starting from rest, such a sediment particle will be moving at 99% of its terminal settling velocity after only 0.002 seconds.

In other words, this means that deviations from the first approximation

$$\mathbf{u}_s = \mathbf{u} + \mathbf{w}_o \tag{4.1.3}$$

are normally very small.

Note that the discussion above is only meant as an order of magnitude estimate. A detailed analysis of situations where the magnitude of $\mathbf{u}_r = \mathbf{u}_s - \mathbf{u}$ changes considerably should include the history term (Basset 1888), but that would preclude analytical solution.

A quantitative discussion of the importance of the Basset term is given in terms of numerical solutions by Hardistry and Lowe (1991).

4.5 SEDIMENT PARTICLES IN ACCELERATED FLOW

4.5.1 Introduction

In the previous section we saw that sediment particles in a resting fluid will attain a relative velocity approximately equal to the terminal settling velocity after only about one hundredth of a second.

However, if the fluid motion is accelerated, deviations of the order εw_o will remain where

$$\varepsilon = \frac{1}{g}\left|\frac{du}{dt}\right| \tag{4.1.1}$$

The nature of these deviations is the subject of the following section.

4.5.2 Small particles in accelerated flow

Most of the important characteristics of sediment motion in accelerated flow are essentially the same for large and small particles. That is, their general nature does not depend on the drag force being linear or non-linear. Consequently, the general nature of these effects is quite adequately described by the linear equation which can be derived from Equation (4.3.11) under the "small-particle-assumptions".

When the particles are so small that the drag force is linear and given by Equation (4.2.4), the equation of motion (4.3.11) is linear. Under this assumption of linearity and with the relative sediment particle velocity written in the form

$$\boldsymbol{u}_r = \boldsymbol{w}_o + \boldsymbol{u}_{r1} \tag{4.5.1}$$

we find the following linear equation for $\boldsymbol{u}_{r1}$

$$\frac{d\boldsymbol{u}_{r1}}{dt} + \alpha g \frac{\boldsymbol{u}_{r1}}{w_o} + \boldsymbol{u}_{r1}\cdot\nabla\boldsymbol{u} = -\alpha\frac{d\boldsymbol{u}}{dt} - \boldsymbol{w}_o\cdot\nabla\boldsymbol{u} \tag{4.5.2}$$

This equation is particularly simple to solve if the flow field may be assumed homogeneous so that the gradient matrix is zero, $\nabla\boldsymbol{u} = \boldsymbol{0}$.

As an example, a solution (Equation 4.5.27, page 176) is given in Section 4.5.4, for settling through homogeneous, oscillatory flow.

4.5.3 Larger particles in homogeneous, accelerated flow

The non-linearity of the drag force for larger ($\gamma < 1$) sediment particles gives rise to a reduction of the settling velocity when the particles settle through an accelerated flow.

To assess the importance of this effect for suspended sediment we must determine its magnitude and that is the subject of the present section.

In order to obtain analytical solutions we shall express the relative sediment particle velocity u_r as a perturbation series in $\varepsilon = \frac{1}{g}\frac{|du|}{dt}$ and formally assume $\varepsilon \ll 1$.

For sediment particles moving in a homogeneous, oscillatory flow, it can then be shown that all effects of order εw_o are analogous to the results for small particles derived from the linear Equation (4.5.2). These effects are all purely oscillatory, so there is no settling velocity reduction of the order εw_o. The settling velocity reduction due to non-linear drag is of the order $\varepsilon^2 w_o$.

We introduce the following dimensionless variables

$$U = u/R\omega$$

$$U_r = u_r/w_o$$

$$T = \omega t \tag{4.5.3}$$

where R is the typical scale of the water motion and ω its radian frequency. Then Equation (4.3.11) for particles in a uniform flow becomes

$$\frac{dU_r}{dT} + \frac{\alpha g}{w_o \omega} U_r^{1-\gamma} U_r = \frac{\alpha g}{w_o \omega} - \alpha \frac{R\omega^2}{w_o \omega}\frac{dU}{dT} \tag{4.5.4}$$

or

$$\frac{dU_r}{dT} + \frac{\alpha g}{w_o \omega} U_r^{1-\gamma} U_r = \frac{\alpha g}{w_o \omega} - \varepsilon \frac{\alpha g}{w_o \omega}\frac{dU}{dT} \tag{4.5.5}$$

where $\varepsilon = R\,\omega^2/g$. We seek solutions to this equation in the form of a perturbation series

$$u_r = w_o + u_{r1} + u_{r2} + \ldots \tag{4.5.6}$$

and the corresponding dimensionless $U_r = u_r/w_o$

$$U_r = U_{ro} + \varepsilon U_{r1} + \varepsilon^2 U_{r2} + \ldots$$

$$U_r = \begin{pmatrix} 0 \\ -1 \end{pmatrix} + \varepsilon \begin{pmatrix} U_{r1} \\ W_{r1} \end{pmatrix} + \varepsilon^2 \begin{pmatrix} U_{r2} \\ W_{r2} \end{pmatrix} + \ldots \tag{4.5.7}$$

Then the magnitude of U_r is given by

$$U_r = \sqrt{(\varepsilon U_{r1} + \varepsilon^2 U_{r2} + \ldots)^2 + (-1 + \varepsilon W_{r1} + \varepsilon^2 W_{r2} + \ldots)^2} \tag{4.5.8}$$

and for the linearisation of the drag term in Equation (4.5.5) we find

$$U_r^{1-\gamma} = 1 - \varepsilon (1-\gamma) W_{r1} + \varepsilon^2 \frac{1-\gamma}{2} U_{r1} + \varepsilon^2 \frac{\gamma^2 - \gamma}{2} W_{r1}^2 - \varepsilon^2 (1-\gamma) W_{r2} + \ldots \tag{4.5.9}$$

after using the binomial expansion

$$(1+x)^p = 1 + px + \frac{p(p-1)}{2} x^2 + \ldots \tag{4.5.10}$$

We can now find a hierarchy of equations of motion corresponding to rising order in ε by inserting the expansion (4.5.9) into Equation (4.5.5), and separating terms of equal order in ε.

Order 1: For the order 1 solution we find the trivial equation

$$\mathbf{0} + \frac{\alpha g}{w_o \omega} \begin{pmatrix} 0 \\ -1 \end{pmatrix} = \frac{\alpha}{w_o \omega} \begin{pmatrix} 0 \\ -g \end{pmatrix} \tag{4.5.11}$$

because we have set the order-1-solution to be w_o

Order ε : For the order ε we find

$$\frac{d}{dT}\begin{pmatrix} U_{r1} \\ W_{r1} \end{pmatrix} + \frac{\alpha g}{w_o \omega}\begin{pmatrix} U_{r1} \\ (2-\gamma)W_{r1} \end{pmatrix} = -\frac{\alpha g}{w_o \omega}\frac{d}{dT}\begin{pmatrix} U \\ W \end{pmatrix} \tag{4.5.12}$$

corresponding to the dimensional

$$\frac{d}{dt}\begin{pmatrix} u_{r1} \\ w_{r1} \end{pmatrix} + \frac{\alpha g}{w_o}\begin{pmatrix} u_{r1} \\ (2-\gamma)\, w_{r1} \end{pmatrix} = -\alpha \frac{d}{dt}\begin{pmatrix} u \\ w \end{pmatrix} \tag{4.5.13}$$

Order ε^2 : For the order ε^2 we find

$$\frac{d}{dT}\begin{pmatrix} U_{r2} \\ W_{r2} \end{pmatrix} + \frac{\alpha g}{w_o \omega}\begin{pmatrix} U_{r2} \\ (2-\gamma)W_2 \end{pmatrix} = \frac{\alpha g}{w_o \omega}\begin{pmatrix} (1-\gamma)U_{r1}\, W_{r1} \\ \frac{1-\gamma}{2}\, U_{r1}^2 + \frac{2-3\gamma+\gamma^2}{2}\, W_{r1}^2 \end{pmatrix} \tag{4.5.14}$$

with the corresponding dimensional equation

$$\frac{d}{dt}\begin{pmatrix} u_{r2} \\ w_{r2} \end{pmatrix} + \frac{\alpha g}{w_o}\begin{pmatrix} u_{r2} \\ (2-\gamma)w_{r2} \end{pmatrix} = \frac{\alpha g}{w_o^2}\begin{pmatrix} (1-\gamma)u_{r1}w_{r1} \\ \frac{1-\gamma}{2}\, u_{r1}^2 + \frac{2-3\gamma+\gamma^2}{2}\, w_{r1}^2 \end{pmatrix} \tag{4.5.15}$$

4.5.4 Settling through a homogeneous, oscillatory flow

Substantial efforts have been made to find out if a homogeneous, oscillatory water motion has any net effect on the settling velocity of sand. One possible cause of a retardation is the non-linearity of the drag force.

Ho (1964) made numerical calculations of this non-linear drag effect, and they compared favourably with experiments. However, no analytical solution was given to provide an easy estimate of the effect.

In the following, we will give a simple analytical solution and show that the settling velocity reduction due to non-linear drag in a homogeneous oscillatory flow is of the order of magnitude $(R\,\omega^2/g)^2 w_o$ where $R\omega^2$ is the typical vertical acceleration of the fluid motion.

Let us consider a flow field similar to the one used by Ho (1964) and others in experiments. The water is oscillated vertically as a rigid body

$$\mathbf{u} = \mathbf{u}(t) = \begin{pmatrix} 0 \\ R\,\omega\cos\omega t \end{pmatrix} \tag{4.5.16}$$

We need only consider the vertical fluid motion

$$w(t) = R\,\omega\cos\omega t \tag{4.5.17}$$

and the corresponding vertical sediment velocity

$$w_s(t) = w(t) - w_o + w_{r1} + w_{r2} + \ldots \tag{4.5.18}$$

To find the first order solution w_{r1}, we use the vertical part of Equation (4.5.13) which becomes, in this case

$$\frac{dw_{r1}}{dt} + \frac{\alpha g}{w_o}(2-\gamma)\,w_{r1} = -\alpha\frac{d}{dt}(R\,\omega\cos\omega t) \tag{4.5.19}$$

$$\frac{dw_{r1}}{dt} + \frac{\alpha g}{w_o}(2-\gamma)\,w_{r1} = \alpha R\,\omega^2\sin\omega t \tag{4.5.20}$$

This equation has the simple harmonic solution

$$w_{r1} = \frac{R\omega^2}{g}\frac{w_o}{2-\gamma}\frac{1}{\sqrt{1+\beta_z^2}}\sin(\omega t - \tan^{-1}\beta_z) \tag{4.5.21}$$

where

$$\beta_z = \frac{w_o\,\omega}{(2-\gamma)\,\alpha\,g} \tag{4.5.22}$$

We note that for sand size particles in water the value of β_z is of the order of magnitude 10^{-2}. Hence, for practical purposes, the expression (4.5.21)may be approximated by

$$w_{r1} \approx \frac{R\omega^2}{g}\frac{w_o}{2-\gamma}\sin\omega t \tag{4.5.23}$$

and hence with reference to Equation (4.5.17)

$$w_{r1} \approx -\frac{w_o}{(2-\gamma)g}\frac{dw}{dt} \tag{4.5.24}$$

Note that the magnitude of w_{r1} relative to the still water settling velocity w_o is the ratio between fluid accelerations and gravity

$$\frac{w_{r1}}{w_o} \approx \frac{1}{g}\left|\frac{du}{dt}\right| = \varepsilon \tag{4.5.25}$$

[(2-γ) is of the order unity].

According to the approximation (4.5.24) the sediment particle velocity, correct to order ε , has the form

$$\mathbf{u}_s = \mathbf{u} + \mathbf{w}_o - \frac{w_o}{(2-\gamma)g}\frac{d\mathbf{u}}{dt} \tag{4.5.26}$$

which we can write as

$$\mathbf{u}_s = \mathbf{u}\,(t-\delta_t) + \mathbf{w}_o \tag{4.5.27}$$

This is in accordance with the time lag hypothesis of Bhattacharya (1971), and shows that his assumed time lag is real and has the value

$$\delta_t = \frac{w_o}{(2-\gamma)g} \tag{4.5.28}$$

or approximately w_o/g, which is of the order *0.001s* to *0.01s* for beach sand.

We note that the solution (4.5.21) has the time average zero so there is no settling velocity reduction of magnitude εw_o.

There is however a net reduction of the settling velocity of order $\varepsilon^2 w_o$ when the drag force is non-linear. We find its magnitude by inserting the first order solution (4.5.21) into the second order equation (4.5.15).

In order to simplify the algebra however, let us write the first order solution (4.5.21) in the form

$$w_{r1} = r_1 \sin \omega t^1 \tag{4.5.29}$$

Then the vertical part of the second order equation reads

$$\frac{dw_{r2}}{dt} + \frac{\alpha g}{w_o}(2-\gamma)\, w_{r2} = \frac{\alpha g}{w_o^2}\frac{2-3\gamma+\gamma^2}{2} r_1^2 \sin^2 \omega t^1 \qquad (4.5.30)$$

or

$$\frac{dw_{r2}}{dt} + \frac{\alpha g}{w_o}(2-\gamma)\, w_{r2} = \frac{\alpha g}{w_o^2}\frac{2-3\gamma+\gamma^2}{2} r_1^2 \frac{1}{2}(1-\cos 2\,\omega t^1) \qquad (4.5.31)$$

We are, at the moment, looking for a net (time-averaged) settling velocity reduction, so we need only consider the time-averaged solution $\overline{w_{r2}}$ corresponding to the time average of Equation (4.5.31). We find

$$\overline{w_{r2}} = \frac{1-\gamma}{4w_o} r_1^2 \qquad (4.5.32)$$

or with the full expression (4.5.21) inserted

$$\overline{w_{r2}} = \frac{1-\gamma}{4(2-\gamma)^2}\left(\frac{R\omega^2}{g}\right)^2 \frac{1}{1+\beta_z^2}\, w_o \qquad (4.5.33)$$

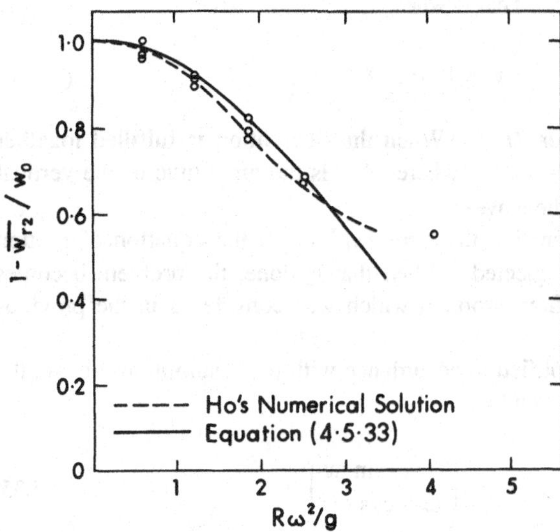

Figure 4.5.1: Settling velocity reduction for a *1mm* brass sphere in vertically oscillating water w_o = *0.39m/s*, ω = *22.7rad/s*. Measurements and numerical solution by Ho (1964).

This shows that the settling velocity reduction due to non-linear drag has the order of magnitude $\varepsilon^2 w_o$. The solution (4.5.33) is compared to measurements and numerical results from Ho (1964) in Figure 4.5.1.

Note that even though the derivation was based on the formal assumption of $\varepsilon \ll 1$, the agreement with Ho's experiments is good up to $\varepsilon \approx 3$.

4.5.5 Sand particles in wave motions

It is of considerable interest with respect to coastal sediment transport modelling to analyse the behaviour of sediment particles in a pure wave motion. That is the purpose of the following Section.

We shall find that deviations from the simple description

$$\boldsymbol{u_S} = \boldsymbol{u} + \boldsymbol{w_O} \tag{4.1.3}$$

are either purely oscillatory, and therefore without a time-averaged effect, or they are for all practical purposes negligibly small.

In most practical cases the water motion above the bottom boundary layer under waves can be considered quasi-uniform as far as the motion of suspended sediment is concerned.

More precisely, this is the case when the distance settled by a sediment particle over one wave period is small compared to the vertical scale on which the wave induced velocities change. That is when

$$k w_O T \ll 1 \tag{4.5.34}$$

where k is the wave number $2\pi/L$. When this condition is fulfilled together with the assumption that $kR_z \ll 1$, where R_z is the amplitude of the vertical water particle motion due to the waves.

These assumptions mean that the term $\boldsymbol{u_r} \cdot \nabla \boldsymbol{u}$ in the equation of motion (4.3.11), page 169, can be neglected. When that is done, the problem becomes analogous to homogeneous flow problem which was considered in the previous section.

For a wave motion simplified in accordance with the assumptions above, the fluid motion near the bed is given by

$$\boldsymbol{u}(t) = \begin{pmatrix} -\omega R_x \sin \omega t \\ \omega R_z \cos \omega t \end{pmatrix} \tag{4.5.35}$$

We now use Equation (4.5.13), page 174, to find the horizontal and vertical components of the first order solution u_{r1}, $[u_s(t) = u(t) + w_o + u_{r1}(t) + \ldots]$. In analogy with the solution (4.5.21) for purely vertical motion, we find

$$u_{r1} = \begin{pmatrix} \dfrac{R_x\,\omega^2}{g} w_o \dfrac{1}{\sqrt{1+\beta_x^2}} \cos\,(\omega t - \tan^{-1}\beta_x) \\ \\ \dfrac{R_z\,\omega^2}{g} \dfrac{w_o}{2-\gamma} \dfrac{1}{\sqrt{1+\beta_z^2}} \sin\,(\omega t - \tan^{-1}\beta_z) \end{pmatrix} \tag{4.5.36}$$

where

$$\beta_x = \frac{w_o\,\omega}{\alpha\, g} \tag{4.5.37}$$

and β_z is given by Equation (4.5.22), page 175.

Note that the time average $\overline{u_{r1}} = 0$. Therefore, as indicated by Figure 4.1.1 page 162, there is no net drift induced by the waves at order εw_o.

The solution is simplified considerably when the particle is so small that $\gamma \approx 1$ in which case we will also have $\beta_x, \beta_z \ll 1$. Then we get

$$u_{r1} \approx \begin{pmatrix} \dfrac{R_x\,\omega^2}{g} w_o \cos\,\omega t \\ \\ \dfrac{R_z\,\omega^2}{g} w_o \sin\,\omega t \end{pmatrix} = -\frac{w_o}{g}\frac{du}{dt} \tag{4.5.38}$$

and hence

$$u_s \approx u(t) + w_o - \frac{w_o}{g}\frac{du}{dt} \tag{4.5.39}$$

$$u_s \approx u\,(t - \frac{w_o}{g}) + w_o \tag{4.5.40}$$

This result proves and quantifies Bhattacharya's (1971) time lag hypothesis for small particles in the two-dimensional case. For larger particles $(\gamma \approx 0)$ the time lag for the vertical component is smaller, namely $w_o/(2-\gamma)g$ instead of w_o/g.

Another instructive interpretation of u_{r1} is given in Figure 4.5.2 from which it is realised that u_{r1} represents a centrifugal effect.

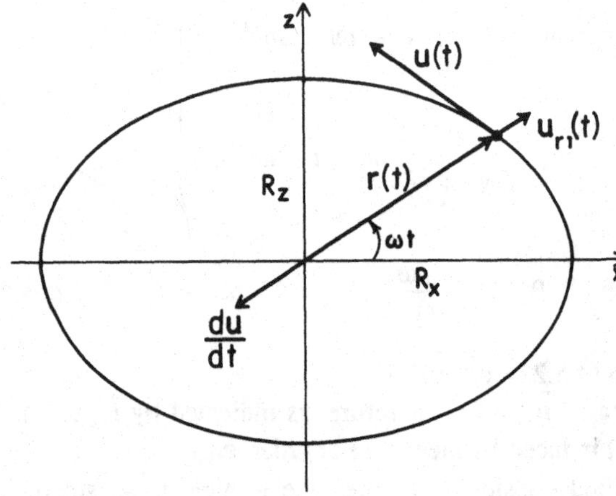

Figure 4.5.2: In a homogeneous flow field where the water particles follow elliptical orbits, the component u_{r1} of the sediment velocity is directed away from the orbit centre and thus represents a centrifugal effect.

For larger sediment particles $(\gamma \approx 0)$ there is a net delay of settling $\overline{w_{r2}}$ due to the non-linearity of the drag force. It can be evaluated by inserting the first order solution (4.5.36) into the second order Equation (4.5.15), page 174, and taking the time average.

The solution is analogous to Equation (4.5.33). Hence, the settling velocity reduction $\overline{w_{r2}}$ is of the order $\varepsilon^2 w_o$.

In most practical cases the reduction is insignificant. For example, for a typical wave motion with vertical semi-excursion $R_z = 0.1m$ and $\omega = 1s^{-1}$ we find from Equation (4.5.33).

$$\overline{w_{r2}} = 6 \cdot 10^{-6}\, w_o \tag{4.5.41}$$

which is a totally insignificant reduction of the settling velocity.

4.6 SEDIMENT TRAPPING BY VORTICES

4.6.1 Introduction

One of the most important mechanisms for entraining and suspending sediment is the trapping of sediment by vortices with horizontal axes. See Figures 4.1.1, page 162, and Figure 4.6.1.

This mechanism was first illustrated experimentally by Tooby et al (1977). The reader is referred to Tooby et al's magnificent photographic illustration. It shows a heavy particle as well as air bubbles trapped on circular paths inside a water filled cylinder which rotates around its horizontal axis.

In the following we shall develop an analytical description of the phenomenon along the same lines as used in the previous sections. That is, the sediment particle velocity is written in the form

$$\mathbf{u}_s = \mathbf{u} + \mathbf{w}_o + \mathbf{u}_{r1} + \mathbf{u}_{r2} + \dots \qquad (4.1.2)$$

$$= \mathbf{u} + w_o\left[\begin{pmatrix}0\\-1\end{pmatrix} + \varepsilon \mathbf{U}_{r1} + \varepsilon^2 \mathbf{U}_{r2} + \dots\right]$$

where $\mathbf{u}$ is the fluid velocity vector and $\mathbf{w}_o$ is the still water settling velocity. The perturbation parameter ε is the ratio between fluid accelerations and gravity

$$\varepsilon = \frac{1}{g}\left|\frac{d\mathbf{u}}{dt}\right| \qquad (4.1.1)$$

It will be shown that the essence of the trapping mechanism is purely kinematic and can be explained by the zero-order solution

$$\mathbf{u}_s = \mathbf{u} + \mathbf{w}_o \qquad (4.6.1)$$

4.6.2 The kinematic trapping effect

In the following we shall see how the zero order solution for the sediment particle velocity

$$\mathbf{u}_s(x,z,t) = \mathbf{u}(x,z,t) + \mathbf{w}_o \qquad (4.6.2)$$

provides adequate approximations to the sediment paths in simple flow structures. That is, on the basis of these approximations we can explain important sediment suspension mechanisms such as the sediment trapping by vortices.

Consider the water motion of a forced vortex with the velocity field

$$\boldsymbol{u}(x,z) = \omega\begin{pmatrix}-z\\x\end{pmatrix} \tag{4.6.3}$$

which is that of a rigid body rotating around the origin with angular velocity ω, see Figure 4.6.1. Applying Equation (4.6.2), we find the approximate sediment particle velocity

$$\boldsymbol{u}_s = \boldsymbol{u} + \boldsymbol{w}_o = \omega\begin{pmatrix}-z\\x\end{pmatrix} + \begin{pmatrix}0\\-w_o\end{pmatrix} = \omega\begin{pmatrix}-z\\x - w_o/\omega\end{pmatrix} \tag{4.6.4}$$

This equation is identical to Equation (4.6.3), which describes the water motion, except for a horizontal shift of magnitude w_o/ω.

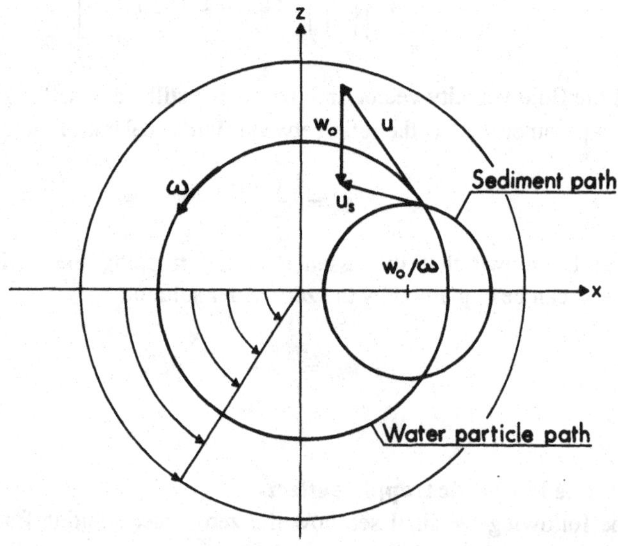

Figure 4.6.1: Sediment particle with settling velocity w_o moving in a forced vortex with velocity field $\boldsymbol{u}$ given by Equation (4.6.3). The particle paths corresponding to $\boldsymbol{u} + \boldsymbol{w}_o$ are circles and thus the time-averaged velocity of a sediment particle will (at this level of approximation) be zero.

Hence, the sediment paths are analogous to the fluid paths except for a horizontal shift of w_o/ω. That is, since all circles around (0, 0) are fluid particle paths, any circle around $(w_o/\omega, 0)$ is a possible sediment path, see Figure 4.6.1.

Note that the sediment paths in a pure, deep water wave motion are not closed, although the water particles move in circles like in the vortex, see Figure 4.1.1, page 162, and Section 4.5.5, page 178 ff.

It is important to understand the difference between wave motion and vortex motion with respect to sediment suspension. It stems from the fact that the wave motion is essentially homogeneous while the vortex motion is not.

Quantitatively, in terms of the equation of motion (4.3.11), page 169, the difference is that the term $\boldsymbol{u_r}\cdot\nabla\boldsymbol{u}$ is approximately $\boldsymbol{0}$ for the wave motion while it is significant for sediment particles inside vortices.

Tooby et al (1977) showed experimentally that small sediment particles in a vortex do, in fact, follow circular paths very closely and only very slowly spiral away from them. This spiralling process is discussed in Section 4.6.3, page 186.

The fact that the settling is strongly delayed in a vortex flow field has been noticed by Reizes (1977), who found the phenomenon in a numerical study. He concluded that the sediment particle must tend to spend more time in the upward moving parts of the flow than in the downward moving parts.

This is indeed the case. The sediment path shown in Figure 4.6.1 lies entirely in the "upward moving" part of the vortex.

For buoyant particles or air bubbles the centre of the "sediment path" would lie on the negative x-axis. Thus, they would spend the majority of the time within downward moving fluid.

The described trapping mechanism will work for all sand grains with settling velocity smaller than the maximum velocity in the vortex.

The next question to be asked is whether the trapping is a feature of the rather unnatural, forced vortex only.

The answer is no for the following reason. In general, the velocity field of a circular vortex with angular velocity ω can be written

$$\boldsymbol{u}(x,z) = \omega F(x^2+z^2)\begin{pmatrix}-z\\ x\end{pmatrix} \tag{4.6.5}$$

and the first approximation for the sediment velocity is then

$$\boldsymbol{u_s} = \boldsymbol{u} + \boldsymbol{w_o} = \omega F(x^2+z^2)\begin{pmatrix}-z\\ x\end{pmatrix} + \begin{pmatrix}0\\ -w_o\end{pmatrix} \tag{4.6.6}$$

For the components u_S and w_S of this velocity field, we therefore have the following symmetries

$$u_S(x,-z) = -u_S(x, z) \tag{4.6.7}$$

and

$$w_S(x,-z) = w_S(x, z) \tag{4.6.8}$$

This means that any particle path which crosses the x-axis twice must be closed. Since, as a result of the symmetry, a particle that has travelled along the curve P_1P_2 in Figure 4.6.2 must travel back to P_1 via the mirror image of P_1P_2.

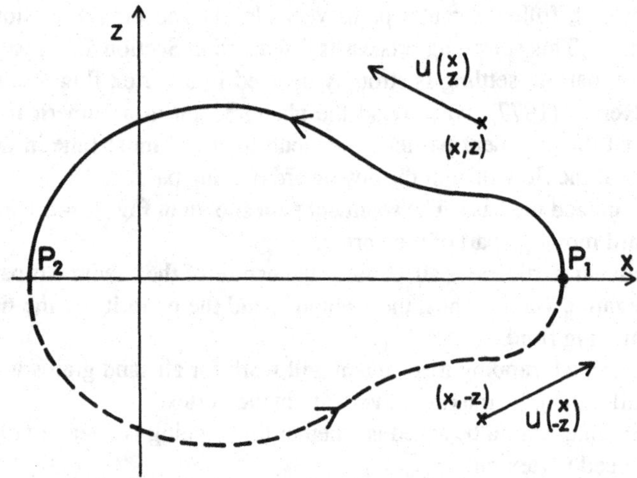

Figure 4.6.2: In a vortex, where the flow field has the form given by Equation (4.6.5), any particle path given by $u_s = u + w_o$ must be closed if it crosses the x-axis twice.

Hence, closed sediment paths, and the corresponding sediment trapping capability, are general features of vortex flow of the general form (4.6.5).

In particular, this includes the irrotational vortex with the fluid velocity field

$$\boldsymbol{u} = \frac{\omega R_o^2}{x^2 + z^2}\begin{pmatrix} -z \\ x \end{pmatrix} \tag{4.6.9}$$

Therefore, the trapping capability of a vortex is not conditional on the flow being rotational.

A fair model of many natural vortices is the Rankine vortex in which the fluid velocity field is given by

$$\boldsymbol{u}(x,z) = \frac{\omega}{1 + (x/R)^2 + (z/R)^2}\begin{pmatrix} -z \\ x \end{pmatrix} \tag{4.6.10}$$

It is characteristic for this vortex model that the core rotates as a rigid body, with the velocity being proportional to the distance from the centre, while $\boldsymbol{u}$ becomes inversely proportional to this distance farther away.

In this vortex a sand grain with settling velocity w_o can theoretically be at rest at the two singular points in Figure 4.6.3. At these two points we have $\boldsymbol{u} + \boldsymbol{w}_s = \boldsymbol{0}$, and their coordinates are $\left(R^2\omega/2w_o \pm \sqrt{(R^2\,\omega\,/\,2w_o)^2 - R^2}\,, 0\right)$. The circle which is shown, is given by the equation

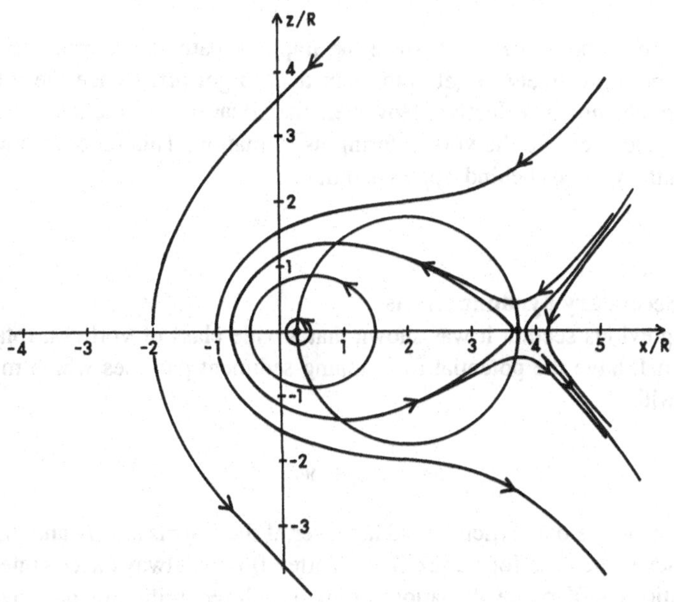

Figure 4.6.3: Particle paths corresponding to $\boldsymbol{u}_s = \boldsymbol{u} + \boldsymbol{w}_o$ in a Rankine vortex with the velocity field (4.6.10). In the inner region, where $\boldsymbol{u}$ is essentially proportional to the distance from the origin, we get closed sediment orbits that are very similar to the circles inside the forced vortex, see Figure 4.6.1.

$$\left(\frac{x}{R}-\frac{\omega R}{2w_o}\right)^2 + \left(\frac{z}{R}\right)^2 = \left(\frac{\omega R}{2w_o}\right)^2 - 1 \qquad (4.6.11)$$

and it is the locus of all points with $w_s = 0$. Sand grains will move upward ($w_s > 0$) in the interior and downward ($w_s < 0$) outside the circle.

Figure 4.6.3 also shows the sediment particle paths corresponding to $\mathbf{u}_s = \mathbf{u} + \mathbf{w}_s$. Some of these are closed and could thus keep sand grains trapped.

Trapping is only possible if the maximum upward water velocity in the vortex exceeds the sediment settling velocity which, for the Rankine vortex, is the case when $w_o < \omega R/2$.

The equation of the sediment path through a point $(x_o, 0)$ is

$$z^2 = -R^2 - x^2 + \left(R^2+x_o^2\right)\exp\left[\frac{2w_o}{\omega R}\frac{x}{R}-\frac{x_o}{R}\right]$$

Now it could be argued that, since the trapping paths are closed, sediment particles are no more likely to get onto them than to get off. Hence the trapping mechanism might not be effective. However, the situation in practice is that the sediment particles get into the vortex during its formation. This process is easy to observe with the vortices behind ripples and dunes.

4.6.3 Secondary dynamic effects

In the previous section it was shown that a wide class of vortices, rotational and irrotational, have the potential for trapping sediment particles which move in accordance with

$$\mathbf{u}_s = \mathbf{u} + \mathbf{w}_o \qquad (4.6.1)$$

This equation is only strictly valid however, if the flow is steady and uniform, and that is never the case for vortex flow. Vortex flow is always accelerated, and the accelerations will cause deviations from the closed sediment particle paths described above. In the following we shall describe these acceleration effects and show that they consist of a spiralling effect and a steady horizontal drift.

Qualitatively the spiralling effect is familiar and easy to understand. The vortex is a centrifuge. Heavy particles will spiral outwards and light particles or air bubbles will spiral inwards towards its orbit centre.

The steady horizontal drift is best understood by considering a sediment particle at rest at the point $(w_o/\omega, 0)$ in a forced vortex with angular velocity ω, see Figure 4.6.1, page 182.

Resting at this position is possible for the sediment particle under the assumption (4.6.1). However, a fluid particle whose orbit passes through this point is travelling on a circular path with radius w_s/ω and radian frequency ω. It is therefore accelerated towards the vortex centre with acceleration $\omega\, w_o$.

The horizontal pressure gradient which provides the fluid particle with this acceleration will tend to push the "resting" sediment particle towards the centre of the vortex. This describes the driving mechanism for the steady horizontal drift which is not only present at the point $(w_o/\omega\,, 0)$ but everywhere in the vortex.

In order to obtain a quantitative description of the drift, consider the sediment particle in Figure 4.6.4 which moves around the circle

$$r_s\,(t) \;=\; \begin{pmatrix} R\cos\omega t + w_o/\omega \\ R\sin\omega t \end{pmatrix} \tag{4.6.12}$$

in a forced vortex in accordance with the zero order solution $\mathbf{u}_s \;=\; \mathbf{u} + \mathbf{w}_o$.

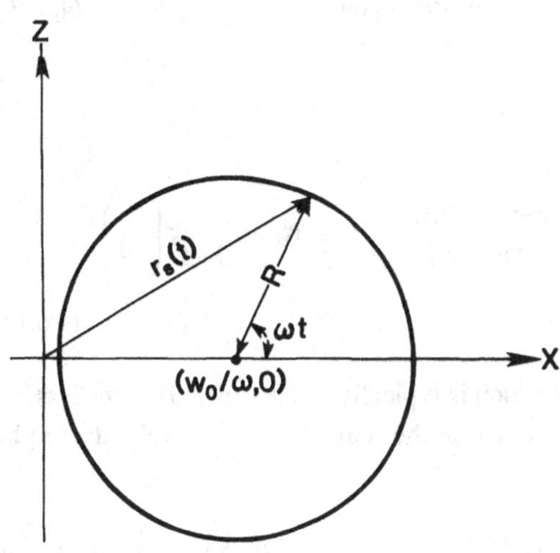

Figure 4.6.4: The acceleration required for the sediment particle to stay on the circular path is directed towards the centre of the circle. However, the pressure gradients in the undisturbed flow are directed towards (0,0). Hence, an excess pressure force will make the particle drift towards (0, 0).

A water particle on the fluid orbit, which passes through the sediment particles position, would be travelling on a circular path with radius $r_S(t)$ and radian frequency ω and hence would have the acceleration

$$\frac{du}{dt} = -\omega^2 r_S(t) \tag{4.6.13}$$

$$\frac{du}{dt} = -\omega^2 R\begin{pmatrix}\cos\omega t\\ \sin\omega t\end{pmatrix} - \begin{pmatrix}w_o\,\omega\\ 0\end{pmatrix} \tag{4.6.14}$$

The velocity field of the forced vortex is given by

$$\begin{pmatrix}u\\ w\end{pmatrix} = \omega\begin{pmatrix}-z\\ x\end{pmatrix} \tag{4.6.15}$$

and with this and Equation (4.6.14) inserted, the equation of motion (4.5.2) for a small particle reads

$$\begin{aligned}\frac{du_{r1}}{dt} + \frac{\alpha g}{w_o}u_{r1} - \omega w_{r1} &= \alpha\omega^2 R\cos\omega t + (\alpha-1)w_o\,\omega\\ \frac{dw_{r1}}{dt} + \frac{\alpha g}{w_o}w_{r1} + \omega u_{r1} &= \alpha\omega^2 R\sin\omega t\end{aligned} \tag{4.6.16}$$

These equations have the solution

$$\begin{pmatrix}u_{r1}\\ w_{r1}\end{pmatrix} = \frac{\dfrac{\omega^2 R}{g}w_o}{\sqrt{1+4\beta^2}}\begin{pmatrix}\cos(\omega t-\tan^{-1}2\beta)\\ \sin(\omega t-\tan^{-1}2\beta)\end{pmatrix} - (1-\alpha)\,w_o\frac{\beta}{1+\beta^2}\begin{pmatrix}1\\ -\beta\end{pmatrix} \tag{4.6.17}$$

where $\beta = w_o\omega/\alpha g$ is a quantity which is typically of the order 10^{-1} or less.

Hence, the order ε contribution u_{r1} to the sediment particle velocity can be closely approximated by

$$\begin{pmatrix}u_{r1}\\ w_{r1}\end{pmatrix} \approx \frac{\omega^2 R}{g}w_o\begin{pmatrix}\cos\omega t\\ \sin\omega t\end{pmatrix} - (1-\alpha)\,w_o\,\beta\begin{pmatrix}1\\ 0\end{pmatrix} \tag{4.6.18}$$

or

$$u_{r1} = \frac{\omega^2}{g} w_o \mathbf{r}(t) - (1-\alpha) w_o \beta \begin{pmatrix} 1 \\ 0 \end{pmatrix} \tag{4.6.19}$$

which shows that the first term represents a centrifugal effect with a spiralling time scale

$$t_s = \left[\frac{1}{R} \frac{dR}{dt} \right]^{-1} = \frac{g}{w_o \omega^2} \tag{4.6.20}$$

while the second term represents a horizontal, steady drift towards the z axis.

4.7 PARTICLES SETTLING THROUGH TURBULENCE

4.7.1 Introduction

In order to understand sediment suspension in turbulent flows it is first necessary to analyse the influence of turbulence on a single particle which is settling through it.

Experiments by Murray (1970) indicate that the turbulence has two main effects on settling particles, which may be quantitatively described in the following way.

Consider an ensemble of many identical particles which are dropped into a water filled jar where a certain level of turbulence is maintained by some kind of stirring, see Figure 4.7.1.

The distances L_i, which the particles settle during the interval t, are measured and the corresponding time averaged velocities $w_i = L_i/t$ are calculated. It may be observed then that the ensemble average $\overline{w_i}$ is less than the still water settling velocity

$$\overline{w_i} < w_o \tag{4.7.1}$$

and that the variance $Var\{w_i\}$ is roughly proportional to: the turbulence intensity $\overline{w^2}$; to the Lagrangian integral scale of the turbulence T_L; and to $1/t$. That is

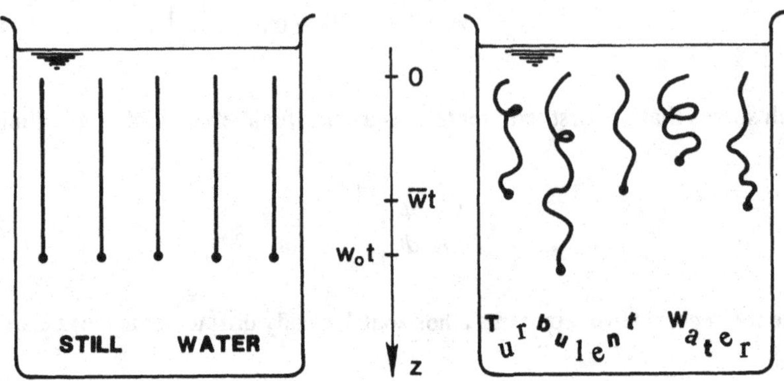

Figure 4.7.1: Sediment particles settling through turbulent water will, on the average, settle more slowly than they would through still water.

$$Var\{w_l\} \propto \overline{w^2}\, T_L/t \qquad \text{for } t > T_L \tag{4.7.2}$$

The second result can be understood in terms of G I Taylor's classical model for diffusion by continuous movement (Taylor 1921), which will be outlined in Section 4.7.3.

The first effect however, which is equally important for the maintenance of sediment suspension, is not as well understood. It is however closely related to the trapping mechanism described in Section 4.6, page 181 ff, by which heavy particles or air bubbles can be trapped for a long time inside a vortex. Obviously, a sediment particle which has been trapped for some time in a few vortices on its way down will be delayed considerably. The same applies to an air bubble which rises through a turbulent flow field.

4.7.2 The loitering effect

It is important to note, that the settling delay described above does not require a specific, highly organised vortex structure. Trapping by vortices is just one manifestation of the general "Loitering effect" which is illustrated in Figure 4.7.2.

The term "Loitering effect" refers to the fact that a settling or rising particle in a steady, non-uniform flow field will spend relatively more time (loiter) in those parts of the flow which move against its natural settling/rising velocity. By this mechanism, a non-uniform flow field with zero spatial mean velocity ($<u> = 0$) will delay the settling of sediment particles and the rising of air bubbles.

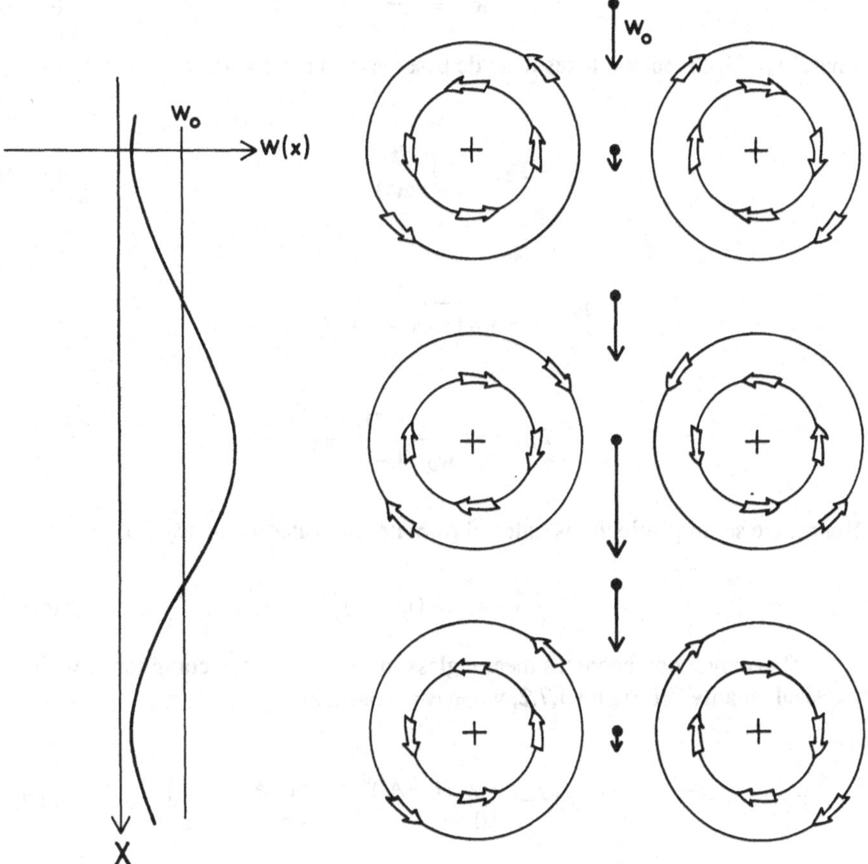

Figure 4.7.2: A sediment particle which settles through this arrangement of vortices will experience a considerable delay compared with settling through still water because it spends more time in the upward moving parts of the fluid than in the downward moving parts. Note that the spatial average of the particle velocity is w_o.

To quantify the effect, consider the special flow pattern in Figure 4.7.2. For simplicity, let the flow structure be such, that the particle velocity varies with z as

$$w_S(z) \;=\; w_O\,(1 + A\cos\frac{\pi z}{D}) \tag{4.7.3}$$

where D is the vortex diameter and $A < 1$. Then the resulting settling velocity is

$$\overline{w_S} \;=\; \frac{2D}{T_{2D}} \tag{4.7.4}$$

where T_{2D} is the time it takes to settle past two complete vortices

$$T_{2D} \;=\; \int_0^{2D} \frac{dz}{w_S(z)} \tag{4.7.5}$$

$$T_{2D} \;=\; \int_0^{2D} \frac{dz}{w_O\,(1+A\cos\pi z/D)}$$

$$T_{2D} \;=\; \frac{2\,D}{w_O\;(1-A^2)^{0.5}}$$

Hence, the settling velocity is reduced from its still water value w_O to

$$\overline{w_S} \;=\; w_O\,(1-A^2)^{0.5} \tag{4.7.6}$$

This expression becomes meaningless for $A > 1$ but in compliance with the physical "reality" of Figure 4.7.2, we may extend it to

$$\overline{w_S} \;=\; F_W(A) \;=\; \begin{Bmatrix} w_O\,(1-A^2)^{0.5} & \text{for } A < 1 \\ 0 & \text{for } A \geq 1 \end{Bmatrix} \tag{4.7.7}$$

This corresponds to the fact that when the maximum upward water velocity along the symmetry line in Figure 4.7.2 exceeds w_O there will be positions where $u + w_O = 0$ at which the particle will stop and could be trapped forever.

In practice, the time limit for trapping is set by the typical lifetime of local flow structures T_E which is called the Eulerian time scale for the turbulence.

4.7.3 Taylor's dispersion model

Taylor (1921) derived a description for the dispersion of a cloud of particles in turbulence. That model has since formed the basis for most research into turbulent diffusion and into some of the more fundamental aspects of turbulence. The main results of Taylor's theory can be summarised as follows.

Consider the one-dimensional case of spreading in the z-direction due to turbulent velocities w with zero mean and variance σ_w^2, $(\overline{w}, \overline{w^2}) = (o, \sigma_w^2)$.

The spread of a cloud of particles, originating at $z=o$ may then be quantified by the ensemble average $\overline{z_t^2}$ or the expected value $E\{z^2\}$ of the positions of these particles after a certain time t.

Noting that

$$\frac{d}{dt} z^2 = 2\, z\, w \qquad (4.7.8)$$

we get

$$E\{z^2(t)\} = E\{ 2 \int_o^t w(t_1) \int_o^{t_1} w(t_2)\, dt_2\, dt_1 \} \qquad (4.7.9)$$

In terms of the Lagrangian autocorrelation coefficient

$$\rho_{ww}(\Delta_t) = E\{w(t_o)\, w(t_o + \Delta_t)\}/\sigma_w^2 \qquad (4.7.10)$$

Equation (4.7.9) can be written as

$$E\{z^2(t)\} = 2\, \sigma_w^2 \int_o^t \int_o^{t_1} \rho_{ww}(t_1 - t_2)\, dt_2\, dt_1 \qquad (4.7.11)$$

The autocorrelation coefficient $\rho_{ww}(\Delta_t)$ is generally expected to have the shape shown in Figure 4.7.3. By definition the value at zero time lag is unity and because $\rho_{ww}(\Delta_t)$ is expected to be a smooth and even function, the derivative at $\Delta_t = 0$ should be zero, $\rho_{ww}'(0) = 0$.

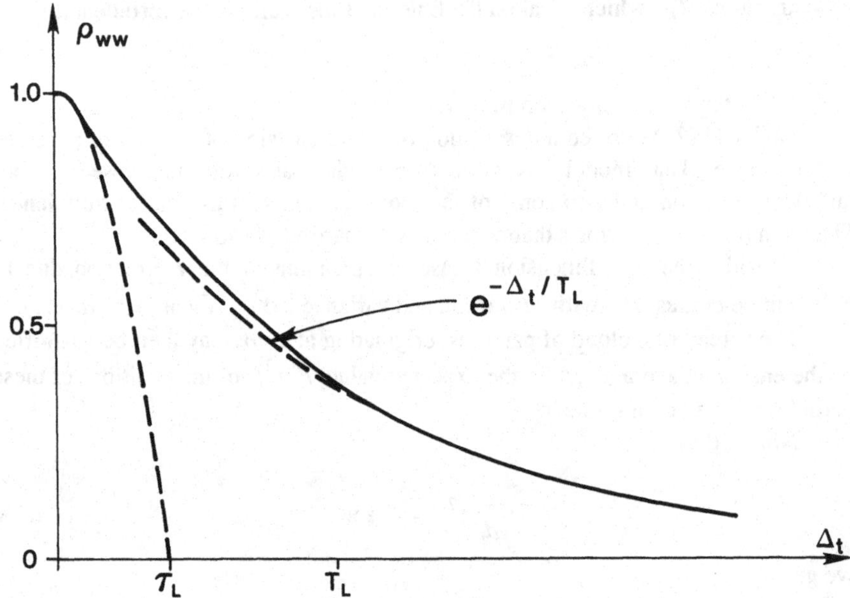

Figure 4.7.3: Typical shape of the Lagrangian autocorrelation coefficient.

Measurements of $\rho_{ww}(\Delta_t)$ are quite scarce, but they tend to confirm the general picture presented above, see e g Snyder & Lumley (1971) or Sato & Yamamoto (1987). Also, the integral

$$T_L = \int_0^\infty \rho_{ww}(t)\, dt \tag{4.7.12}$$

normally exists and it is called the Lagrangian integral scale.
The short time scale τ_L is called the Lagrangian microscale and is defined by

$$\tau_L = \sqrt{-2/\rho_{ww}''(0)} \tag{4.7.13}$$

In terms of these parameters, two asymptotic approximations for the spread given by Equation (4.7.11) can be derived, which are valid for relatively short and relatively long times respectively

$$E\{z^2\} \approx \sigma_w^2 t^2 \qquad \text{for } t < \tau_L \tag{4.7.14}$$

$$E\{z^2\} \approx 2\sigma_w^2 T_L t \qquad \text{for } t > T_L \tag{4.7.15}$$

Since, for $t_1 < \tau_L$, the inner integral in Equation (4.7.11) is approximately equal to t_1, while for large t_1, it is approximately T_L.

Considering a heavy particle with settling velocity w_o we may, as a first approximation, use the approximation $w_s \approx w - w_o$ for its vertical velocity. The accuracy of this approximation was discussed in Section 4.5.

Using this simple superposition law, Taylor's dispersion model can be used to estimate the variance of the settling velocities of the particles settling through the turbulence in Figure 4.7.1. If the settling time is large compared to the Integral time scale, we can use the approximation (4.7.15) and get

$$\mathrm{Var}\{w_i\} = E\{(z - w_o t)^2\}/t^2 \approx 2\sigma_w^2 T_L/t \tag{4.7.16}$$

Thus Taylor's model enables us to model the variance of the measured average velocities, w_i/t from the experiment in Figure 4.7.1. However, it says nothing about a reduction of the ensemble average settling velocity, $\overline{w_i}$ due to trapping in vortices or loitering in general.

To predict a delay, a model must in some way capture the trapping by vortices or the loitering effect. To do this, some account must be taken of the flow structure.

Taylor did not explicitly consider details of the flow structure because his approach was purely Lagrangian i e, $w = w(r_o, t)$. That is, after noting the particle's starting position, there is no further consideration of position. Time is the only independent variable.

4.7.4 Dispersion with loitering effect

Taylor (1921) developed the concept of diffusion by continuous movements in analogy with a simple, discrete random walk model where the correlation between the velocities w_i and w_{i+1} in successive steps was given by

$$\rho = \exp[-\delta_t/T_L] \tag{4.7.17}$$

where δ_t is the time step.

The model which is developed in the following, is designed to predict the loitering effect for non-neutral particles. The necessary difference from Taylor's model, which enables this, is that the correlation between velocities in successive steps must depend on the magnitude $|u_{i-1}|$ of the instantaneous velocity.

The correlation formula (4.7.17) corresponds to the relation

$$E\{|\delta_w|\} = E\{|\frac{dw}{dt}|\}\,\delta_t = \frac{\sigma_w}{T_L}\,\delta_t \tag{4.7.18}$$

In order to capture the loitering effect we must consider the velocity derivative in its Eulerian form

$$\frac{dw}{dt} = \frac{\partial w}{\partial t} + u\frac{\partial w}{\partial x} + v\frac{\partial w}{\partial y} + w\frac{\partial w}{\partial z} \tag{4.7.19}$$

and then in analogy with Equation (4.7.18) write

$$E\{\,|\delta_w|\,\} = E\left\{\left|\left(\frac{\partial w}{\partial t} + u\frac{\partial w}{\partial x} + v\frac{\partial w}{\partial y} + w\frac{\partial w}{\partial z}\right)\right|\right\}\delta_t = \frac{\sigma_w}{T_L}\,\delta_t \tag{4.7.20}$$

for the unconditional expected value of the absolute velocity increment.

Wishing to model the loitering effect, the essence of which is that small velocities correspond to slow changes of velocity, see Figure 4.7.2, we consider the conditional expected value of $|\delta_w|$

$$E\{\,|\delta_w|\;|\;u_{i-1}\} = E\left\{\left|\left(\frac{\partial w}{\partial t} + u_{i-1}\frac{\partial w}{\partial x} + v_{i-1}\frac{\partial w}{\partial y} + w_{i-1}\frac{\partial w}{\partial z}\right)\right|\;\Big|\;u_{i-1},v_{i-1},w_{i-1}\right\}\delta_t \tag{4.7.21}$$

The value of this expression will depend on the possible correlations between the velocity components (u,v,w) and the partial derivatives $(\frac{\partial w}{\partial t}, \frac{\partial w}{\partial x}, \frac{\partial w}{\partial y}, \frac{\partial w}{\partial z})$. However, knowing nothing about these correlations it seems reasonable, as the simplest option, to assume zero correlation so that

$$E\{\,|\delta_w|\;|u_{i-1}\} \;=$$

$$\delta_t\,[(E\{\frac{\partial w}{\partial t}\})^2 + (u_{i-1}\,E\{\frac{\partial w}{\partial x}\})^2 + (v_{i-1}\,E\{\frac{\partial w}{\partial y}\})^2 + (w_{i-1}\,E\{\frac{\partial w}{\partial z}\})^2\,]^{0.5} \tag{4.7.22}$$

After applying the Eulerian equivalents of (4.7.18):

$$E\{\,|\frac{\partial w}{\partial t}|\,\} \;=\; \frac{\sigma_w}{T_E} \tag{4.7.23}$$

and

$$E\{|\frac{\partial w}{\partial x}|\} \;=\; E\{|\frac{\partial w}{\partial y}|\} \;=\; E\{|\frac{\partial w}{\partial z}|\} \;=\; \frac{\sigma_w}{L_E} \tag{4.7.24}$$

we find

$$E\{|\delta_w|\;u_{i-1}\} \;=\; \frac{\delta_t}{T_E}(\,1 + A_E^2\,[\,(\frac{u_{i-1}}{\sigma_w})^2 + (\frac{v_{i-1}}{\sigma_w})^2 + (\frac{w_{i-1}}{\sigma_w})^2\,]\,)^{0.5} \tag{4.7.25}$$

where the parameter A_E which measures the time scale against the spatial scale of the turbulence is given by

$$A_E \;=\; \frac{\sigma_w\,T_E}{L_E} \tag{4.7.26}$$

Based on Equation (4.7.25), we see that the replacement for Taylor's constant correlation

$$\rho \;=\; \exp\,[-\delta_t/T_L] \tag{4.7.17}$$

is (under the assumption 4.7.22)

$$\rho_{i-1} \;=\; \exp\Big\{-\frac{\delta_t}{T_E}(\,1 + A_E^2\,[\,(\frac{u_{i-1}}{\sigma_w})^2 + (\frac{v_{i-1}}{\sigma_w})^2 + (\frac{w_{i-1}}{\sigma_w})^2\,]\,)^{0.5}\Big\} \tag{4.7.27}$$

For a non-neutral particle with still water settling velocity w_o which moves in accordance with $w_s = w - w_o$, the analogous expression is

$$\rho_{i-1} = \exp\left\{ -\frac{\delta_t}{T_E}\left(1 + A_E^2 \left[\left(\frac{u_{i-1}}{\sigma_w}\right)^2 + \left(\frac{v_{i-1}}{\sigma_w}\right)^2 + \left(\frac{w_{i-1} - w_o}{\sigma_w}\right)^2 \right] \right)^{0.5} \right\} \quad (4.7.28)$$

The last term in this expression generates the loitering effect because it will be small if w_o is nearly balanced by w_{i-1} and hence ρ_{i-1} will then be closer to unity.

The loitering effect expressed by Equation (4.7.28) is much weaker for three dimensional flow than it would be for purely vertical flow. This is because of the first two terms inside the square bracket whose randomness will smooth things out. However, the loitering effect may still be significant.

Figure 4.7.4 shows results of a numerical simulation for a particle with still water settling velocity w_o settling through three-dimensional turbulence with $\overline{u^2} = \overline{v^2} = \overline{w^2} = \sigma^2$. The model is described by

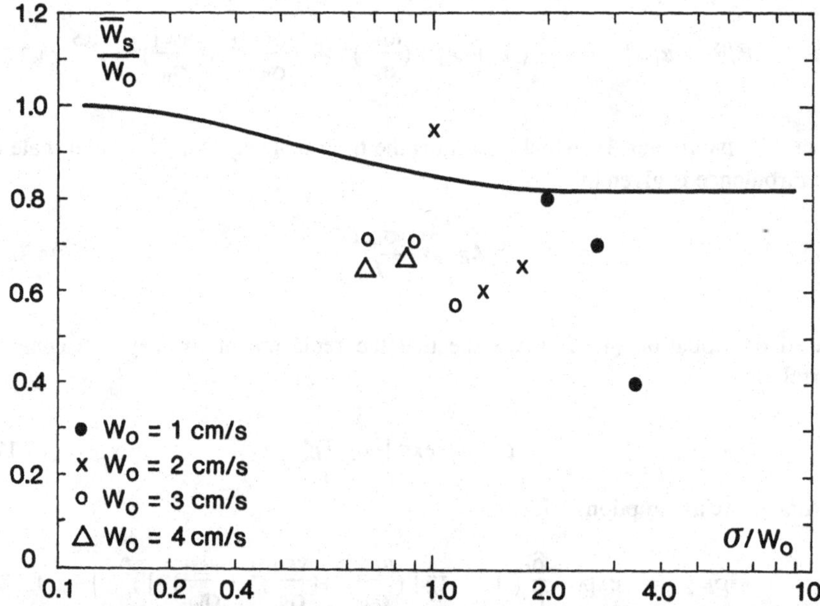

Figure 4.7.4: Measured settling velocities from grid turbulence by Murray (1970) and simulated values based on the random walk model described by Equations (4.7.28), (4.7.29), and (4.7.30) with $A_E = 1$.

$$u_i = \rho_{i-1}\, u_{i-1} + \sigma\sqrt{1-\rho_{i-1}^2}\;\zeta_{u,i}$$

$$v_i = \rho_{i-1}\, v_{i-1} + \sigma\sqrt{1-\rho_{i-1}^2}\;\zeta_{v,i}$$

$$w_i = \rho_{i-1}\, w_{i-1} + \sigma\sqrt{1-\rho_{i-1}^2}\;\zeta_{w,i} \qquad (4.7.29)$$

and

$$w_{s,i} = w_i - w_o \qquad (4.7.30)$$

where $(\zeta_u,\zeta_v,\zeta_w)$ are independent, normalised, normal variates and ρ_{i-1} is calculated from Equation (4.7.28).

The model predicts a settling velocity reduction which depends on the scale ratio A_E, as well as on the relative strength σ/w_o of the turbulence.

For $A_E \geq 1$ and $\sigma \geq w_o$ the model predicts a settling velocity reduction of about 20%, which is generally less than what was measured by Murray (1970).

The fact that Murray measured greater delays is probably due to his turbulence containing an element of organised vortices rather than being totally random, and thus creating the possibility of occasional trapping. This is likely to be the case for most types of natural turbulence.

Murray suggested that the observed delays might be due to non-linearity of the drag force, but he also showed in his Figure 5, that with the typical eddy radius assumed equal to the grid bar diameter (*1cm*), a frequency of *25-40Hz* would be required in order to generate settling delays of the order 40%.

Such high frequencies seem incompatible with the recorded rms-velocities which were only a few centimetres per second.

The analytical expression (4.5.33) for the non-linear drag effect gives for $R = 1cm$ and $\omega = \sigma_{max}/[1cm] = 3.4\, rad/s$, a reduction of only $10^{-5} w_o$. Thus the effect of possible non-linear drag is insignificant.

In the special case of $A_E = 0$ which corresponds to $L_E \rightarrow \infty$, the model described above becomes identical to Taylor's model, and there is no delay. We may call this case the strongly homogeneous case because it corresponds to a situation where the turbulence changes in step everywhere so to speak. The magnitude of the velocity increment is then independant of the instantaneous velocity magnitude. (Taylor showed that in order for the turbulence to be a stationary stochastic process, the velocities and the velocity increments must be uncorrelated: $\rho(w, \frac{dw}{dt}) = 0$, but this does not automatically mean that the absolute values must be uncorrelated).

On the other hand when $A_E > 0$ there is a positive correlation between $|u_{i-1}|$ and $|u_i - u_{i-1}|$ in the model above. For $A_E = 1$ we find

$$\rho(|u_{i-1}|, |w_i - w_{i-1}|) \approx 0.13 > 0 \tag{4.7.31}$$

The actual value of this correlation coefficient for various types of natural turbulence is presently unknown.

This means that the validity of the assumption (4.7.22) is basically unknown. If the correlation turned out to be zero for some flow, it would mean that there was no settling delay due to loitering in this flow.

If the correlation turned out to be negative, a random turbulence field would tend to enhance the settling velocity.

CHAPTER 5

SEDIMENT SUSPENSIONS

5.1 INTRODUCTION

5.1.1 Sediment concentrations and the transport rate

A considerable portion of the sediment transport in coastal areas is due to sediment which moves in suspension. It is therefore necessary to develop models for sediment suspensions so that concentration distributions can be calculated. These may in turn be combined with the sediment velocity distributions to give the transport rate. The sediment transport rate is usually given in the form

$$Q(t) = \int_0^D c(z,t)\, u_s(z,t)\, dz \tag{5.1.1}$$

for the x-direction and correspondingly for the y-direction.

The suspended sediment concentration $c(z,t)$ is, in the following, a dimensionless quantity calculated as the volume of solid sediment divided by the total volume of water-sediment mixture. The horizontal velocity $u_s(z,t)$ of a sediment particle can generally be assumed equal to the horizontal velocity of the immediately surrounding fluid, see Section 4.5.5, page 178.

The sediment concentration being dimensionless, the transport rate per unit width $Q(t)$, defined by Equation (5.1.1), has the dimension of L^2T^{-1} and the corresponding SI units of m^2/s.

5.1.2 Mixing length and sediment diffusivity

Since the sediment is heavier than water, it has a natural tendency to sink and settle out unless a compensating upward sediment flux is somehow created to balance the settling rate cw_s. The vertical sediment velocity w_s is practically equal to the local, vertical fluid velocity minus the still water settling velocity, $w_s \approx w - w_o$, see Section 4.5.5, page 178.

Stationary sediment suspensions do exist. Hence, the upward fluxes which balance settling must be present. They occur in turbulent flows because water parcels which travel upward through a given plane generally contain larger sediment concentrations than water parcels which travel downward through the same plane. In other words, the upward sediment flux required to balance the settling rate in a steady situation, is generated by vertical mixing. This type of process is called *gradient diffusion*.

Random vertical mixing can generate an upward sediment flux provided the average sediment concentration decreases with height above the bed.

To obtain a quantitative description of gradient diffusion, consider the simple mixing process which is illustrated by Figure 5.1.1.

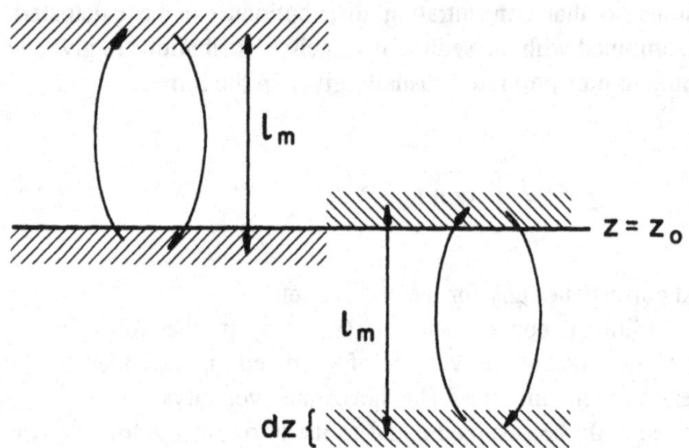

Figure 5.1.1: The process of gradient diffusion may be quantified in terms of the depicted exchange process.

Assume that the sediment concentration varies linearly with z and that mixing is provided by exchanging layers of thickness dz and separation l_m once every t seconds.

Since the concentration difference between such a pair of exchanging layers is $l_m \frac{dc}{dz}$, each exchange corresponds to transport of the volume $-\,dx\,dy\,dz\,l_m \frac{dc}{dz}$ upward through an area $dx\,dy$ of an intermediate plane, say $z = z_o$. The minus is due to the fact that, if $\frac{dc}{dz}$ is positive, the exchange results in a net downward transport of sediment. In most natural cases, $\frac{dc}{dz}$ is negative and the exchange process provides the upward transport required to balance settling.

Summation of the contributions from all exchanges which intersect the plane $z = z_0$, gives a total transport of $-dx\,dy\,l_m^2\frac{dc}{dz}$ through the area $dx\,dy$.

With the exchanges occurring at the rate of one every t seconds, the vertical transport rate per unit area is therefore

$$q_z = -\frac{l_m^2}{t}\frac{dc}{dz} \qquad (5.1.2)$$

This is generally written in terms of a sediment diffusivity ε_s as

$$q_z = -\varepsilon_s\frac{dc}{dz} \qquad (5.1.3)$$

corresponding to

$$\varepsilon_s = \frac{l_m^2}{t} \qquad (5.1.4)$$

Thus, sediment diffusivity has the dimension of L^2T^{-1}, like viscosity, and the SI units are m^2/s.

The equation which expresses a time-averaged balance between settling and the diffusive flux, given by Equation (5.1.3), reads

$$w_o\bar{c} + \varepsilon_s\frac{d\bar{c}}{dz} = 0 \qquad (5.1.5)$$

To obtain values of the sediment diffusivity from measured concentrations, under the assumption of a pure diffusion process, Equation (5.1.5) may be alternatively written as

$$\varepsilon_s = \frac{-w_o\bar{c}}{\frac{d\bar{c}}{dz}} \qquad (5.1.6)$$

Gradient diffusion can be used to describe the upward sediment flux if the mixing length l_m is small compared to the overall height of the concentration profile. However, if the two lengths are of the same order of magnitude, a different model must be applied. See the discussion in Section 5.4.2, page 233 for details.

Nevertheless, gradient diffusion has been used exclusively for the modelling of suspended sediment distributions since the Nineteen Thirties without critical assessment of its validity as a description of the actual physical process. Worrying experimental data like those of Coleman (1970) and of Nielsen (1983) have generally been bypassed by ad hoc modifications to the sediment diffusivity, see Figure 5.1.2.

Coleman's data show that, in order to model the distribution of suspended sediment in steady, open channel flows as pure gradient diffusion, the sediment diffusivity ε_s must generally be of a very different magnitude than the eddy viscosity ν_t, ($\nu_t \approx \kappa u_* z (1- z/D)$ for $z < D/2$). Furthermore, the magnitude of ε_s must be a strongly increasing function of the relative settling velocity w_o/u_*.

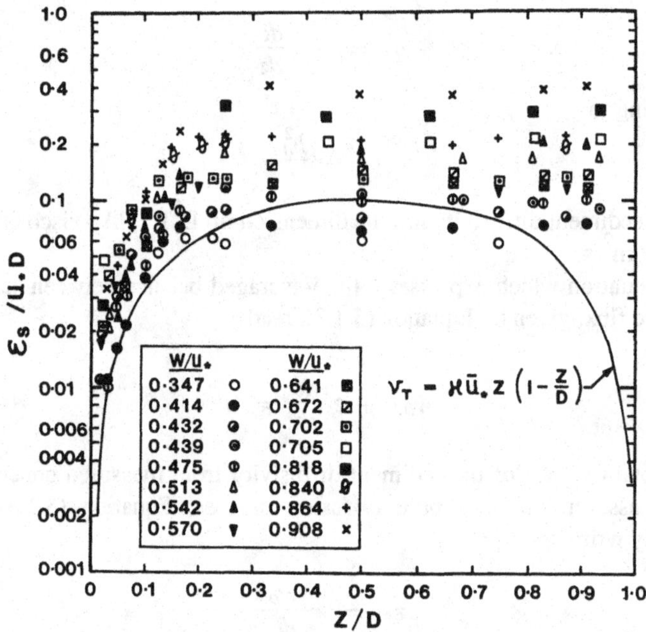

Figure 5.1.2: Sediment diffusivities derived through Equation (5.1.6) from measured concentration profiles $\bar{c}(z)$ under the assumption of pure gradient diffusion. Different sediment sizes (different values of the relative settling velocity w_o/u_*) give very different values for the diffusivity ε_s. Data from Coleman (1970).

The data of Nielsen (1983) and McFetridge & Nielsen (1985) from oscillatory flow over ripples (Figures 5.2.11 and 5.2.12, p 222-223) give further evidence that pure gradient diffusion is inadequate as a model of suspended sediment distributions. In order to explain their measurements, the magnitude of ε_s would have to be different for different grain sizes, as for Coleman's measurements. More importantly, however, the distribution of $\varepsilon_s(z)$ would have to be significantly different for different sand sizes in the same flow.

This would be very unsatisfactory. Using gradient diffusion as a model makes sense only if the same diffusivity can be used for all particles and if this diffusivity is closely related to the diffusivity of momentum, i e, the eddy viscosity.

In recognition of the increasing weight of experimental evidence against pure gradient diffusion as the sole distribution mechanism, Section 5.4 is devoted to a new modelling framework which acknowledges the presence of *large scale* or *convective* mixing mechanisms.

The new, combined convection-diffusion model is discussed qualitatively in general terms and developed quantitatively for the special case of pure, non-breaking waves over rippled beds. For that case, the new model leads to improved agreement with observations, but the details of the model outlined in Section 5.4.8 should not be seen as definitive. As more detailed data become available, the details of the model should be adjusted as appropriate.

The main point at present is, that a combined convection-diffusion model with the general features outlined in Section 5.4.5 has the potential of vastly improved modelling of sediment suspensions, including aspects which could not possibly be accounted for by pure gradient diffusion.

It is hoped that the combined convection-diffusion approach will make it possible to reconcile the concepts of sediment diffusivity and eddy viscosity, so that $\varepsilon_s \approx \nu_t$. This is not possible under the assumption of pure gradient diffusion, as illustrated by Coleman's data in Figure 5.1.2.

Before the development of the new model, an illustrated introduction to suspended sediment distributions in coastal areas is given in Section 5.2, page 206.

5.1.3 The bottom boundary condition

The description of the physical processes at the bottom end of the suspended sediment distribution presents many difficult problems. In mathematical terms that is. The specification of the bottom boundary condition for the differential equations, which describe the distribution of suspended sediment, present no lesser problems than the formulation of the equations themselves.

For flows that are both steady and uniform, it is reasonable to assume instantaneous equilibrium between near-bed sediment concentrations and the flow conditions (represented for example by the Shields parameter). However, this is not possible if the flow is unsteady or non-uniform. In unsteady or non-uniform flows, the near-bed concentrations are often considerable when the instantaneous bed shear stress is zero because sand is arriving from above.

The alternative approach, which will be discussed in detail in Section 5.3 (page 222), is to consider entrainment and deposition separately and to try and specify the pickup rate $p(t)$ in terms of the instantaneous, local flow parameters. It is acknowledged, however, that a complete description of the sediment pickup process under waves may not become available for some time.

5.2 THE NATURE OF SEDIMENT SUSPENSIONS

5.2.1 What is suspended sediment

In Section 2.3 the bed-load was defined as that part of the total sediment load which is supported by intergranular forces while the suspended load is carried by fluid drag. The same criterion will be used here to distinguish suspended load from bed-load.

The wash load is considered as part of the suspended load. It consists of fine sediment which is not in equilibrium with sediment in the local bed because the capacity of the flow to entrain these fine particles exceeds their settling rate at the bed. Thus, wash load is, in a sense, similar to water vapour over a dry or drying surface. Wash load sediment may be carried to the coastal environment by rivers, but the general conditions on beaches are such that the wash load will move offshore and get deposited there. Hence, beaches away from river entrances are generally free from wash load. The water looks clear apart from distinct clouds of suspended sand.

At elevations of more than a centimetre or so above the bed, the suspended concentration equals the total sediment concentration and is therefore directly measurable. Close to the bed however, where a considerable part of the moving sediment is bed-load, it is difficult to get information about the suspended part of the sediment concentrations.

To obtain suspended sediment concentrations very close to the bed is a challenge to the present measuring techniques and is, to some extent, a matter of definition. The measurable quantity is the total concentration which, with

Bagnold's (1956) definitions, may be partly bed-load and partly suspended load near the bed. Thus, after measuring the total concentrations near the bed, the problem of separating bed-load and suspended load remains.

As an example, this separation will be attempted with the total concentration measurements by Horikawa et al (1982), using Bagnold's definitions.

Bagnold's definition of bed-load can be written as

$$\rho\,(s-1)\,g \int_z^{\infty} c_b(z_1)\,dz_1 \;=\; \sigma_e(z) \qquad (5.2.1)$$

where $\sigma_e(z)$ is the dispersive stress which supports the bed-load that is travelling above the level z.

Solving this equation with respect to the bed-load concentration gives

$$c_b \;=\; \frac{-1}{\rho\,(s-1)\,g}\,\frac{d\sigma_e}{dz} \qquad (5.2.2)$$

Consider Case 1-1 of Horikawa et al (1982) at the phase *90* degrees which corresponds to maximum free stream velocity. The general experimental conditions are described by $(A, T, d, s) = (0.72m, 3.6s, 0.2mm, 2.65)$.

The linear sediment concentration λ, derived from the total, volumetric concentration in accordance with the definition (2.1.1) page 96, is of the order unity, and the velocity gradient is of the order $100s^{-1}$ close to the bed. Hence, the Bagnold number, defined by Equation (2.1.2), page 97, is about *20*. Therefore, the relevant formula for the dispersive stress according to Bagnold (1954) is

$$\sigma_e \;=\; 1.3\,(1+\lambda)(1+\lambda/2)\,\rho\,\nu\,\frac{du}{dz} \qquad (5.2.3)$$

See Section 2.1.2 pp 95-98.

By applying Equations (5.2.2) and (5.2.3) to the measured values of λ and du/dz we can estimate the bed-load concentration c_b. Subtracting this from the total concentration then gives an estimate of the suspended concentration ($c_s = c_{total} - c_b$). The distributions of total, bed-load and suspended concentrations thus derived are shown in Figure 5.2.1.

It can be seen that the suspended concentration is indistinguishable from the total concentration for $z > 5\,mm$ and, that the distribution of $c_s(z,t)$ is nearly exponential in the range $5mm < z < 25mm$. The nominal value $c_s(o)$ obtained by extrapolating the exponential part of the distribution to the level of the undisturbed bed ($z=o$) is of the order $0.1c_{max}$.

Between the outer, "exponential region" of pure suspension and the level of the undisturbed bed, the concentration gradients are considerably greater. Through this layer, which for these experiments has a thickness of five millimetres, the total concentrations drop by a factor *100.*

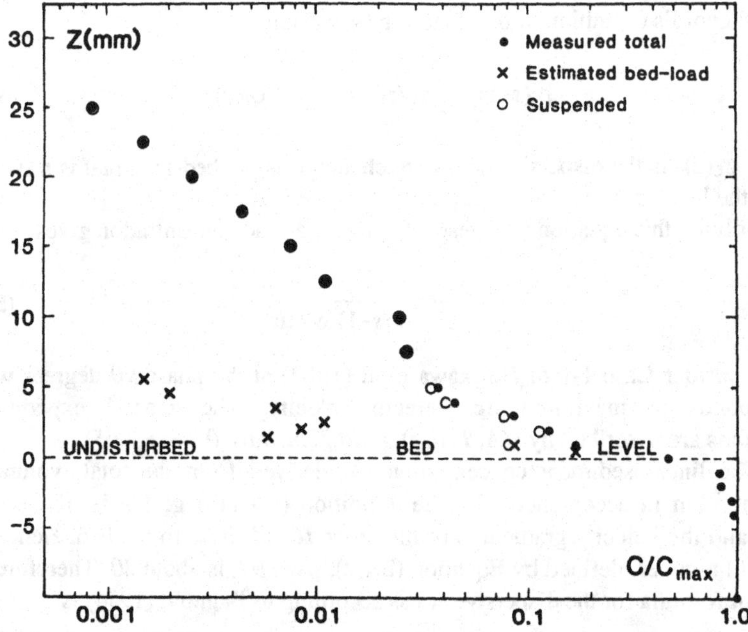

Figure 5.2.1: Measured values of the total, relative sediment concentrations and estimated values of the contributions $c_b(z,t)$ and $c_s(z,t)$ in oscillatory sheet flow. Data from Horikawa et al (1982), Case 1-1.

In the moving layer below the level of the undisturbed bed, the concentrations decrease rather slowly with elevation, from c_{max} in the stationary bed to between *50%* and *80%* of c_{max} at the level of the undisturbed bed.

5.2.2 Concentration measurements

Many different devices are now available for measuring suspended sediment concentrations.

The oldest and simplest type of device is the suction sampler. These have the advantage of being simple and reliable and their output is easy to interpret. The concentrations, determined from drying and weighing the captured samples, are independent of such factors as grain size, shape, colour and grading. These factors influence the outputs of other devices to varying extents.

When using suction samplers, consideration must be given to nozzle orientation and intake velocity because these may influence the amount of sand that gets captured. If the suction speed is low compared to the ambient flow and to the sediment settling velocity, and if the grains have to turn sharp corners to get in, the larger grains will tend to escape.

For steady flow the criterion is simple. The intake velocity should match the flow velocity with respect to both direction and magnitude. This is of course not possible in a wave motion so compromises must be made and corrections may be required.

Bosman et al (1987) investigated the efficiency of suction samplers in oscillatory flows. They found that for nozzles perpendicular to the main flow oscillation the capture rate increased with increasing intake velocity up to $u_{\text{intake}} \approx A\,\omega$. For higher intake velocities, the capture rate was fairly constant at around *0.8*.

In order to capture instantaneous suspended sediment concentrations, many other types of concentration gauges have been developed over recent years. Most of these rely on the absorption or back-scatter of sound (Jansen, 1978), or electromagnetic radiation (Sleath 1982, Downing et al 1981). The output of these types of instruments depends strongly on such sediment characteristics as grain size, shape, grading and colour which vary with both location and height. Precise, quantitative information is therefore difficult to derive. Nevertheless, even qualitative information such as correlations between velocities and concentrations is valuable at the present state of the art.

The complications mentioned above are less severe for devices which are based on X-ray or γ-ray absorption. The absorption of these types of radiation depends almost exclusively on the density of the absorbing medium.

Also the dependence of the electric conductivity on sediment concentration has been used to measure sediment concentrations, see e g Horikawa et al 1982.

5.2.3 Concentration time series

The concentration of suspended sediment is in general a function of all three space coordinates and of time $c = c(x,y,z,t)$, and the amount of detail involved in a complete description is overwhelming. It is therefore appropriate to start with less detailed descriptions.

Theoretical considerations often deal with the horizontally averaged quantity $c(z,t)$. It is, however, important to acknowledge the difference between this theoretical quantity and conveniently measurable quantities such as the average concentration along a sensor beam or over a small measuring volume. While the time series of the real horizontal average $c(z,t)$ may be fairly well behaved, the time series of the measured quantities may look very noisy due to the "*spotted carpet effect*". That is, since natural sediment concentrations are horizontally non-uniform, the "carpet of concentrations" at a certain level $c(x,y,z_o,t)$ is spotted and these spots are convected past the sensor. This may result in much high frequency variation which is irrelevant to the modelling of the horizontal average $c(z,t)$.

Figure 5.2.2 shows suspended sediment concentrations under irregular waves measured with γ-absorption along a *30cm* shore parallel path. The bed was covered by rounded megaripples which do not shed vortices regularly as vortex ripples do. The measured concentrations seem random to a great extent although some correlation obviously exists with the measured near-bed velocities.

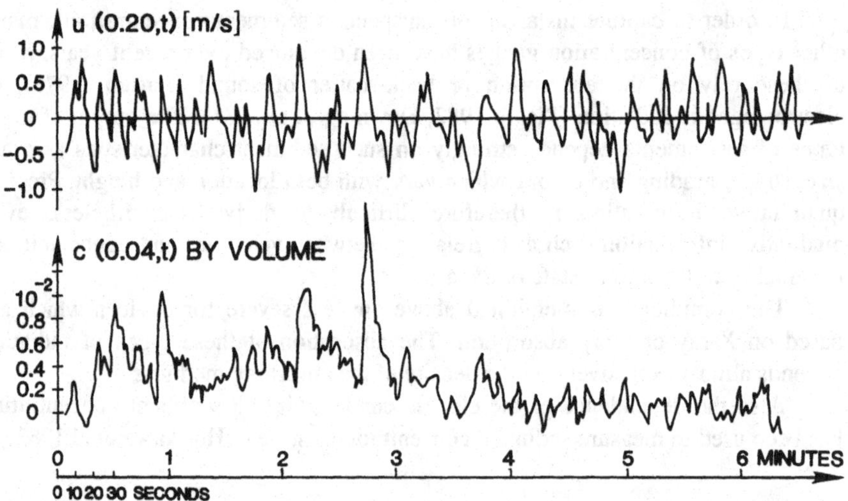

Figure 5.2.2: Field measurements of suspended sediment concentrations *4cm* above irregular megaripples. During the first three minutes, several concentration events occurred. During the last three minutes, the concentrations dropped to the "background level" even though the wave conditions were essentially unchanged. The difference is due to subtle changes in the bed topography in the vicinity of the probe. Data from Nielsen (1984).

Sometimes, a certain area of a sand bed with megaripples will generate large suspended sediment concentrations for a given set of flow conditions. At other times, it may not.

Thus, individual entrainment events are not entirely predictable in terms of the free stream velocity history. They result from complicated flow instabilities which also depend on subtle details of the micro-topography.

Nadaoka et al (1988) described an important mechanism for the creation of large clouds of suspended sediment in the outer surf zone, where the bed is commonly covered by megaripples. They attributed the formation of these clouds to obliquely descending vortices which are able to lift considerable amounts of sand into suspension after they reach the bed. Some of the large concentration peaks in Figure 5.2.2 may be associated with this type of sediment clouds.

The concentration variations in Figure 5.2.2 result not only from clouds of sand rising from the bed or settling from above, but also from clouds being convected horizontally past the probe.

More details about concentration time series under irregular waves are given by, for example, Hanes & Huntley (1986) and Vincent & Green (1990).

The picture is somewhat less confused if the wave motion is regular, especially under sheet-flow where the conditions are essentially constant in a

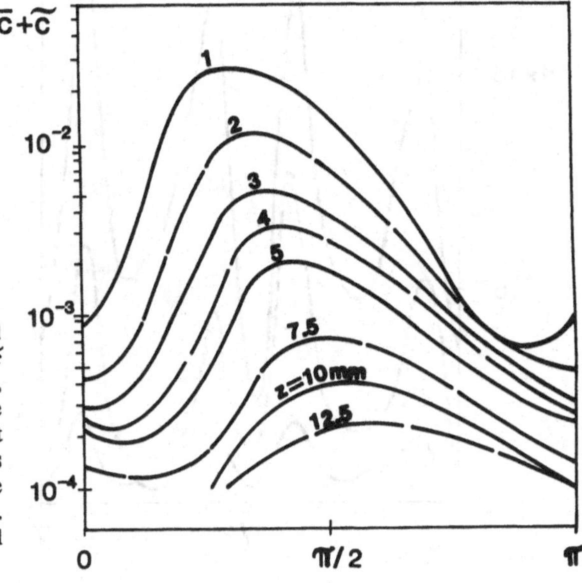

Figure 5.2.3: Phase-averaged sediment concentrations $\bar{c}+\tilde{c}$ over a flat bed under regular, symmetric oscillatory flow. One peak occurs each half-period at all levels. Near the bed, it occurs shortly before the peak of the free stream velocity ($\omega t = \pi/2$). Data from Horikawa et al (1982).

horizontal plane, $c \approx c(z,t)$. Such conditions were investigated by Horikawa et al (1982) and Staub et al (1984). An example of the smoothed, phase-averaged behaviour is shown in Figure 5.2.3. Pronounced concentration peaks are present at all levels. Near the bed they occur fairly shortly before the maximum free stream velocity (at $\omega t = \pi/2$). At higher elevations they occur progressively later.

Nakato et al (1977) and Sleath (1982) investigated $c(x,z,t)$ over vortex ripples under regular, oscillatory flow conditions. Over vortex ripples the concentration distribution is dominated by dense, travelling sediment clouds. These are formed when sand is swept over the ripple crest and into the lee vortex. Subsequently, they are transported in a fairly coherent form by the lee vortices which are released at the time of free stream reversal. See page 160.

At each ripple crest there will be two vortices released each period but the phase-averaged concentration records at most levels show more than two peaks. See Figure 5.2.4.

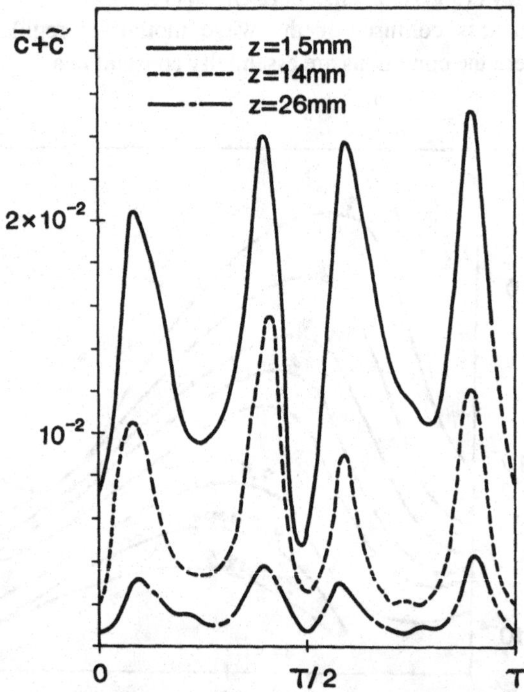

Figure 5.2.4: Suspended sediment concentrations over vortex ripples in a symmetric, regular oscillatory flow. In contrast to the situation over a flat bed (Figure 5.2.3), the phase-averaged concentrations show more than two peaks per period. Data from Nakato et al (1977).

This is because the sand clouds remain identifiable for a considerable part of a wave period. Hence, other sand clouds than the latest one generated at the nearest ripple are also making an organised imprint. In some cases a sensor may also see the same sediment cloud twice.

When the sand bed has high-profile features like vortex ripples, the suspended sediment concentrations cannot be expected to be horizontally uniform and the velocity field will likewise be non-uniform in a horizontal plane.

Consequently, the time-averaged vertical flux of sediment $\overline{w_s c}$ will vary greatly in a horizontal plane. In fact, the upward transport of sediment over vortex ripples occurs through quite narrow spatial corridors which originate near the ripple crest and follow the typical upward paths of the released lee vortices, see Figure 5.2.5.

The existence of special corridors for upward sediment movement is very obvious from the observations of Nakato et al (1977). They found that the time-averaged, vertical sediment transport rate $\overline{c\, w_s}$ was clearly different from zero almost everywhere in the two vertical sections which they studied. One above the

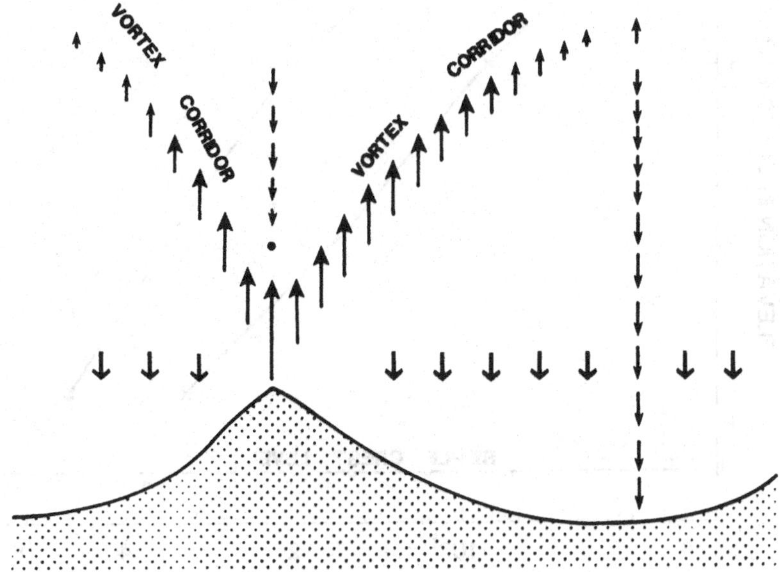

Figure 5.2.5: The upward transport of sand over vortex ripples is very organised. The sand is carried upwards by the lee vortices which are released at the free stream velocity reversals. The arrows in this figure indicate the local, time-averaged, vertical sediment flux $\overline{c\, w_s}$.

ripple crest and one above the trough. In general, they found that $\overline{w_S c}$ was negative everywhere in their two vertical sections except very close to the ripple crest, see Figure 5.2.5. At the same time, the stationarity of the situation obviously requires zero net vertical transport on the average (averaged over time and over any horizontal plane).

Nakato et al (1977) estimated the average upward sediment flux as $(\overline{w} - w_O)\overline{c} + \overline{\tilde{w}\,\tilde{c}}$ in accordance with the assumption that the sediment velocity equals the fluid velocity plus the still water settling velocity of the sediment $\mathbf{u}_S = \mathbf{u} + \mathbf{w}_O$. The limitations of this assumption are insignificant for the present purpose, see Sections 4.5 and 4.6 in general and Equation (4.5.26) page 176 for a simple estimate.

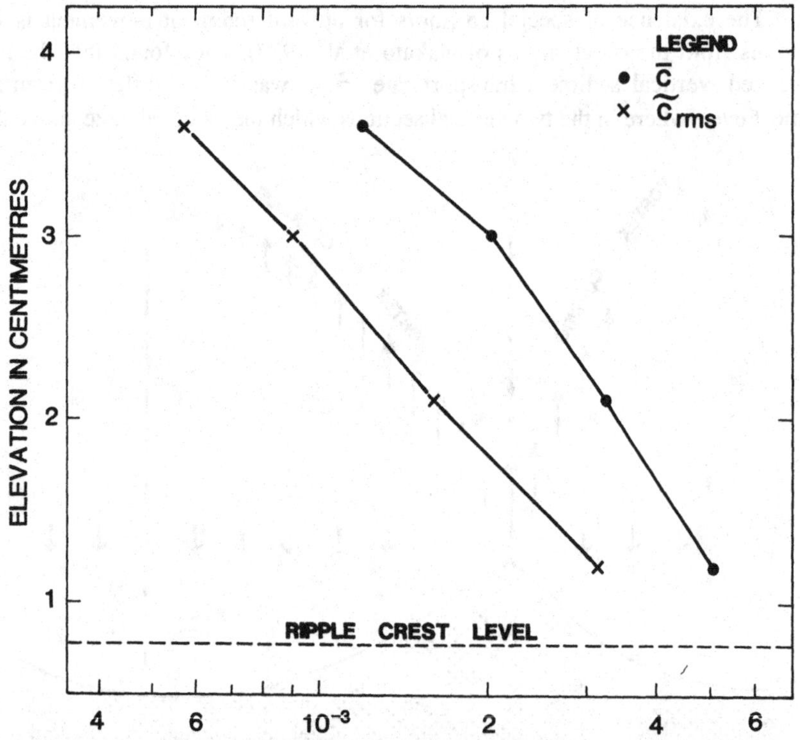

Figure 5.2.6: Vertical distributions of the time-averaged concentration $\overline{c}$ and the root-mean-square of the periodic $\tilde{c}_{rms}$. The values are obtained by averaging the values given by Nakato et al (1977) for two vertical sections, one over the ripple crest and one over the trough.

In relation to the discussion of sediment distribution models (Section 5.4, page 233) it is of interest to compare the decay with increasing elevation of the steady concentration component $\overline{c}(z)$ and of the periodic component $\tilde{c}(z,t)$. In purely convective descriptions, the decay rates are essentially the same. Diffusion models will, however, predict significantly faster decay for the oscillatory component, compare Equation (5.4.21) page 240 and Equation (5.4.38) page 244.

Figure 5.2.6 shows a comparison of these decay rates based on the average of the crest-section values and the trough-section values of $\overline{c}$ and $\tilde{c}_{rms}$ from Nakato et al (1977), Test 2. It can be seen that the decay rates are essentially identical in agreement with the obviously convective nature of the sediment entrainment process over vortex ripples.

5.2.4 Time-averaged sediment concentrations

It is comparatively easy to measure time-averaged sediment concentrations at "a point" for example with a simple suction sampler. Consequently, there is a substantial amount of experimental data available for $\overline{c}(x,z)$.

Given a flat sand bed, the time-averaged sediment concentrations should not vary much in a horizontal plane.

For rippled beds, however, there are subtle differences between the time-averaged concentration profiles above the ripple crests and over the troughs. Nielsen (1979) found that the crest sections had larger concentrations close to the bed while the trough sections had larger concentrations higher up. As a result, the total amount of suspended material in each vertical section was fairly constant.

Figure 5.2.7 shows four different $\overline{c}(z)$-profiles measured over vortex ripples in a large oscillating water tunnel.

This sequence of four $\overline{c}(z)$-profiles with fixed sediment size show the typical change of profile shape with increasing wave period. That is, for the shortest wave period, the profile is upward convex.

For the two intermediate periods, the profile is practically linear over the first 4 ripple heights $(0 < z/\eta < 4)$. Hence, this major part of the profile is appropriately described by the simple exponential relationship

$$\overline{c}(z) = C_o\, e^{-z/L_s} \tag{5.2.4}$$

For the longest wave period, the $\overline{c}(z)$-profile is upward concave from the start.

A similar sequence of profiles is seen in Figure 5.2.8, for time-averaged concentrations over flat beds.

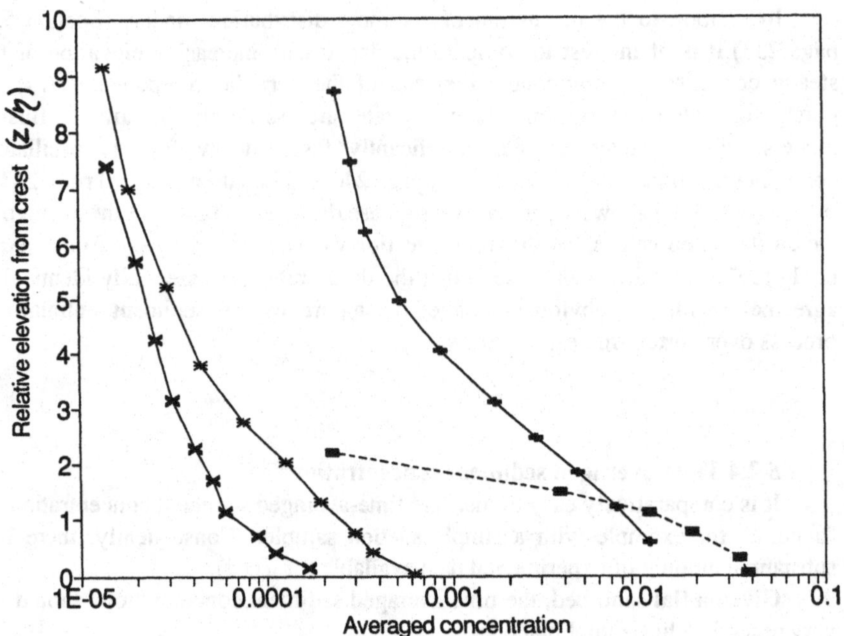

Figure 5.2.7: Time-averaged sediment concentrations $\overline{c}(z)$ measured over vortex ripples in an oscillating water tunnel. In all cases the sand size was 0.2mm. Legend: *square* : *T=1s, Aω = 0.5m/s*; + : *T* = 2*s, Aω* = 0.5m/s; * : T=4s, *Aω* = 0.3m/s; x: T=10s, *Aω* = 0.3m/s. Data from Bosman (1982) and Delft Hydraulics (1989).

A similar, gradual change in concentration profile shape with increasing T for fixed d, which is seen in Figures 5.2.7 and 5.2.8 is also seen in Figure 5.2.12. However, in Figure 5.2.12, where different sand sizes in the same flow are considered, the wave period is obviously fixed and the change from upward convex to upward concave happens with increasing settling velocity.

These profile changes are predicted by the combined convection-diffusion model developed in Section 5.4.5, page 245. The dimensionless parameter which determines the profile shape is found to be the dimensionless settling velocity $w_o L/\varepsilon_s$, where w_o is the sediment settling velocity, L is the vertical scale of the convective mixing process, and ε_s is the sediment diffusivity.

The dimensionless settling $w_o L/\varepsilon_s$ velocity is found to be roughly proportional to $w_o T/\sqrt{A r}$, where r is the bed roughness, see page 254 for

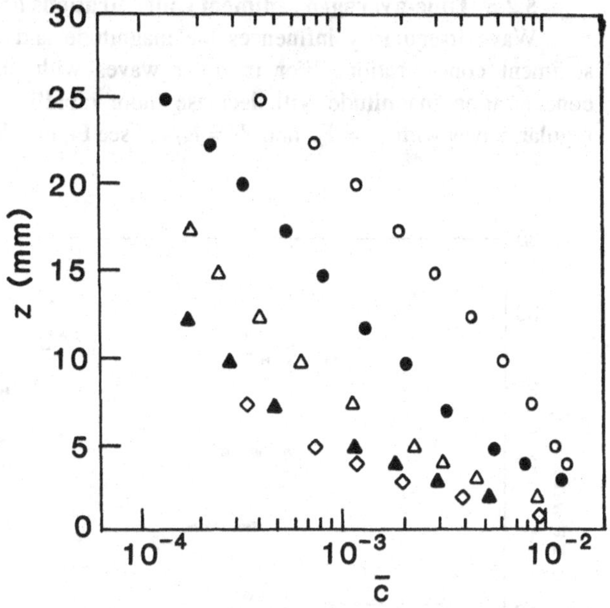

Figure 5.2.8: $\bar{c}(z)$-profiles over flat beds in an oscillating water tunnel. While $(A,d) \equiv (0.72m,\ 0.2mm)$ in all tests, the period was varied from *3.6s* to *6s*. As the wave period increases, the profiles become flatter and change from being upward convex to being upward concave. Data from Horikawa et al (1982).

details. Since the shape parameter $w_o T/\sqrt{A\,r}$ is proportional to both w_o and T, the same change in the shape of the $\bar{c}(z)$-profile is observed when the settling velocity is increased and when the wave period is increased while the other parameters are kept constant. See Figures 5.2.7, 5.2.8 and 5.2.12.

The simple exponential model

$$\bar{c}(z) = C_o\, e^{-z/L_s} \tag{5.2.4}$$

which is a reasonable description for $\bar{c}(z)$ over rippled beds within a fairly wide, intermediate range of the shape parameter $w_o L/\varepsilon_s$, has been studied in considerable detail by Nielsen (1979, 1984, 1986, 1990). The reference concentration C_o [$=\bar{c}(o)$] can be estimated from Equation (5.3.10), page 228, and the vertical scale L_s is closely related to the ripple height η for sharp crested ripples. Nielsen (1990) suggested

$$L_s = \begin{cases} 0.075\,\dfrac{A\omega}{w_o}\,\eta & \text{for } \dfrac{A\omega}{w_o} < 18 \\ 1.4\,\eta & \text{for } \dfrac{A\omega}{w_o} > 18 \end{cases} \tag{5.2.5}$$

5.2.5 Time-averaged sediment concentrations under irregular waves

Wave irregularity influences the magnitude and distribution of suspended sediment concentrations. For irregular waves with parameters (T_p, H_{rms}) the concentration magnitude will decrease more rapidly with elevation than under regular waves with $T = T_p$ and $H = H_{rms}$, see Figure 5.2.9.

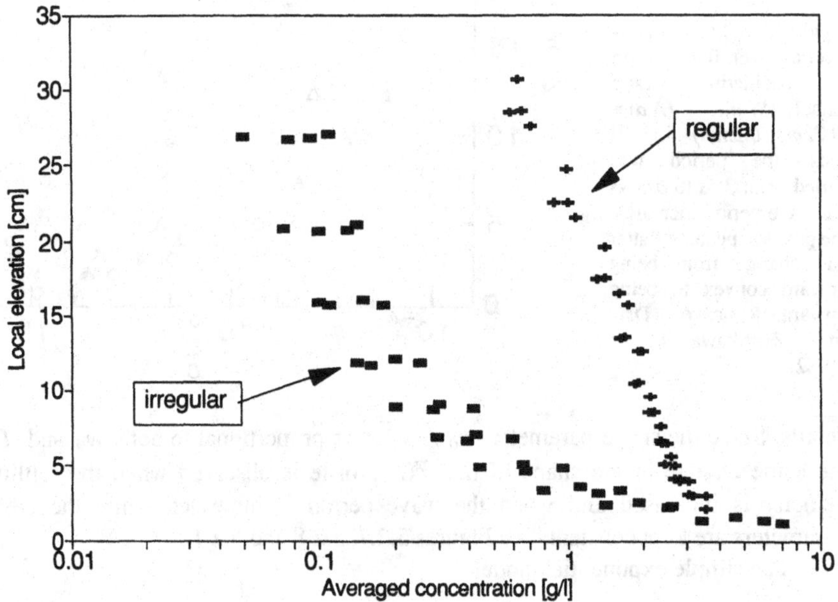

Figure 5.2.9: Time-averaged concentration profiles from regular and irregular oscillatory flows with the same d_{50} , T_p and $u_{\infty,rms}$, data from Delft Hydraulics (1989), oscillating water tunnel.

The main reason for the difference in concentration profiles is the difference in ripple size. Thus, for the two cases shown in Figure 5.2.9, the regular flow formed ripples with an average height of *15cm*, while the ripples in the irregular flow had an average height of only *4cm*.

As a consequence of the bed roughness (~ ripple height) being smaller, the shape parameter $w_o T/\sqrt{A r}$ will be larger for the irregular waves and the profile shape therefore tends to be more upward concave.

5.2.6 Time-averaged concentration profiles under breaking waves

Wave breaking has profound and highly diverse effects on the suspended sediment concentrations. However, except for extreme cases of plunger jets hitting the bed, the pickup rate at the bed and hence the sediment concentrations very close to the bed will be unaffected by the breaking. The main effect of the turbulence from wave breaking is a vertical stretching of the sediment concentration profiles. See Figure 5.2.10.

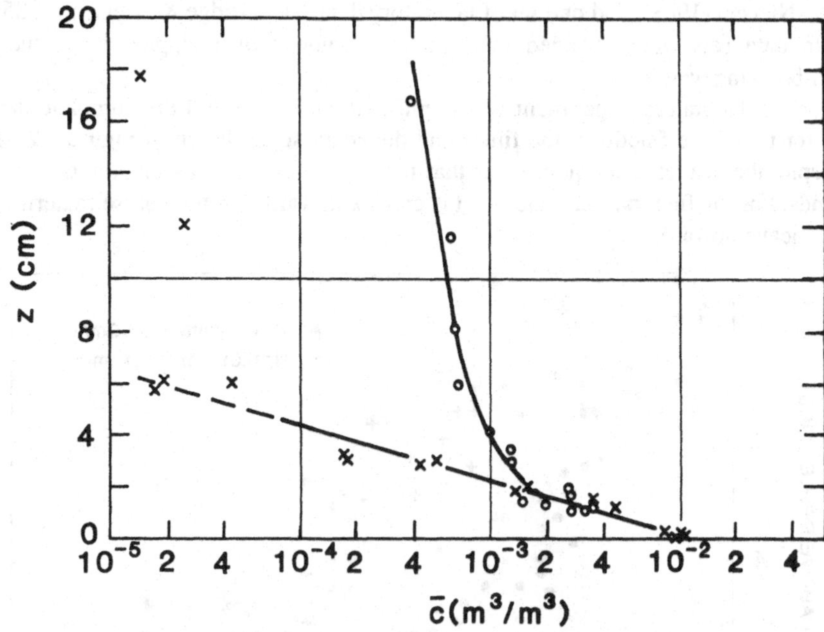

Figure 5.2.10: Time-averaged concentration profiles under non-breaking (x) waves and under spilling breakers (o) of the same height. Wave flume data from Nielsen (1979).

With respect to the modelling of sediment concentration profiles under breaking waves, it may be possible to treat the case of gently spilling breakers as "simply" a case of increased diffusivity compared to the non-breaking case. For plunging breakers however, the distribution process seems to be dominated by large scale convection, with large amounts of sediment travelling straight to the surface with the entrained air behind the plunging breakers. See Figure 6.5.4, page 290 and Nielsen (1984).

5.2.7 Different sand sizes suspended in the same flow

It is of interest to examine concentration profiles for different sand sizes suspended in the same flow. Similarities and differences between such profiles provide information about the mechanisms which distribute the sediment. A couple of such mechanisms, namely the classical gradient diffusion and a simple convective process will be discussed in detail in Sections 5.4.2 through 5.4.4.

Two comprehensive data sets with concentration distributions of different sand sizes in the same flow, are presently known to the writer. One set of field data, Nielsen (1983), and one set of laboratory data, McFetridge & Nielsen (1985). Both data sets were obtained with suction samplers over rippled beds under non-breaking waves.

The laboratory experiment was duplicated nine times and the complete data set for two sieve fractions (the finest and the coarsest) is shown in Figure 5.2.11. Despite the scatter, it is quite clear that the two sieve fractions display different trends. For the fine material, the trend is convex upward. For the coarse material it is concave upward.

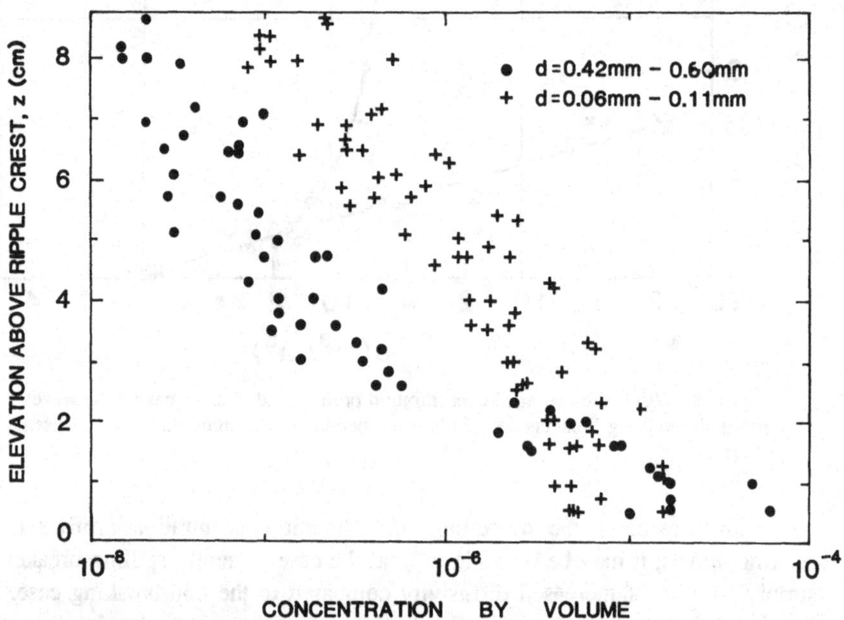

Figure 5.2.11: Concentration profiles for two sieve fractions of sand suspended in the same flow (waves over natural ripples). The profiles are not similar. The trend for the fine sand is convex upward while the trend for the coarse sand is concave upward. $(h, T, H, d_{50}, \lambda, \eta)$ = *(0.3m, 1.51s, 0.13m, 0.19mm, 0.078m, 0.011m)*.

The difference is significant because it shows that the entrainment process can neither be described as purely diffusive or as purely convective. The details behind this conclusion will be given in Sections 5.4.3 to 5.4.5, page 238 ff.

Figure 5.2.12 shows the hand-drawn trends of the data from McFetridge & Nielsen (1985) for all six sieve fractions. The curves show a continuous transition from the upward convex for the fine sand fractions to the upward concave for the coarse ones. The same transition is shown by the field data of Nielsen (1983) in Figure 5.4.7, page 256.

This transition can be explained in terms of the distribution mechanism being partly diffusive and partly convective, see Section 5.3.4, page 245 ff. The diffusive characteristics are displayed by the finer fractions while the coarse fractions are more influenced by the convective mechanisms. The parameter which describes the profile shape change is $w_0\, T/\sqrt{A\, r}$, see also page 216, and Section 5.4.8, page 252.

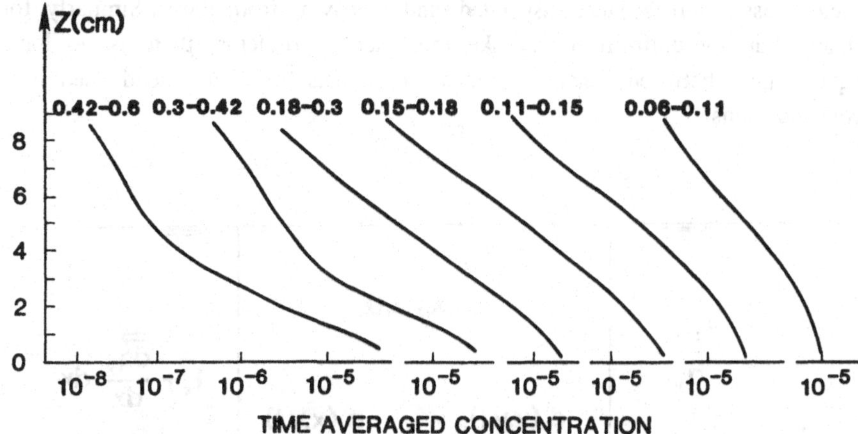

Figure 5.2.12: Hand-drawn concentration profiles for the data from McFetridge & Nielsen (1985). The curves show a continuous transition from upward convex for fine sand to upward concave for coarse sand. The numbers on the curves indicate the grain size interval in millimetres.

Different grain sizes are not picked up in direct proportion to their abundance in the undisturbed bed. Small particles have a greater chance of becoming suspended than large particles. Details of this phenomenon are discussed in Section 5.3.7, page 230.

5.3 PICKUP FUNCTIONS

5.3.1 Introduction

At the fundamental level, the modelling of suspended sediment transport presents two challenges. One is to describe the mechanisms which distribute the sand through the water column. The other is to describe the pickup (entrainment) process at the bed. The present section deals with the latter process.

The traditional approach to sediment suspension modelling in steady, uniform flows has been to assume equilibrium between bed shear stresses and near-bed sediment concentrations. Under this assumption, it should be possible to relate sediment transport rates directly to the bed shear stress. It is however, fairly obvious that this equilibrium assumption is unrealistic if the flow is either non-uniform or unsteady.

In an unsteady situation, where the bed shear stress varies with time, the near-bed sediment concentrations may well be considerable at times when the bed shear stress is zero because suspended sand is arriving from above. Similarly, for steady but non-uniform flows, like most scour problems, there is no local equilibrium between shear stresses and near-bed suspended sediment concentrations.

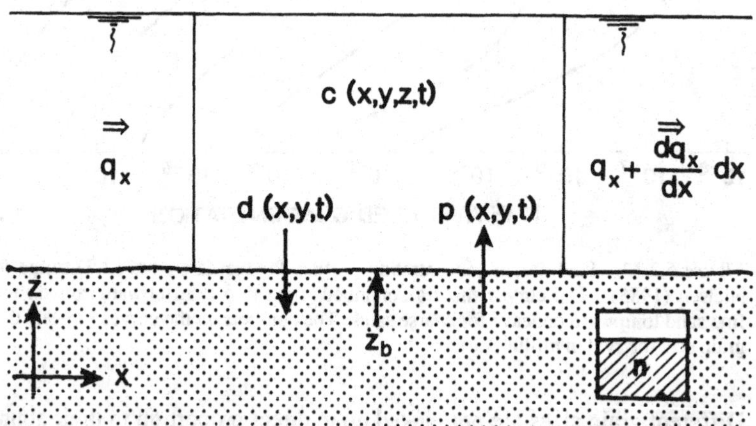

Figure 5.3.1: The imbalances between local pickup rate $p(x,y,t)$ and deposition rate $d(x,y,t) = w_o\, c(x,y,z_r,t)$ through the near bed reference level z_r generate the changes in bed level.

Thus, unsteady and non-uniform flows call for a different modelling approach, capable of dealing with non-equilibrium situations. The alternative approach, which has been suggested by Nielsen et al (1978) for oscillatory flows, and by van Rijn (1984) for steady flows, is to consider sediment entrainment and deposition as independent processes.

The deposition rate down through a level z_r is then taken as $c(x,y,z_r,t)\, w_o$. The reference level z_r is usually taken as the bed level, corresponding to $z_r = o$, but other choices are equally possible.

The instantaneous rate at which sand is picked up through the level z_r is given in terms of a pickup function $p(x,y,t)$. The pickup function is a non-negative function with the dimension of sediment flux, i e, sediment concentration times velocity. Similarly, for the deposition rate $d(x,y,t) = c(x,y,z_r,t)\, w_o$, see Figure 5.3.1. The SI units for both pickup functions and deposition rates are thus m/s.

The rate of morphological change quantified as the rate of bed level change dz_b/dt is related to the pickup and deposition rates and to the horizontal sediment fluxes q_x and q_y through the conservation of sediment expressed by the conservation equation

$$n\frac{\partial z_b}{\partial t} = d - p = -\int_o^D \left(\frac{\partial c}{\partial t} + \frac{\partial q_x}{\partial x} + \frac{\partial q_y}{\partial y}\right) dz \qquad (5.3.1)$$

where n is the solids volume contained in a unit volume of bed material.

5.3.2 Pickup functions in steady flows

Van Rijn (1984) performed experiments on scour rates in steady flows and on the development of the sediment concentration profile in a situation similar to Figure 5.3.2. From the experiments, the pickup function was determined directly for a wide range of flow velocities and grain sizes ($0.13mm < d < 1.5mm$, $0.06 < \theta' < 1.0$).

Van Rijn recommended the following formula for the pickup function in steady flow

$$\frac{p}{\sqrt{(s-1)gd}} = 0.00033\left(\frac{\theta' - \theta_c}{\theta_c}\right)^{1.5}\left(\frac{(s-1)\, g\, d^3}{\nu^2}\right)^{0.1} \qquad (5.3.2)$$

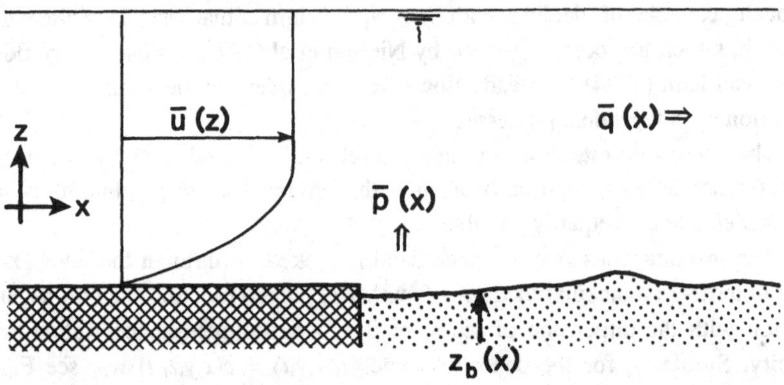

Figure 5.3.2: Van Rijn (1984) studied pickup functions in steady flow via the scour rates and the build up of the concentration profile over a sand bed following a fixed concrete bed.

Van Rijn compared his experimental results to previous pickup functions which are implicit in some earlier work on steady bed-load transport. He found that with the exception of that of Fernandez-Luque (1974) these implicitly assumed pickup functions did not match his data at all. The dependence of the pickup function $p(t)$ on the bed shear stress was generally too weak.

5.3.3 Pickup functions in unsteady flows

For the estimation of pickup functions in unsteady flows we have two options at present. The first, is to adapt van Rijn's expression (5.3.2) or a similar steady flow formula. This requires some knowledge about the effective Shields parameter $\theta'(t)$ as a function of time. The other option is to infer the pickup function from measured near-bed sediment concentrations in unsteady flows.

The time-averaged pickup function, $\bar{p}$ may be inferred from the time-averaged near-bed sediment concentration C_o by the following argument. The averaged deposition rate is assumed to be $w_o C_o$ and, if the process is stationary, the average pickup rate must equal the average deposition rate, i e,

$$\bar{p} = w_o C_o \qquad (5.3.3)$$

A considerable amount of information is available about C_o $[=\bar{c}(o)]$ under waves, see e g Nielsen (1986).

Alternatively, we can try to derive an expression for $p(t)$ from van Rijn's formula (5.3.2) (or a similar steady-flow formula) if we assume that this steady-flow formula can be applied instantaneously to an unsteady flow. With a slight rearrangement van Rijn's formula reads:

$$p(t) = 0.00033 \left(\frac{\theta'(t) - \theta_c}{\theta_c} \right)^{1.5} \frac{(s-1)^{0.6}\, g^{0.6}\, d^{0.8}}{\nu^{0.2}} \quad \text{for } \theta' > \theta_c \qquad (5.3.4)$$

In order to apply this formula it is, of course, necessary to know how the effective Shields parameter $\theta'(t)$ varies with time. Attempts to estimate $p(t)$ for three special cases are presented in the following sections.

5.3.4 Pickup functions for sine waves over flat beds

For sine waves over flat beds (sheet-flow), it might be justified to make the assumption that the effective Shields parameter $\theta'(t)$ is simple harmonic with amplitude $\theta_{2.5}$ and a phase shift φ_τ relative to the free stream velocity

$$\theta'(t) \approx \theta_{2.5} \cos(\omega t + \varphi_\tau) \qquad (5.3.5)$$

and to apply the general value $\theta_c \approx 0.05$ for the critical Shields parameter. This model for $\theta'(t)$ was found to work well for estimation of bed-load transport rates over flat beds in Section 2.4.4, page 121.

Hence, we shall try and insert these expressions for $\theta'(t)$ and θ_c into van Rijn's pickup formula (5.3.4) and compare the result with experimental data.

For the purpose of comparing van Rijn's formula (5.3.4) with $\bar{p}$*–values* determined from concentration measurements via Equation (5.3.3), we note that the last factor in Equation (5.3.4) is almost proportional to the settling velocity within the relevant range of quartz grain sizes. Thus, the expression

$$w_o \approx 0.4 \frac{(s-1)^{0.6} g^{0.6} d^{0.8}}{\nu^{0.2}}$$

is within twenty percent of Gibbs et al's (1971) expression (4.2.6) for quartz spheres in the diameter range $0.3mm < d < 1.5mm$.

Inserting these approximations for $\theta'(t)$ and w_o into van Rijn's formula (5.3.4) we find

$$p(t) \approx 0.017\, w_o\, (\theta_{2.5} - \theta_c)^{1.5}\, |\cos^3(\omega t + \varphi_\tau)| \qquad \text{for } \theta_{2.5} >> \theta_c \tag{5.3.6}$$

Then, noting that $\overline{|\cos \omega t|^3} = \frac{4}{3\pi}$ we find the time-average

$$\bar{p} \approx 0.007\, w_o\, (\theta_{2.5} - 0.05)^{1.5} \tag{5.3.7}$$

Combined with Equation (5.3.3), this formula corresponds to

$$C_o \approx 0.007\, (\theta_{2.5} - 0.05)^{1.5} \tag{5.3.8}$$

which is shown by Figure 5.3.3 to agree reasonably with measurements from oscillatory sheet flow.

Thus, quasi-steady application of van Rijn's pickup function with the instantaneous, effective Shields parameter given by Equation (5.3.5) leads to reasonable agreement with observed C_o-values for oscillatory sheet-flow.

5.3.5 Pickup functions for irregular waves over flat beds

Not much is known at present about sediment concentrations and corresponding pickup rates under irregular waves of arbitrary shape. However, if the bed is flat (no sharp crested ripples), it seems reasonable to apply the quasi-steady approach outlined in the previous section. That is, instead of the "sine wave expression" (5.3.6) we might use

$$p(t_n) \approx \begin{cases} 0.017\, w_o\, |\theta'(t_n) - \theta_c|^{1.5} & \text{for } \theta'(t_n) > \theta_c \\ 0 & \text{for } \theta'(t_n) \le \theta_c \end{cases} \tag{5.3.9}$$

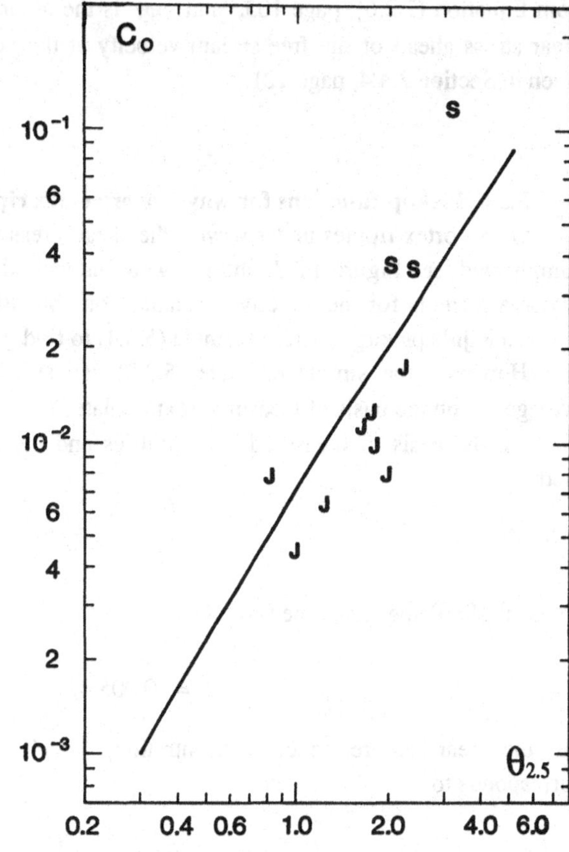

Figure 5.3.3: Equation (5.3.8) compared to data from oscillatory sheet flow by Horikawa et al (1982), (J) and by Staub et al (1984), (S).

with the instantaneous, effective Shields parameter, at time step t_n estimated by

$$\theta'(t_n) = \frac{\frac{1}{2} f_{2.5} A_{rms}}{(s-1)\, g\, d} \left(\cos\varphi_\tau\, \omega_p\, u_\infty(t_n) + \sin\varphi_\tau \frac{u_\infty(t_{n+1}) - u_\infty(t_{n-1})}{2\,\delta_t}\right) \tag{2.4.15}$$

on the basis of measured or simulated values of the free stream velocity $u_\infty(t)$. ω_p is the spectral peak angular frequency for the velocity record and $A_{rms} = \sqrt{2}\,(u_\infty)_{rms}/\omega_p$. The grain roughness friction factor $f_{2.5}$ is calculated from Equation (2.2.6), page 105, and φ_τ is the assumed phase shift of the bed shear stress ahead of the free stream velocity at the peak frequency. Details are given in Section 2.4.4, page 121.

5.3.6 Pickup functions for waves over vortex ripples

Over vortex ripples under waves, the shear stress variation with time is very complicated, see Figure 1.2.3, and not well understood. It is therefore difficult to devise a formula for the effective, instantaneous Shields parameter $\theta'(t)$ to insert into van Rijn's pickup function formula (5.3.4) to find $p(t)$ for rippled beds.

However, the simple estimate (5.3.3) can still be used to find the time average $\bar{p}$ on the basis of measured (extrapolated) C_o-values.

On the basis of measured $\bar{c}(z)$-profiles and the simple exponential profile model

$$\bar{c}(z) = C_o\, e^{-z/L_s} \tag{5.2.4}$$

Nielsén (1986) suggested the formula

$$C_o = 0.005\, \theta_r^3 \tag{5.3.10}$$

for the near-bed reference concentration, which through Equation (5.3.3) corresponds to

$$\bar{p} = 0.005\, w_o\, \theta_r^3 \tag{5.3.11}$$

These formulae for C_o and $\bar{p}$ apply reasonably well for both rippled and flat beds, see Nielsen 1986. The modified effective Shields parameter θ_r is given by

$$\theta_r = \frac{\theta_{2.5}}{(1 - \pi\,\eta/\lambda)^2} \tag{5.3.12}$$

where the grain roughness Shields parameter $\theta_{2.5}$ is calculated as described in Section 2.2.5, page 105. The correction factor $(1 - \pi\,\eta/\lambda)^2$ is the square of the velocity correction suggested by Du Toit & Sleath (1981) for the flow enhancement near the crest of vortex ripples.

For irregular waves, Equations (5.3.10) and (5.3.11) may be used with the regular wave height H replaced by H_{rms} or, correspondingly with $A\omega$ replaced by $\sqrt{2}\,(u_\infty)_{rms}$ in the calculation of $\theta_{2.5}$.

With respect to the time dependence of $p(t)$ over rippled beds, Nielsen (1988) noted that most of the resulting sediment transport over vortex ripples occurs above the ripple crest level and that the input of sand into this domain happens virtually as instantaneous puffs at the time of free stream velocity reversals. Thus, the pickup function must vary with time as shown qualitatively in Figure 5.3.4. It has two distinct peaks at the times t^d and t^u where the free stream velocity changes direction. These are the times when the lee vortices with their clouds of sand move upwards into the main flow.

Quantitatively, these pickup functions may be described in terms of functions of the form

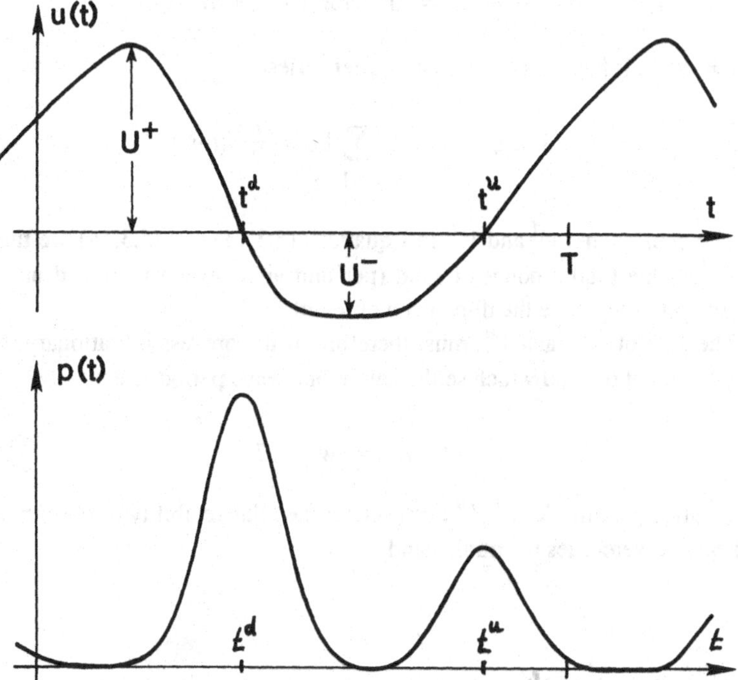

Figure 5.3.4: The sediment pickup function $p(t)$ under regular waves has two peaks per wave period. For rippled beds the peaks are very sharp and correspond to the release of the lee vortices. For flat beds (sheet-flow) the peaks are flatter and occur closer to the extreme velocity in either direction, see Figures 5.2.3 and 5.2.4.

$$p(t) \;=\; V^d \frac{\cos^{2m}\frac{\omega}{2}(t-t^d)}{\cos^{2m}\frac{\omega}{2}t} + V^u \frac{\cos^{2m}\frac{\omega}{2}(t-t^u)}{\cos^{2m}\frac{\omega}{2}t} \qquad (5.3.13)$$

where m is a positive integer, see Nielsen (1979). Alternatively, a train of delta functions can be used

$$p(t) \;=\; V^d\,\delta_T(t-t^d) + V^u\,\delta_T(t-t^u) \qquad (5.3.14)$$

The periodic delta function $\delta_T(t-t_o)$ has the dimension T^{-1}. It is periodic with period T and is further specified by

$$\delta_T(t-t_o) \;=\; 0 \quad \text{for } t \neq t_o + nT \qquad (5.3.15)$$

and $\overline{\delta} = 1s^{-1}$. It has the convenient Fourier series

$$\delta_T(t-t_o) \;=\; 1 + \sum_{n=1}^{\infty} 2\cos\frac{2n\pi}{T}(t-t_o) \qquad (5.3.16)$$

The coefficients V^d and V^u in Equations (5.3.13) and (5.3.14) are then seen to represent the total amount of sand (per unit area) which is picked up in each pickup event. They have the dimension of length.

The sum of V^d and V^u must therefore, if the process is stationary, balance the total amount of sand which settles out in one wave period , i e

$$V^d + V^u \;=\; w_o\,C_o\,T \qquad (5.3.17)$$

Their relative magnitude V^d/V^u must reflect the relative ability of the forward and the backward velocities to entrain sand.

5.3.7 Selective pickup from graded sand beds

Natural sand beds always consist of a mixture of different sand sizes with correspondingly different settling velocities. Hence, measured near-bed concentrations and pickup rates are generally compounded by contributions from the different grain sizes which are present in the bed.

For most practical purposes, it is sufficient to consider the bulk quantities (total concentrations and total pickup rates), and try to model them in terms of the median sediment parameters. However, it is sometimes of interest to consider the concentrations and pickup rates for the individual size fractions, for example in relation to sediment sorting across a beach profile.

Intuitively, one would expect that the finer fractions of a graded sand bed would have a greater tendency to go into suspension than the coarser fractions. This expectation is confirmed by the data in Figure 5.3.5.

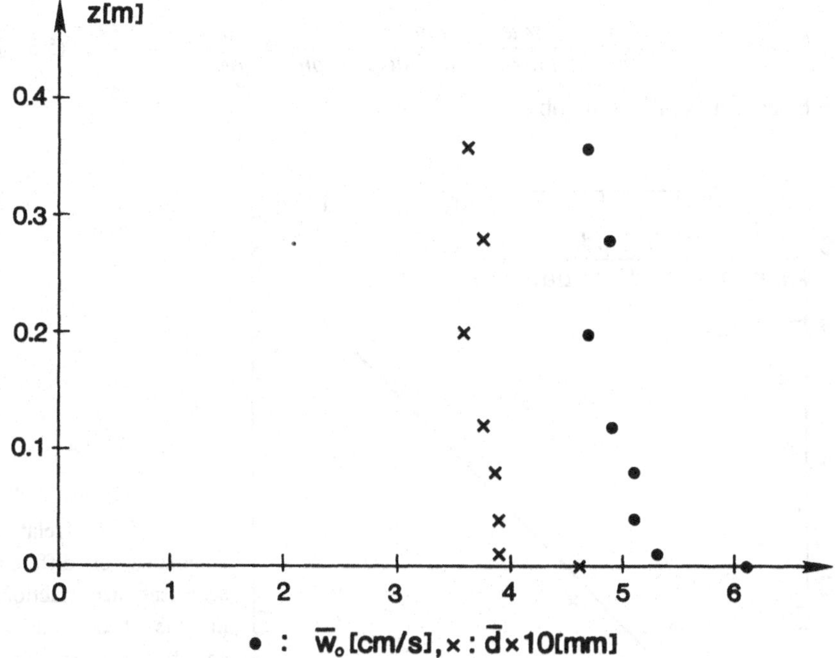

Figure 5.3.5: Variation of the average grain size and the average settling velocity with elevation above the ripple crest. The abrupt decrease in both quantities from $z = 0$ to $z = 0.01m$ indicates that most of the sorting happens in the pickup process. After that, the size distribution remains fairly constant. Data from Nielsen (1983).

Noting that the Shields parameter, see Equation (2.2.2), p 103, is proportional to d^{-1} for a given bed shear stress. One might expect the relative pickup rate for sand with size d to be proportional to d_{50}/d. It might also be argued that, the pickup rate of a sand fraction with settling velocity w should be proportional to u_*/w. Hence, the pickup rate relative to that of the median size should be proportional to w_{50}/w. On the other hand, small particles are sometimes sheltered against the eroding effect of the flow by larger grains (the armouring effect), and this makes things more complicated.

There is precious little information available about the selective entrainment of different sand sizes under waves. However, the data in Figure 5.3.6 indicate that the simple formula

$$\frac{\textit{fraction in bed}}{\textit{fraction in near–bed suspension}} \approx \frac{d}{d_{50}} \tag{5.3.18}$$

can be used as a rule of thumb.

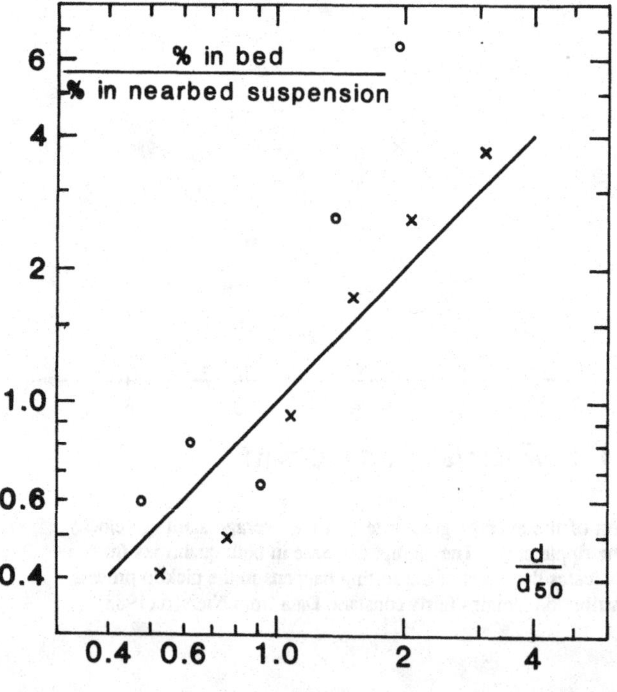

Figure 5.3.6: relative abundance of different sediment size fractions in the bed and in near-bed suspension. Rippled beds under waves, Nielsen (1983) (x), and McFetridge & Nielsen (1985) (o).

Equation (5.3.18) corresponds to the time-averaged pickup rate for sand with parameters (d,w) being given by

$$\bar{p}(d,w) = \frac{d_{50}}{d} * \textit{fraction in bed} * \bar{p}(d_{50}) \tag{5.3.19}$$

where $\bar{p}(d_{50})$ is given by a suitable formula for the bulk pickup rate. For example Equation (5.3.11), page 228.

5.4 SUSPENDED SEDIMENT DISTRIBUTION MODELS

5.4.1 Introduction

In the previous section the processes of sediment pickup and deposition were considered, i e the processes by which sediment goes from a state of rest to a state of movement and vice versa. The present section deals with the processes which take the sand upwards from the immediate vicinity of the bed.

These processes are under one referred to as distribution processes but two distinct categories are defined and analysed in detail. These are convective processes and gradient diffusion.

A quantitative framework is suggested for describing convective entrainment. The characteristic behaviour of sediment concentrations $c(z,t)$ resulting from this model are compared to that resulting from the gradient diffusion description.

It is shown that natural entrainment processes, in general, contain elements of both convection and diffusion. Therefore, a combined convection-diffusion description for the distribution of suspended sediment is developed in Sections 5.4.5 through 5.4.9.

5.4.2 Convection or diffusion

The process of sediment distribution may be considered as an orderly convective process or as a disorganised, diffusive process or, most commonly, as a combination of these two.

In terms of the mixing length l_m, defined in Section 5.1.2, p 201, the distinction between convective and diffusive processes is made as follows. If the mixing length is large compared to the overall scale of the sediment distribution, the process is convective. Conversely, if the mixing length is small compared to

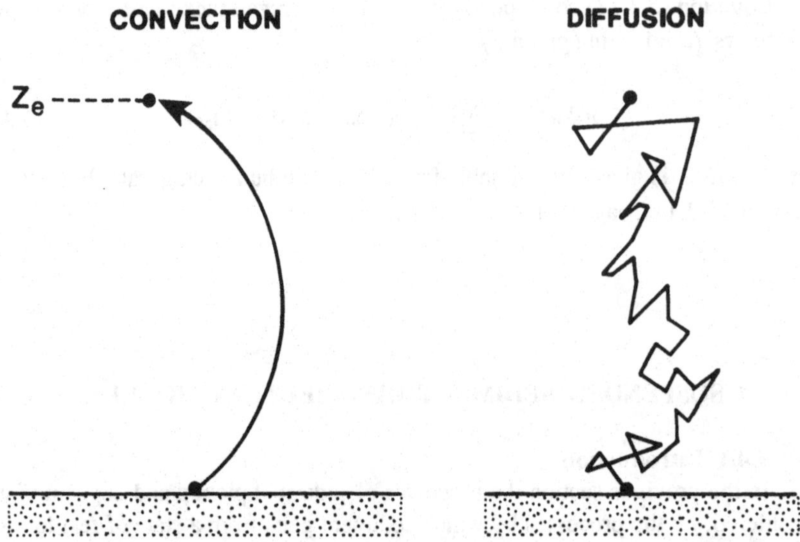

Figure 5.4.1: In a convective process the distance of interest is travelled in a smooth, organised way while in diffusive processes it is covered in a large number of more or less random steps.

the overall scale, the process may be described as diffusive, see Figure 5.4.1.

For both types of processes, and indeed for combined convection-diffusion processes, the conventional means of describing concentration profiles of suspended sediment is the conservation equation for the volume of sediment which may be written as

$$\frac{\partial c}{\partial t} = -\, div\,(\mathbf{u}_s\, c) \tag{5.4.1}$$

which expresses that a divergence of the sediment flux field $\mathbf{q} = c\,\mathbf{u}_s$ must result in a change of the local sediment concentration.

For the sake of simplicity, we shall in the following, consider a horizontally uniform sediment concentration field $c = c(z,t)$ and a correspondingly uniform sediment velocity field $\mathbf{u}_s = \mathbf{u}_s(z,t)$. The conservation equation (5.4.1) is then simplified to

$$\frac{\partial c}{\partial t} = -\frac{d}{dz}(c w_s) = -\frac{dq_z}{dz} \tag{5.4.2}$$

where q_z denotes the total sediment flux (transport rate per unit area) in the z-direction.

It is generally considered too complicated to describe the vertical sediment velocity w_S in detail in a turbulent flow. So instead, a broader approach is generally applied to the sediment flux. In a horizontally averaged description, the vertical sediment flux q_z is considered to consist of a downward component $-w_O c$ due to gravitational settling, and an upward flux which can be of convective (subscript C) or diffusive (subscript D) nature or a combination of both. The total, vertical sediment flux is thus written as

$$q_z = -w_O\, c + q_D + q_C \qquad (5.4.3)$$

and hence, the conservation equation (5.4.2) is written

$$\frac{\partial c}{\partial t} = w_O \frac{\partial c}{\partial z} - \frac{\partial q_D}{\partial z} - \frac{\partial q_C}{\partial z} \qquad (5.4.4)$$

see Figure 5.4.2.

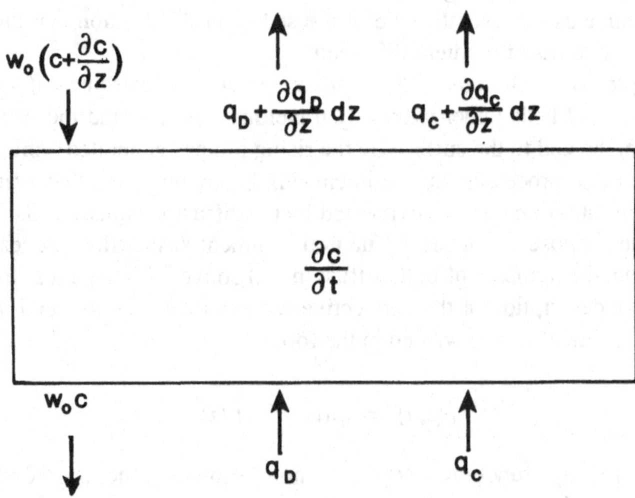

Figure 5.4.2: For a horizontally uniform situation, conservation of sediment is expressed in terms of the three vertical flux components: The settling flux $-w_O c$; the diffusive flux q_D; and the convective flux q_C.

Diffusive sediment flux is generally described in terms of gradient diffusion

$$q_D = -\varepsilon_S \nabla c \tag{5.4.5}$$

i e, the diffusive flux vector is directed from larger towards smaller concentrations. It is proportional to the concentration gradient vector ∇c and to the sediment diffusivity ε_S. The diffusivity has dimension length squared per unit time (or velocity times length), the same as kinematic viscosity and eddy viscosity which can both be considered as diffusivities of momentum. The diffusivity may be expressed in terms of a typical mixing length as outlined in Section 5.1.2, p 201.

The concept of gradient diffusion was developed in connection with the kinetic theory of gasses and statistical mechanics where macroscopic distances are covered in many random steps between which the velocity changes due to collisions. In that case the diffusivity has a clear physical meaning. It is simply the typical step length (the mixing length) squared divided by the typical time between collisions. Thus, diffusion is a useful concept when the motion considered is of this random walk nature.

However, gradient diffusion cannot describe details of a mixing process on a smaller scale than the mixing length. Processes where the mixing length is of the same magnitude as the overall scale of the sediment distribution can therefore not be described in terms of gradient diffusion.

Examples of such *convective* processes are the entrainment of sediment from rippled sand beds under waves by travelling vortices, and the lifting of sand straight from the bed to the surface by the rising plumes generated behind plunging breakers. In these processes, the sediment flux is obviously not necessarily related to the concentration gradient as expressed by the diffusion equation (5.4.5).

For the purpose of modelling natural sediment suspension processes, which generally contain elements of both diffusion and convection, we must first develop a quantitative description of the convective sediment flux. To this end, the vertical convective sediment flux is written in the form

$$q_C(z,t) = p\left(t - \frac{z}{w_C}\right) F(z) \tag{5.4.6}$$

where the pickup function $p(t)$ is a non-negative function describing the instantaneous pickup rate at the bed. The pickup function has the dimension of sediment flux (velocity times concentration) while w_C is the average vertical velocity with which the sand is convected upwards, see Figure 5.4.3.

Quantitatively, the sediment convection velocity w_C may be assumed similar to the speed w_t with which bursts of turbulence travel upward from the bed.

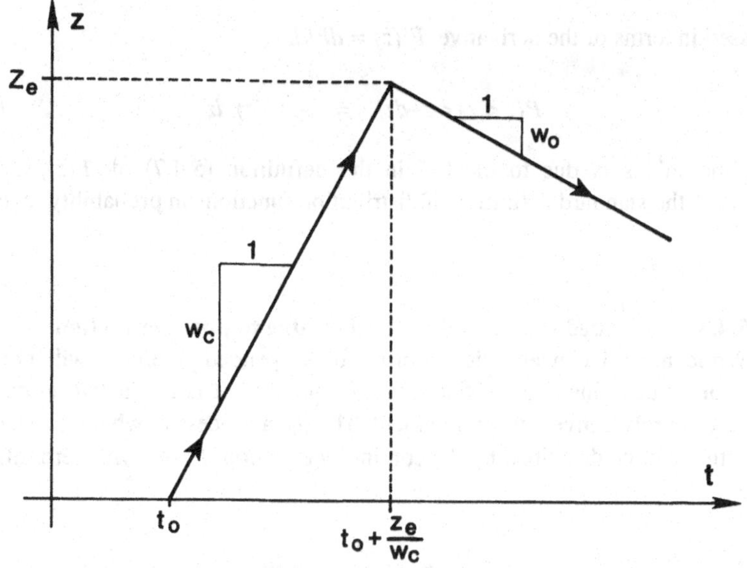

Figure 5.4.3: In the simple convection model considered here, a sediment particle which is picked up from the bed at time t_o will travel upwards with speed w_c until it reaches its entrainment level z_e at time $t_o + z_e/w_c$. After that, it is assumed to settle with its still water settling velocity w_o.

Sleath (1987) found that in oscillatory boundary layers

$$w_c \approx w_t \approx \frac{\omega\, \delta_{.05}}{2.27} \tag{1.2.3}$$

see Section 1.2.3, page 19.

The dimensionless *convective distribution function* $F(z)$ in Equation (3.4.6) determines the fraction of the entrained sand which travels (convectively) beyond the level z. Thus, $F(z)$ is a probability function expressing the probability that the entrainment level z_e reached by a given particle is higher than z

$$P\{z_e > z\} = F(z) \tag{5.4.7}$$

Hence $F(z)$ must be a positive function with $F(o) = 1$ if the process is assumed purely convective, and in general $F(z) \to o$ for $z \to \infty$. Correspondingly, the probability that a given particle is entrained to the level z is expressed in terms of the derivative $F'(z) = dF/dz$

$$P\{z < z_e < z+dz\} = -F'(z)\,dz \qquad (5.4.8)$$

where the minus is due to the ">" in the definition (5.4.7) of $F(z)$, which is opposite to the standard definition of distribution functions in probability texts.

5.4.3 Suspended sediment distributions due to pure convection

While natural concentration profiles of suspended sediment will normally result from a combined convection-diffusion process, it is instructive to consider the case of purely convective entrainment. That is, a process in which the sediment distribution can be described by the continuity equation (5.4.4) with zero diffusive flux, $q_D = 0$

$$\frac{\partial c}{\partial t} = w_o \frac{\partial c}{\partial z} - \frac{\partial q_c}{\partial z} \qquad (5.4.9)$$

and with the convective flux described in accordance with Equation (5.4.6) and Figure 5.4.3. With the expression (5.4.6) inserted for q_C Equation (5.4.9) becomes

$$\frac{\partial c}{\partial t} = w_o \frac{\partial c}{\partial z} + \frac{1}{w_c} p'(t - \frac{z}{w_c})\, F(z) - p(t - \frac{z}{w_c})\, F'(z) \qquad (5.4.10)$$

where the brief notation $p' = \frac{dp}{dt}$ and $F' = \frac{dF}{dz}$ has been used for the derivatives.

The time-average of this equation, which describes the time-averaged concentration profile $\bar{c}(z)$, is considerably simpler

$$w_o \frac{d\bar{c}}{dz} = \bar{p}\, F'(z) \qquad (5.4.11)$$

and is easily integrated to yield

$$w_o\, \bar{c}(z) - \bar{p}\, F(z) = const \qquad (5.4.12)$$

The constant of integration is found by considering this equation at the bed level, $z = o$

$$w_o \overline{c}(o) - F(o)\overline{p} = w_o C_o - \overline{p} = n\frac{\overline{dz_b}}{dt} \qquad (5.4.13)$$

where w_oC_o is the time-averaged rate of deposition , see Figure 5.3.1, page 222. Hence, for a stationary situation with $dz_b/dt = 0$, we get

$$w_o \overline{c}(o) = \overline{p} \qquad (5.3.3)$$

That is, the average pickup rate $\overline{p}$ must balance the average settling rate at the bed.

For such a stationary situation, Equation (5.4.12) takes the form

$$w_o \overline{c}(z) - \overline{p}F(z) = 0 \qquad (5.4.14)$$

which, with $w_o\,\overline{c}(o)$ substituted for $\overline{p}$, shows that the relative concentrations at different levels, i e the concentration profile shape, is the same for all settling velocities, w_o, and given directly by the convective distribution function $F(z)$

$$\frac{\overline{c}\,(z, w_o)}{\overline{c}\,(o, w_o)} \equiv \frac{\overline{c}(z)}{\overline{c}(o)} = F(z) \qquad (5.4.15)$$

Solutions to the time-dependent Equation (5.4.10) are, in general, somewhat more complicated. However, since Equation (5.4.10) is linear, and since most coastal sediment transport problems can be described in terms of Fourier series, it is natural to consider the solution which corresponds to a simple harmonic pickup function with amplitude P_n and angular frequency ω

$$p_n(t) = P_n\, e^{in\omega t} \qquad (5.4.16)$$

which generates the convective flux

$$q_{cn}(z,t) = P_n\, e^{in\omega(t-z/w_c)}\, F(z) \qquad (5.4.17)$$

upward through the level z in accordance with Equation (5.4.8).

Assuming that the solution has the form

$$c_n(z,t) = C_n f_n(z) e^{in\omega t} \tag{5.4.18}$$

with $f_n(o) = 1$, we find the following differential equation for $f_n(z)$

$$-w_o C_n f_n' + in\omega C_n f_n = \frac{P_n}{w_c} in\omega e^{-in\omega z/w_c} F(z) - P_n e^{-in\omega z/w_c} F'(z) \tag{5.4.19}$$

This equation is linear with constant coefficients. It can therefore be solved analytically for most realistic functions $F(z)$, but it has particularly simple solutions if the distribution function $F(z)$ is an exponential

$$F(z) = e^{-z/L} \tag{5.4.20}$$

In this case the solution is

$$c_n(z,t) = C_n e^{-\beta_n z/L} e^{in\omega t} \tag{5.4.21}$$

where

$$C_n = \frac{P_n}{w_o} \frac{1 + in\omega L/w_c}{1 + in\omega L/w_c + in\omega L/w_o} \tag{5.4.22}$$

and

$$\beta_n = 1 + \frac{in\omega L}{w_c} \tag{5.4.23}$$

The fact that C_n is complex with a small negative argument shows that the concentration $c_n(o,t)$ at the bed lags behind the pickup function p_n. The imaginary part of β_n makes this phase lag grow with distance from the bed. In terms of real-valued functions, the solution (5.4.21)-(5.4.23) can be written

$$c_n(z,t) = |C_n| e^{-z/L} \cos[n\omega(t - z/w_c) + \text{Arg}\{C_n\}] \tag{5.4.24}$$

This solution, which corresponds to the pure convection process, differs from the pure gradient diffusion solution, which will be discussed below, in two important respects. Firstly, the magnitude of all harmonic concentration components decay as $e^{-z/L}$ since $\text{Re}\{\beta_n\} = 1$ for all ω. Secondly, the time lag $z/w_c\text{-Arg}\{C_n\}/n\,\omega$ grows at the same rate with z for all frequencies.

The expression (5.4.24) shows that, when the distribution process is convective, a concentration peak will travel upwards with the convection speed w_c

as expected. Hence, w_C can be inferred from measurements of $c(z,t)$. Such measurements were made by Homma et al (1965) and Nakato et al (1977) over rippled beds, and by Horikawa et al (1982) and Staub et al (1984) under sheet-flow conditions.

5.4.4 Sediment concentrations due to pure gradient diffusion

For a pure diffusion process (zero convective flux), the conservation equation (5.4.4), page 235, for suspended sediment, becomes

$$\frac{\partial c}{\partial t} = w_O\frac{\partial c}{\partial z} - \frac{\partial q_D}{\partial z} \tag{5.4.25}$$

and hence, with the diffusive sediment flux given as

$$q_D = -\varepsilon_s\frac{\partial c}{\partial z} \tag{5.4.26}$$

we get

$$\frac{\partial c}{\partial t} = w_O\frac{\partial c}{\partial z} + \frac{\partial}{\partial z}(\varepsilon_s\frac{\partial c}{\partial z}) \tag{5.4.27}$$

For the steady concentration component $\overline{c}(z)$ Equation (5.4.27) becomes

$$w_O\frac{\partial \overline{c}}{\partial z} + \frac{\partial}{\partial z}(\varepsilon_s\frac{\partial \overline{c}}{\partial z}) = 0 \tag{5.4.28}$$

This time-averaged equation can immediately be integrated once to yield

$$w_O\overline{c} + \varepsilon_s\frac{d\overline{c}}{dz} = 0 \tag{5.4.29}$$

since the constant of integration is zero when there is no sediment flux at infinity.

The time-averaged Equation (5.4.29) can also be written as

$$\frac{d}{dz}\ln\overline{c} = -\frac{w_O}{\varepsilon_s} \tag{5.4.30}$$

and is therefore seen to have the general solution

$$\bar{c}(z) = \bar{c}(o)\, e^{-w_o \int_o^z \frac{dz}{\varepsilon_s}} \tag{5.4.31}$$

The constant $\bar{c}(o)$ i e, the concentration at the bed, may be given in terms of the time-averaged bed shear stress if the flow is uniform.

Alternatively, if the bed-level is stationary $(\frac{\overline{dz_b}}{dt} = 0)$, $\bar{c}(o)$ can be expressed in terms of the time averaged pickup rate $\bar{p}$, since the deposition rate $w_o\,\bar{c}(o)$ must be balanced by $\bar{p}$. See Figure 5.3.1. This leads to the expression

$$\bar{c}(z) = \frac{\bar{p}}{w_o}\, e^{-w_o \int_o^z \frac{dz}{\varepsilon_s}} \tag{5.4.32}$$

for the time-averaged sediment concentration.

For graded sediments this shows that, in contrast to the pure convection solution (5.4.21)-(5.4.23), the shape of the $\bar{c}$-profile depends strongly on the settling velocity in a pure gradient diffusion process. The concentrations of coarse sand decay faster with elevation than the concentrations of fine sand. Still, the profiles for different grain sizes in the same flow are similar in the sense expressed by Equation (5.4.30). See Figure 5.4.4.

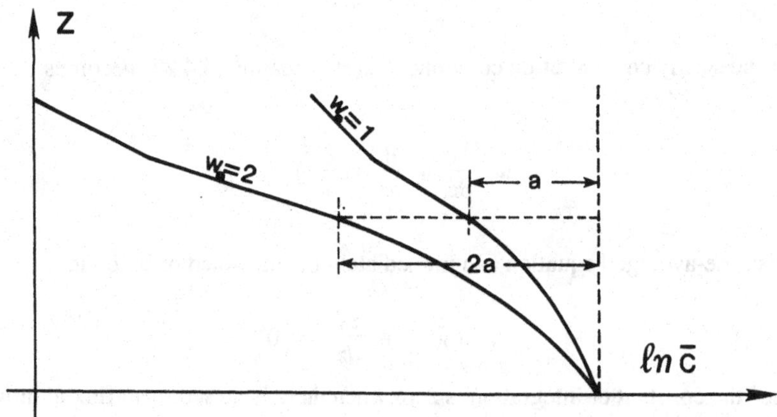

Figure 5.4.4: In a pure gradient diffusion process, the time-averaged concentration profiles for different grain sizes are different but similar in the sense described by Equation (5.4.30). That is, they are stretched horizontally in proportion to the settling velocity w.

The traditional approach, from steady flow, of prescribing the sediment concentration at the bed level as a function of the bed shear stress, does not work for time-dependent suspension problems, because there is no instantaneous equilibrium between bed shear stress and suspended sediment concentration. The concentration at the bed may well be large when the stress is zero because dense clouds of sediment are settling from above.

The alternative is to express the upward sediment flux at the bed which, in the diffusion terminology is written as $-\varepsilon_s \frac{dc}{dz}$, in terms of a pickup function

$$-\varepsilon_s \frac{dc}{dz} = p(t) \qquad \text{for} \quad z=0 \tag{5.4.33}$$

as suggested by Nielsen et al (1978).

As for the pure convection case considered in Section 5.4.3, page 238, the general form of the time-dependent sediment concentrations $c(z,t)$, in a pure gradient diffusion process, can be investigated by considering a Fourier component

$$c_n(z,t) = C_n f_n(z)\, e^{in\omega t} \tag{5.4.34}$$

of $c(z,t)$, which is generated by the Fourier component

$$p_n(t) = P_n\, e^{in\omega t} \tag{5.4.35}$$

of the pickup function.

This leads to the following ordinary, linear differential equation $f_n(z)$

$$f_n'' + \left(\frac{w_o + \varepsilon_s'}{\varepsilon_s}\right) f_n' - \frac{in\omega}{\varepsilon_s} f_n = 0 \tag{5.4.36}$$

which corresponds to Equation (5.4.19), page 240, for the pure convection problem. In order to get an impression of similarities and differences between the two descriptions, consider the special case of constant diffusivity. For constant diffusivity, Equation (5.4.36) is reduced to

$$f_n'' + \frac{w_o}{\varepsilon_s} f_n' - \frac{in\omega}{\varepsilon_s} f_n = 0 \tag{5.4.37}$$

With the boundary condition (5.4.33) and the simple harmonic pickup function

(5.4.35), Equation (5.4.37) has the solution

$$c_n(z,t) \;=\; \frac{P_n}{w_o\,\alpha_n}\, e^{-\alpha_n z w_o/\varepsilon_s}\, e^{in\omega t} \tag{5.4.38}$$

where

$$\alpha_n \;=\; \frac{1}{2} + \sqrt{\frac{1}{4} + \frac{in\omega\varepsilon_s}{w_o^2}} \tag{5.4.39}$$

The nature of the complex coefficient α_n is illustrated by Figure 5.4.5. In terms of real-valued functions, this pure gradient diffusion solution may be written as

$$c_n(z,t) \;=\; \frac{P_n}{w_o|\alpha_n|}\, e^{-\mathrm{Re}\{\alpha_n\} w_o z/\varepsilon_s}\, \cos(n\omega t - \mathrm{Im}\{\alpha_n\}\frac{w_o}{\varepsilon_s} z - \mathrm{Arg}\{\alpha_n\}) \tag{5.4.40}$$

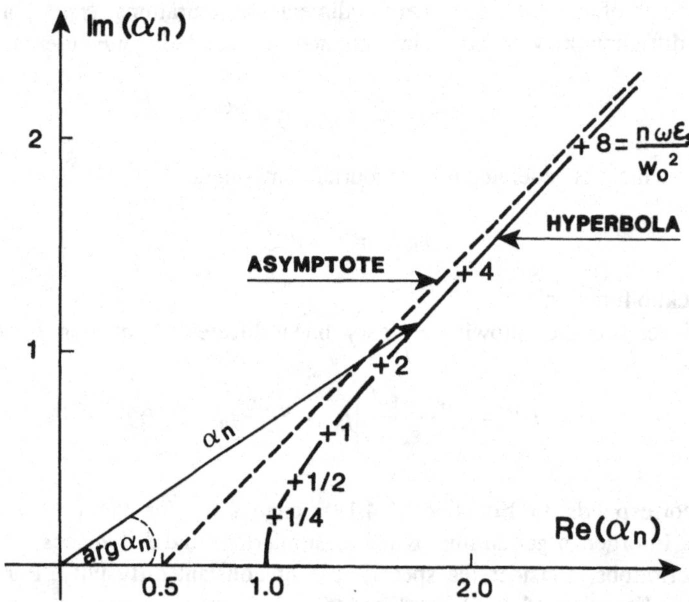

Figure 5.4.5: The variation of the complex coefficient α_n of the pure diffusion solution. It differs from the corresponding β_n (Equation (5.4.23)) of the pure convection solution in that Re{α_n} is a function of ω and, Im{a_n} is not proportional to $n\omega$. This means that the different harmonic components of the diffusion solution decay at different rates with z, and their time lags grow at different rates with z.

from which we see that, in contrast with the convection solution, the rate of decay with elevation of individual Fourier components $c_n(z,t)$ increases with frequency. In the pure convection solution (5.4.24), all Fourier components of $c(z,t)$ decay at the same rate.

Furthermore, the time lags of $c_n(z,t)$ relative to the pickup function grow at different rates with z. In the pure convection solution, all the time lags grow at the same rate, namely, z/w_c. This means that in a convection-dominated process, concentration peaks will show up with very similar shape in records from different elevations. In a diffusion-dominated process, the peaks will become blurred more rapidly.

The length scale L of the convection solution is replaced by ε_s/w_o in the pure gradient diffusion solution.

5.4.5 Sediment concentrations due to combined convection-diffusion

Natural sediment suspension processes contain elements of both convection and diffusion. This is obvious from physical inspection of the processes. Most sediment distribution processes involve mixing mechanisms with mixing lengths l_m (see Section 5.1.2, p 201) of quite different magnitudes. For some of the mechanisms the typical mixing length is small compared to the overall scale of the sediment distribution. For others the mixing length is relatively small.

When l_m is small compared to the overall scale of the suspended sediment distribution, gradient diffusion may provide a satisfactory description. However, when l_m is relatively large, a different approach is required. Recall that the term "convective" is used in the present text for processes where the mixing length is not small compared to the overall distribution scale.

Thus, convective sediment entrainment in rivers is exemplified by the "boils" of sediment laden water which sometimes appear at the surface. They are generated by travelling vortices formed in the lee of dune crests , see e g Jackson (1976). Due to the trapping mechanism described in Section 4.6, page 181, these vortices can carry suspended sediment directly from the bed to the surface.

In the surf zone, the air which is entrained by plunging breakers will generate strong upward flows which can carry large amounts of suspended sand directly to the surface, see Figure 6.5.4, page 290, and Nielsen 1984.

The travelling vortices which carry suspended sand in a fairly organised manner over rippled beds under waves are also an example of convective entrainment, see page 160.

At the same time it is fairly obvious that in all of these environments, small scale turbulence is present. The role of the small scale turbulence is to smooth concentration differences by gradient diffusion.

The combined nature of the entrainment process is also evident from measured time-averaged concentration profiles of different grain sizes in the same flow. As discussed in Section 5.2.3, page 209, the distributions of different sand sizes are not always similar, as they should be in any one of the pure processes, see Equation (5.4.15) page 239 and Figure 5.4.4 page 242.

For a combined convection-diffusion process, the behaviour of the horizontally averaged concentrations *c(z,t)* can be described by the conservation equation (5.4.4) page 235 with both convective and diffusive fluxes included and given by Equations (5.4.6) page 238, and Equation (5.4.26) page 241 respectively

$$\frac{\partial c}{\partial t} = w_o \frac{\partial c}{\partial z} + \frac{1}{w_c} p'(t - \frac{z}{w_c}) F(z) - p(t - \frac{z}{w_c}) F'(z) + \frac{\partial}{\partial z}(\varepsilon_s \frac{\partial c}{\partial z}) \tag{5.4.41}$$

By taking time-averages, this equation is reduced to

$$w_o \frac{\partial \bar{c}}{\partial z} - \bar{p} F'(z) + \frac{\partial}{\partial z}(\varepsilon_s \frac{\partial \bar{c}}{\partial z}) = 0 \tag{5.4.42}$$

which describes the time-averaged suspended sediment concentrations in a horizontally uniform, combined convection-diffusion process. This equation can immediately be integrated once with respect to z

$$w_o \bar{c} - \bar{p} F(z) + \varepsilon_s \frac{d\bar{c}}{dz} = \text{const} \tag{5.4.43}$$

where the constant of integration must be zero if all concentrations and concentration gradients are assumed to vanish far from the bed. In that case the equation can be written

$$w_o \bar{c} + \varepsilon_s \frac{d\bar{c}}{dz} = \bar{p} F(z) \tag{5.4.44}$$

By comparing this equation with the corresponding time-averaged Equation (5.4.29) page 241 for the pure gradient diffusion case, we see that the homogeneous version of the combined Equation (5.4.44) is the time-averaged diffusion equation. Hence, the pure gradient diffusion solution is always part of the combined solution (with a suitable multiplying factor).

We shall now turn to the bottom boundary condition for the suspended sediment distribution in the combined convection-diffusion process.

The bottom boundary condition in the combined convection-diffusion system would logically have to be a combination of

$$w_o\, c(o,t) \;-\; p(t)\, F(o) \;=\; n \frac{\partial z_b}{\partial t}$$

from the pure convection case, and

$$w_o\, c(o,t) \;+\; \varepsilon_s \frac{\partial c}{\partial z}\Big|_{z=o} \;=\; n \frac{\partial z_b}{\partial t}$$

from the pure gradient diffusion case. That is

$$w_o\, c(o,t) \;-\; p(t)\, F(o) \;+\; \varepsilon_s \frac{\partial c}{\partial z}\Big|_{z=o} \;=\; n \frac{\partial z_b}{\partial t}$$

The second and the third terms on the left-hand side are minus the convective and diffusive fluxes upward from the bed. Together, these terms amount to the total upward sediment flux from the bed, i e , to the pickup rate

$$p(t)\, F(o) \;-\; \varepsilon_s \frac{\partial c}{\partial z}\Big|_{z=o} \;=\; p(t) \tag{5.4.45}$$

or

$$-\,\varepsilon_s \frac{\partial c}{\partial z}\Big|_{z=o} \;=\; [1 - F(o)\,]\, p(t) \tag{5.4.46}$$

Thus, for a given pickup rate $p(t)$, the concentration solution depends on the chosen value $F(0)$ of the convective distribution function at the bed. In the following we shall mostly be using $F(0) = 1$ which through Equation (5.4.46) corresponds

$$-\,\varepsilon_s \frac{\partial c}{\partial z}\Big|_{z=o} \;=\; 0 \tag{5.4.47}$$

That is, the diffusive sediment flux is assumed to vanish at the bed. If the diffusivity ε_s tends smoothly towards zero at the bed, then the diffusive flux vanishes smoothly. However, it vanishes abruptly if ε_s is not tending towards zero at the bed. Thus, the general validity of the choice of $F(0) = 0$ is perhaps debatable. It is a convenient choice however, and it leads to reasonable agreement with measurements, see Example 5.4.2, page 256.

For the time-averaged concentration we always have the simple bottom boundary condition

$$w_o\,\overline{c}(o) - \overline{p} = n\left(\frac{\partial z_b}{\partial t}\right)$$

see Figure 5.3.1, page 222 which for a stationary situation with no net erosion or deposition becomes

$$w_o\,\overline{c}(o) - \overline{p} = 0 \tag{5.3.3}$$

5.4.6 Time-averaged concentrations in combined convection-diffusion

The particular solution to the time-averaged conservation Equation (5.4.43), which takes the value $\overline{c}(z_r)$ at the reference level z_r can be written

$$\overline{c}(z) = e^{-G(z)}\left(\int_{z_r}^{z}\frac{\overline{p}}{\varepsilon_s(\zeta)}F(\zeta)\,e^{G(\zeta)}\,d\zeta + \overline{c}(z_r)\right) \tag{5.4.48}$$

or, with $\overline{c}(z_r) = \overline{p}/w_o$

$$\overline{c}(z) = \frac{\overline{p}}{w_o}\,e^{-G(z)}\left(\int_{z_r}^{z}\frac{w_o}{\varepsilon_s(\zeta)}F(\zeta)\,e^{G(\zeta)}\,d\zeta + 1\right) \tag{5.4.49}$$

where the function $G(z)$ can be any integral of $w_o/\varepsilon_s(z)$

$$G(z) = \int\frac{w_o}{\varepsilon_s(z)}dz \tag{5.4.50}$$

as long as the same function is used both inside and outside the brackets.

The first term inside the brackets of Equation (5.4.49) grows very rapidly with the settling velocity w_o. Thus, while the pure diffusion solution, which corresponds to the last term, may dominate for the finer sand fractions, the first term will become dominant for the coarser sand fractions. The first term of the solution accounts for the convective entrainment, and for vanishing diffusivity, it tends towards the pure convection solution

$$\bar{c}(z) = \bar{c}(o)\, F(z) = \frac{\bar{p}}{w_o} F(z) \qquad (5.4.15)$$

This may not be immediately obvious from the form of the solution (5.4.49). It will however become evident from the Examples 5.4.1 and 5.4.3.

The fact that the shape of the convective distribution function $F(z)$ can be expected to be displayed by the coarsest sand through Equation (5.4.15) is important. It shows that information about $F(z)$ can be extracted directly from measured concentration distributions of the coarsest sand fractions.

***Example 5.4.1*: A simple case of combined convection-diffusion**

The following example illustrates the roles of convection and diffusion in the combined system with $F(z_r) = F(o) = 1$ in general. In particular, it illustrates the asymptotic behaviour of the combined convection-diffusion distribution model in the limits of negligible and dominant diffusion.

Consider the time-averaged solution (5.4.49) for the simple situation where the convective entrainment function is a simple exponential with length scale L

$$F(z) = e^{-z/L}$$

and where the diffusivity ε_s is a constant. In this case, the integral (5.4.50) may be taken as

$$G(z) = \frac{w_o}{\varepsilon_s} z$$

and the time-averaged solution (5.4.49) with $z_r=0$ reads

$$\bar{c}(z) = \frac{\bar{p}}{w_o} e^{-w_o z/\varepsilon_s} \left(\frac{w_o}{\varepsilon_s} \int_0^z e^{-\zeta/L} e^{w_o \zeta/\varepsilon_s} d\zeta + 1 \right)$$

which can be rewritten as

$$\bar{c}(z) = \frac{\bar{p}}{w_o}\left\{\frac{1}{1-\frac{\varepsilon_s}{w_oL}}\,e^{-z/L} + \left(1 - \frac{1}{1-\frac{\varepsilon_s}{w_oL}}\right) e^{-w_oz/\varepsilon_s}\right\} \tag{5.4.51}$$

This concentration distribution is shown in Figure 5.4.6 for selected values of the relative diffusivity ε_s/w_oL.

It can be seen that the solution tends towards the pure convection solution

$$\frac{\bar{c}(z)}{C_o} = F(z) \tag{5.4.15}$$

for vanishing relative diffusivity, $\varepsilon_s/w_oL \to 0$

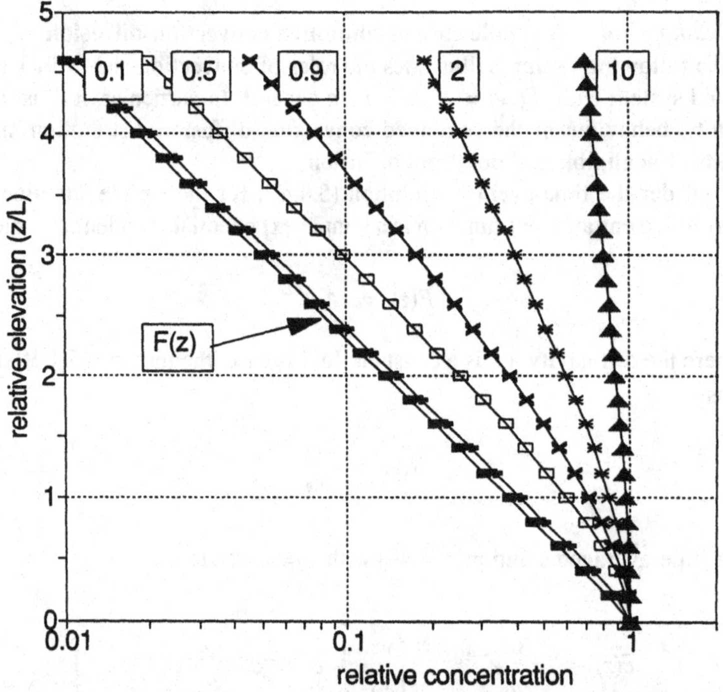

Figure 5.4.6: Relative time-averaged sediment concentrations corresponding to the combined convection-diffusion solution (5.4.49) for different values of the relative diffusivity ε_s/w_oL. For small values of ε_s/w_oL the solution approaches the pure convection solution (5.4.15).

5.4.7 Time-dependent concentrations in combined convection-diffusion

This section contains a comparison of the solution to the time dependent Equation (5.4.41), page 246, for the combined convection-diffusion process with the cases of pure convection and pure diffusion considered in Sections 5.4.3. and 5.4.4. All of the processes have been assumed horizontally uniform for simplicity.

Consider the combined convection-diffusion equation (5.4.41) for the special case of a constant sediment diffusivity ε_s and a simple exponential convective distribution function $F(z) = e^{-z/L}$. Applying the simple harmonic pickup function $p_n(t) = P_n e^{in\omega t}$, and assuming that the concentration solution has the form

$$c_n(z,t) = f_n(z)\, e^{in\omega t} \tag{5.4.52}$$

we find the following differential equation for the complex function $f_n(z)$

$$f_n'' + \frac{w_o}{\varepsilon_s} f_n' - \frac{in\omega}{\varepsilon_s} f_n = \frac{-\beta_n P_n}{L\,\varepsilon_s} e^{-\beta_n z/L} \tag{5.4.53}$$

where, as for the pure convection solution (p 240)

$$\beta_n = [1 + \frac{in\omega L}{w_c}] \tag{5.4.23}$$

Note that the homogeneous version of Equation (5.4.53) is identical to the corresponding Equation (5.4.37) for the pure diffusion problem which has the solution (5.4.38) - (5.4.39), p 244. Hence, the complete solution to Equation (5.4.53) for combined convection-diffusion is a constant times the diffusion solution plus a particular integral to Equation (5.4.53). It is easily verified that

$$f_n(z) = B_n\, e^{-\beta_n z/L} \tag{5.4.54}$$

with

$$B_n = \frac{P_n}{w_o} \frac{1}{1 - \frac{\beta_n \varepsilon_s}{w_o L} + i \frac{n\omega L}{\beta_n w_o}} \tag{5.4.55}$$

is a solution to Equation (5.4.53). Thus, the complete solution is

$$c_n(z,t) = \left(A_n\, e^{-\alpha_n w_o z/\varepsilon_s} + B_n\, e^{-\beta_n z/L}\right) e^{in\omega t} \tag{5.4.56}$$

where the constant A_n is determined by the boundary condition (5.4.46), p 247.

5.4.8 Determining ε_s, w_c and $F(z)$ for combined convection-diffusion

In order to make use of the combined convection-diffusion model, for predictive purposes, it is of course, necessary to be able to predict the sediment diffusivity ε_s, the typical convection velocity w_c, and the convective distribution function $F(z)$. The physical meanings of the convective process parameters w_c and $F(z)$ are explained in connection with Figure 5.4.3, p 237.

The prediction of all three parameters is a major task because a large number of cases would need to be given special consideration. That is, suspended sediment distributions under waves may well be different from those in steady flows and wave breaking is likely to have a profound effect. It is also possible that the unsteady components of $c(z,t)$ correspond to a different diffusivity than $\bar{c}(z)$ just like the steady and unsteady flow components of a combined wave-current flow seem to feel different eddy viscosities, see Section 1.5.3.

For the special case of pure non-breaking waves it seems reasonable to expect that ε_s, w_c and $F(z)$ will all be closely related to the bottom boundary layer parameters. It turns out that a simple model based on this philosophy is in fair agreement with the available $\bar{c}(z)$-data for flat beds as well as rippled beds.

With respect to the sediment diffusivity ε_s, it seems reasonable to relate it to some eddy viscosity. The phrase "the eddy viscosity" is consciously avoided because of the complicated nature of the eddy viscosity concept relating to oscillatory boundary layers and combined wave-current flows. See Sections 1.2.8, 1.2.9 (pp 31-40) and Section 1.5.3 p 68.

A priori, a few expressions for ε_s might be suggested, e g, Equations (1.2.43), (1.2.44) and (1.3.14). However, in view of the experimental evidence, as discussed below, the most pragmatic choice at the moment seems to be

$$\varepsilon_s = 4\nu_1 = 4\,[0.5\,\omega\,z_1^2] = 4\,[0.5\,\omega\,(0.09\sqrt{r\,A}\,)^2] \approx 0.016\,\omega\,r\,A \tag{5.4.57}$$

The factor 4 corresponds to the factor of approximately 4 between the diffusivities of time-averaged momentum $(\bar{u})$ and of oscillatory momentum $(\tilde{u})$ in a wave-dominated boundary layer. See Equation (1.2.38) and Figure 1.5.6, page 70. The eddy viscosity expression applied in Equation (5.4.57) is Equation (1.3.14), page 52. This eddy viscosity estimate should be a reasonable choice for the large relative roughness values $(0.085<r/A<1.15)$ which are generally exhibited by sand beds under waves. See Sections 1.3.3, p 42, and 3.6.4, p 152.

The vertical speed of sediment convection w_c should be closely related to the speed w_t with which boundary layer turbulence propagates upwards in oscillatory boundary layers. Sleath (1987) found, from purely oscillatory flow that

$$w_t \approx \frac{\omega\,\delta_{.05}}{2.27} \tag{1.2.3}$$

where $\delta_{.05}$ is the boundary layer thickness which corresponds to a dimensionless velocity defect $|D|$ of 0.05 at the top of the boundary layer.

Assuming the simple boundary layer structure which corresponds to constant eddy viscosity, the defect magnitude $|D|$ decays exponentially in accordance with Equation (1.3.7). Then Equation (1.2.3), for the convection velocity can be written

$$w_c = w_t \approx \frac{\omega\,3\,z_1}{2.27} \approx 0.12\,\omega\sqrt{r\,A} \tag{5.4.58}$$

where the factor *"3"* comes from $0.05 \approx e^{-3}$ and the boundary layer length scale z_1 is found from $z_1 = 0.09\sqrt{r\,A}$, see Equation (1.3.10), page 47.

It should also be possible to observe actual values of the sediment convection velocity w_c as the speed of upward propagating concentration peaks. Provided of course, the convective (the second) term of the solution (5.4.56) is dominant.

Remaining to be modelled is the convective distribution function $F(z)$. Assuming a close link between advection of boundary layer turbulence and of suspended sediment, it should be possible to derive information about $F(z)$ from measured turbulence intensity distributions. The vortex trapping described in Section 4.6 (pp 181-189) provides the close link because it enables the trapped sediment to travel with the turbulence (the vortices). However, the derivation of $F(z)$ from turbulence intensity distributions offers the extra complication of modelling the decay of turbulence.

Therefore, a simpler and more directly empirical approach will be applied in the following. The general nature of the sediment distribution function $F(z)$ will be inferred directly from the distributions of coarse, suspended sediment. In this context, the term "coarse" corresponds to large values of the relative settling velocity, $w* = \frac{w_o L}{\varepsilon_s} >> 1$, where L is the vertical scale associated with $F(z)$. See Example 5.4.1, page 249 ff.

As discussed in relation to the time-averaged convection-diffusion solution (5.4.48), p 248, and as indicated by Example 5.4.1, the distribution of relative concentrations for very coarse suspended sand ($w* >> 1$) is essentially $F(z)$ as for the pure convection case.

$$\frac{\bar{c}(z)}{\bar{c}(o)} \approx F(z) \qquad \text{for} \quad \frac{w_o L}{\varepsilon_s} >> 1 \tag{5.4.59}$$

Hence, considering the $\bar{c}$-distribution for the coarsest sand in Figure 5.2.12 we see that, in this semi-logarithmic presentation, $F(z)$ has an upward concave shape which is typical of functions of the form $(1+z/L)^{-n}$ where L is an appropriate vertical scale. Furthermore, $F(z)$ is seen to decrease by roughly a factor 10^3 through nine centimetres of height for those experimental conditions.

It seems natural to chose $L = z_1$, where z_1 is the equivalent of the laminar Stokes length in an oscillatory boundary layer with constant eddy viscosity, see Equation (1.3.7), page 46. For the experiments reported in Figure 5.2.12, page 221, z_1 was approximately three millimetres. Hence, the elevation of $9cm$ corresponds to $z/L = z/z_1 \approx 30$ and a power of -2 is thus required to give the observed concentration decrease of the order 10^{-3} over $9cm$. Hence we may suggest

$$F(z) = (1 + z/z_1)^{-2} \qquad (5.4.60)$$

Although the derivation above is very case specific, the expressions (5.4.57) and (5.4.60) lead to fairly reasonable predictions of $\bar{c}(z)$ when inserted into the convection-diffusion solution (5.4.49). This is shown for large vortex ripples and for flat beds (sheet-flow), in Examples 5.4.2 and 5.4.3 in the following section.

Nevertheless, the expressions (5.4.57) - (5.4.60) should only be seen as first estimates of ε_s and $F(z)$ and improved versions should be sought as soon as the necessary data become available. For this purpose, $\bar{c}(z)$-data of the type presented in Figure 5.4.7 are essential. That is, separate concentration profiles for the different sand fractions. No data of this type are presently available from sheet-flow conditions or from long period waves over fairly small ripples.

Note that when the diffusivity is proportional to ωz_1^2, as in Equation (5.4.57) and when the vertical scale L is proportional to z_1 as assumed in the derivation of Equation (5.4.60). Then, the relative settling velocity or profile shape parameter $w^* = w_o L/\varepsilon_s$ becomes proportional to the wave period $w^* \sim w_o/(\omega z_1) \sim w_o T/\sqrt{rA}$ as observed in Figures 5.2.7 and 5.2.8, pp 216-217.

5.4.9: Convection-diffusion modelling of $\bar{c}(z)$ undernon-breaking waves

The time-averaged concentration profiles under non-breaking waves over both vortex ripples and flat beds can be successfully modelled by the combined convection-diffusion model which was developed in Section 5.4.6. That is, with appropriate choices of the time-averaged pickup function $\bar{p}$, the sediment diffusivity ε_s, and the convective distribution function $F(z)$, the concentration magnitude and the general profile shape are reasonably predicted.

Quantitatively, the combined convection-diffusion model consists of the following formulae from the previous sections:

$$\bar{c}(z) = \frac{\bar{p}}{w_o} e^{-G(z)} \left(\int_{z_r}^{z} \frac{w_o}{\varepsilon_s(\zeta)} F(\zeta)\, e^{G(\zeta)}\, d\zeta + 1 \right) \tag{5.4.49}$$

see page 248, where $G(z) = \int \frac{w_o\, dz}{\varepsilon_s}$ and the reference level z_r is generally chosen to be the bed level, i e, $z_r = 0$. The elevation z is measured from the ripple crest level if the bed is rippled. The time-averaged pickup function is predicted by

$$\bar{p} = w_o C_o = 0.005\, w_o\, \theta_r^3 \tag{5.3.11}$$

see page 228. The sediment diffusivity is given by

$$\varepsilon_s \equiv 4\,[0.5\,\omega\, z_1^2] = 4\,[0.5\,\omega\,(0.09\sqrt{r\,A}\,)^2] \approx 0.016\,\omega\, r\, A \tag{5.4.57}$$

see page 252, and the convective distribution function by

$$F(z) = \frac{1}{(1 + z/z_1)^2} \tag{5.4.60}$$

see page 254. The optimal choice of the bed roughness r seems to be

$$r = 8\,\eta^2/\lambda + 5\,\theta_{2.5}\, d_{50} \tag{3.6.14}$$

(page 159). If Equation (3.6.13) is used instead, for sheet flow conditions, the resulting values of $z_1 = 0.09\sqrt{r\,A}$ seem to be too large to be used with Equation (5.4.60).

With the expressions (5.4.57) and (5.4.60) for ε_s and $F(z)$, and with the reference level taken at the ripple crest level, $z_r = 0$, the time-averaged solution (5.4.49) becomes

$$\bar{c}(z) = \frac{\bar{p}}{w_o} e^{-w_o z/\varepsilon_s} \left(\frac{w_o}{\varepsilon_s} \int_0^z \frac{e^{\,w_o \zeta/\varepsilon_s}}{(1 + \zeta/z_1)^2}\, d\zeta + 1 \right) \tag{5.4.61}$$

or, with the dimensionless elevation $\xi = \zeta/z_1$ and the dimensionless settling velocity $w* = w_o z_1/\varepsilon_s$

$$\bar{c}(z) = \frac{\bar{p}}{w_o} e^{-w_* z/z_1} \left(w_* \int_0^{z/z_1} \frac{e^{w_* \xi}}{(1+\xi)^2} d\xi + 1 \right) \tag{5.4.62}$$

***Example 5.4.2:* $\bar{c}(z)$ over vortex ripples**

In this example, the combined convection-diffusion model is applied to the suspended sediment distribution under non-breaking waves over a rippled sand bed.

Consider the field situation studied by Nielsen (1983) and described in further detail as Tests 57-60 by Nielsen (1984). The measured concentration profiles for the different sediment sieve fractions are shown in Figure 5.4.7.

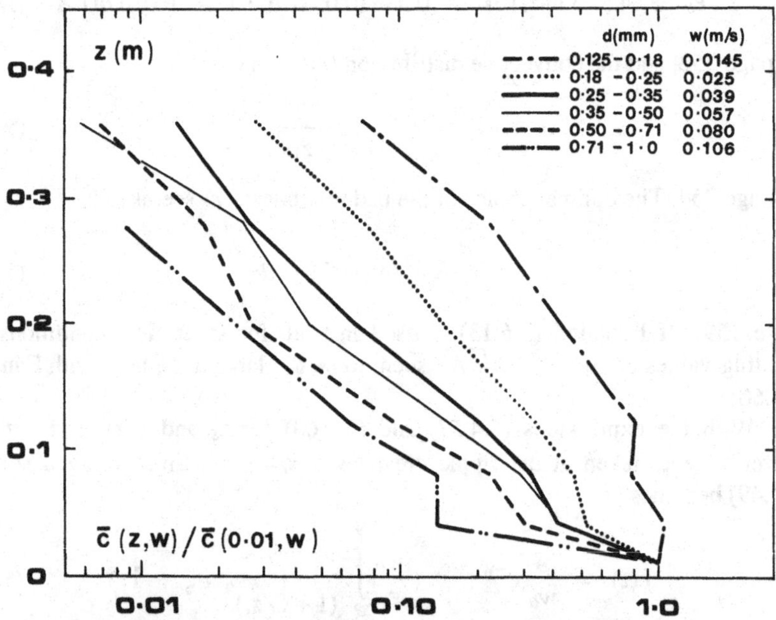

Figure 5.4.7: Time-averaged concentration profiles for different sand size fractions over vortex ripples under non-breaking waves. Field data from Nielsen (1983).

The distributions of the different grain sizes show the characteristic behaviour which calls for a combined convection-diffusion description rather than the traditional, pure gradient diffusion description. That is, the profiles plotted with z against $\ln \overline{c}$ do not show the similarity predicted by the pure diffusion solution, see Figure 5.4.4, p 242. The profiles for fine sand tend to be upward convex while the profiles for coarse sand are upward concave. This is the same pattern as shown, by the laboratory measurements in Figures 5.2.11 and 5.2.12, pp 220-221.

The relative (the factor $\overline{p}(w_o)/w_o$ omitted) concentrations corresponding to the combined convection-diffusion solution (5.4.61) are shown in Figure 5.4.8 for a few values of the dimensionless settling velocity $w* = w_o z_1/\varepsilon_s$.

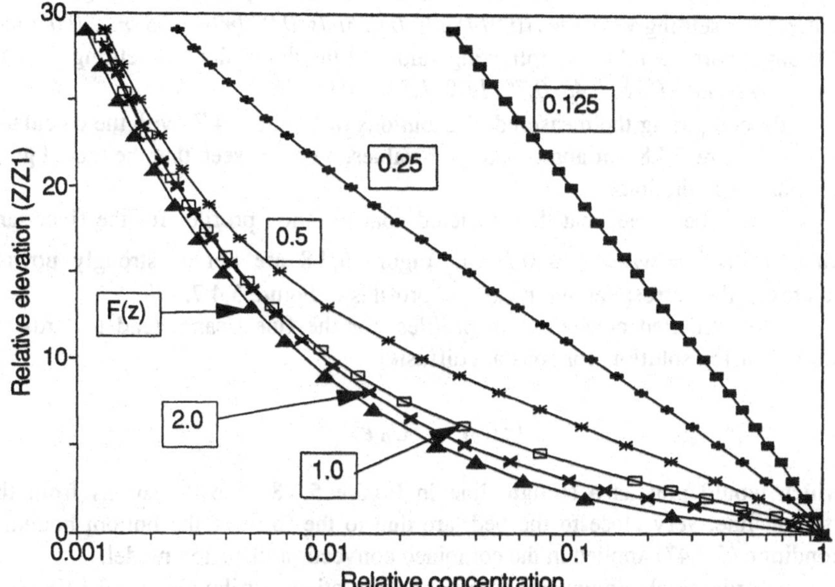

Figure 5.4.8: Relative sediment concentrations calculated from the combined convection-diffusion model, Equation (5.4.61). The numbers on the curves correspond to the dimensionless settling velocity $w* = w_o z_1/\varepsilon_s$.

It can be seen, in qualitative agreement with the data in Figure 5.4.7, that the concentration profiles for the finer sand are upward convex while those for the coarser sand are upward concave.

In detail, the calculations behind the $\overline{c}(z)$-profiles in Figure 5.4.8 are as follows. Based on the measured wave and sediment parameters $(T, D, H_{rms}, d_{50}, \eta, \lambda)$ = (7.2*s*, 1.3*m*, 0.34*m*, 0.46*mm*, 0.09*m*, 0.53*m*) and linear wave theory, the following near-bed flow parameters $A = 0.52m$, $A\omega = 0.65m/s$ are found, corresponding to "the rms wave". The hydraulic roughness is then calculated in accordance with Equation (3.6.14), page 159, which gives

$$r = 0.12m$$

and hence, $z_1 = 0.09\sqrt{rA} = 0.023m$. The sediment diffusivity is then calculated from (5.4.57), $\varepsilon_s \equiv 4 \cdot 0.5\,\omega\, z_1^2 \approx 1.2 \cdot 10^{-3} m^2/s$.

With these values of (z_1, ε_s), the sand fractions represented in Figure 5.4.7, which have settling velocities *(0.0145m/s, 0.025m/s, 0.039m/s, 0.057m/s, 0.080m/s, 0.106m/s)* correspond to the following values of the dimensionless settling velocity $w* = w_o z_1/\varepsilon_s$: *(0.28, 0.48, 0.75, 1.09, 1.53, 2.03)*.

By comparing the measured distributions in Figure 5.4.7 with the calculated ones in Figure 5.4.8 for appropriate $w*$-values, it can be seen that the model gives reasonable predictions.

It may be noted, that the predicted concentration profiles for the finer sand fractions $(w^* = w_o z_1/\varepsilon_s \leq 0.25)$ in Figure 5.4.8 are not as strongly upward convex as the corresponding, measured profiles in Figure 5.4.7.

The predicted concentration profiles, for the finest sand, tend towards the pure diffusion solution, for constant diffusivity

$$\overline{c}(z, w_o) = C_o\, e^{-w_o z/\varepsilon_s}$$

which would plot as a straight line in Figure 5.4.8. The deviations from this straight line, very close to the bed, are due to the form of the bottom boundary condition (5.4.47) applied in the combined convection-diffusion model.

In order to obtain upward convex concentration distributions for the fine sand fractions, it is necessary to assume a diffusivity distribution for which the pure diffusion solution (5.4.32) is upward convex. This is obtained when $\varepsilon_s(z)$ is a decreasing function of z. There are good physical reasons for suggesting that the sediment diffusivity is a decreasing function of z over ripples under non-breaking waves, but this fine tuning of the model will not be attempted at the present stage.

At present, the main point is, that the combined convection-diffusion model is capable of explaining the qualitative difference between the distributions of different grain sizes suspended in the same flow. That is not possible within the framework of either pure gradient diffusion or pure convection.

The distribution of total concentrations (the "sum" of the size fraction profiles discussed above) over a bed of fairly well graded sediment is not necessarily itself given by Equation (5.4.61) with any particular value of w^*. Nevertheless, it is often attempted to predict the distribution of total concentrations with a formula like Equation (5.4.61) and using the mean (or median) settling velocity of the bed sediment.

For the case corresponding to the data in Figure 5.4.7, that leads to the total $\bar{c}(z)$-profile which is shown in Figure 5.4.9. The reference concentration $C_o = \bar{c}(0) = \bar{p}/w_o$ is calculated from Equation (5.3.10), page 228.

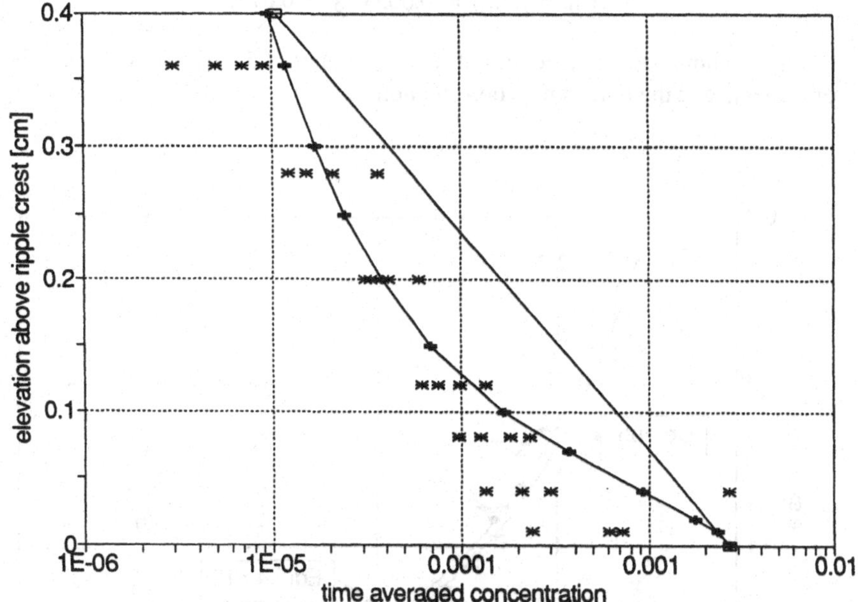

Figure 5.4.9: Measured (*) total concentrations compared to the distribution (5.4.61), page 255, calculated for the mean sediment size ($\bar{d}, \overline{w_o}$) = (0.46mm, 0.061m/s) of the bed sediment (+). The straight line corresponds to the simple exponential model given by Equations (5.2.4), (5.3.10) and (5.2.5) page 217.

Example 5.4.3 **$\bar{c}(z)$ from combined convection-diffusion over a flat bed**

As an example of sediment suspension in purely oscillatory flow under sheet-flow conditions consider Test 26 of Delft Hydraulics (1989).

The experiment was performed in a large oscillating water tunnel, and the flow and sediment parameters were $(T, A\omega, d_{50}, \overline{w_o})$ = (*8s, 1.50m/s, 0.21mm, 0.033m/s*) corresponding to a grain roughness Shields parameter of $\theta_{2.5} = 2.43 >> 1$. Hence, the bedforms which were present $(\eta, \lambda) = (0.17m, 2.5m)$ would have been rounded megaripples rather than sharp crested vortex ripples.

The boundary layer parameters and the concentration profile therefore are calculated as for a flat bed. Then we find the hydraulic roughness $r = 0.0026m$ from Equation (3.6.14), page 159, leading to $z_1 = 0.09\sqrt{rA} = 0.0063m$, and the diffusivity $\varepsilon_s = 2\,\omega\, z_1^2 = 8.2{\cdot}10^{-5} m^2/s$, see Equation (5.4.57), p 252. The reference concentration is found from Equation (5.3.10), p 228

$$C_o = \overline{p}/w_o = 0.005\, \theta_{2.5}^3 = 0.072.$$

The combined convection-diffusion solution (5.4.61) based on these parameters, and the pure convection solution

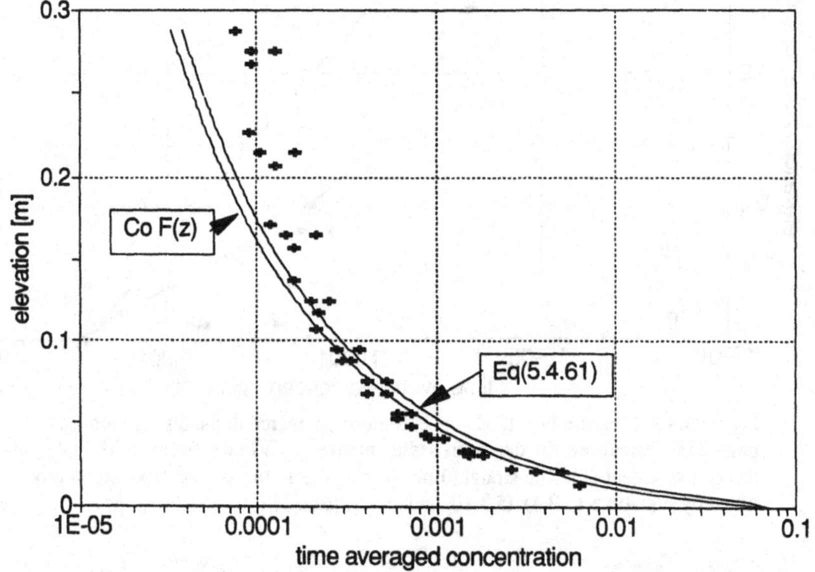

Figure 5.4.10: Measured $\overline{c}(z)$-values from oscillatory sheet-flow compared to the pure convection solution (5.4.15) and to the combined convection-diffusion solution (5.4.61). Data from Delft Hydraulics (1989), test T26.

$$\bar{c}(z) = C_o F(z) = \frac{C_o}{(1 + z/z_1)^2} \qquad (5.4.15)$$

are compared with the measured concentrations in Figure 5.4.10.

The agreement is reasonable considering that the used expressions for the diffusivity ε_s and for the convective distribution function $F(z)$ (Equations 5.4.57 and 5.4.60) are the same as applied in the previous example for very different flow conditions. These formulae were derived in Section 5.4.8 from concentration data measured over vortex ripples. Nevertheless, these expressions for ε_s and $F(z)$ are temporary. Improvements should be sought when detailed measurements of $\bar{c}(z)$ for different sand sizes in the same oscillatory sheet-flow become available.

The difference between the combined convection-diffusion solution (5.4.61) and the much simpler, pure convection solution (5.4.12) is fairly small for this case (Figure 5.4.10). This corresponds to the fairly large value of the dimensionless settling velocity $w* = \frac{w_o z_1}{\varepsilon_s} = 2.5$, compare with the sequence of concentration profiles in Figure 5.4.8.

Assuming that the diffusivity and the convective distribution function can be estimated from Equations (5.4.57) and (5.4.60) as above for oscillatory sheet-flow, it seems likely that the relative settling velocity will be quite large

$$w* = \frac{w_o z_1}{\varepsilon_s} \gtrsim 2$$

for most sheet-flow conditions. Hence, using the pure convection solution (5.4.12) instead of the complete convection-diffusion solution (5.4.61), may be acceptable for sheet-flow under non-breaking waves in general.

It should be remembered however, that the use of Equations (5.4.57) through (5.4.60) for sheet-flow conditions has not been verified in detail. The formulae were derived from measurements over rippled beds. The verification will have to be done in relation to measured concentration distributions of different sand sizes in the same oscillatory sheet-flow, when such data become available.

Figure 6.1.1: The modelling of coastal sediment transport processes still presents many interesting, unresolved problems. The challenges of the swash zone are particularly daunting.

CHAPTER 6

SEDIMENT TRANSPORT MODELS

6.1 INTRODUCTION

Natural beaches are constantly changing in response to changes in wind and wave conditions and sometimes also in response to human interference, e g the construction of breakwaters and groynes.

Such changes of the beach morphology are usually modelled iteratively with each iteration involving three main steps. First the main flow pattern is calculated on the basis of the incoming waves and the existing topography. Secondly, the local sediment transport rates are calculated from the main flow pattern and the sediment characteristics. Thirdly, the rates of morphological change are calculated from the local sediment transport rates.The focus of the present text is the second step in this modelling process.

Chapters 1 through 5 contain a discussion of the basic sediment transport mechanisms for simplified topographical conditions and with the hydrodynamic conditions, basically the free stream velocity $u_\infty(t)$ taken for granted. The aim of the present chapter is to provide a selection of manageable sediment transport models each applicable to a certain combination of flow type and bed topography. These simple models can then be used as building blocks in comprehensive sediment transport models.

6.2 TRANSPORT MODELS ARE IN ESSENCE OF TWO VARIETIES

Sediment transport models are, in essence, of two kinds. Namely, *cu*-integral models, and particle trajectory models. These two types are described in the following.

First the traditional *cu*-integral (concentration times velocity integral) type of models. The sediment transport rate, through a unit width of a vertical plane perpendicular to the x,u-direction, can be calculated as

$$Q(t) = \int_{z=0}^{D} c(z,t)\, u_s(z,t)\, dt \tag{6.2.1}$$

where $c(z,t)$ is the local, instantaneous sediment concentration and $u_s(z,t)$ is the instantaneous, horizontal sediment velocity. In most cases we are mainly interested in the time-averaged transport rate, $\overline{Q}$ which, for a periodic process can be calculated as

$$\overline{Q} = \frac{1}{T}\int_{t}^{t+T}\int_{z=0}^{D} c(z,t)\, u(z,t)\, dz\, dt \tag{6.2.2}$$

where T is the wave period. In terms of the steady, periodic and random components, this may be written as

$$\overline{Q} = \int_{z=0}^{D} (\overline{c}\,\overline{u} + \overline{\tilde{c}\,\tilde{u}} + \overline{c'u'})\, dz \tag{6.2.3}$$

see Equation (1.1.17), page 11.

The last term $\overline{c'u'}$ in this expression is generally expected to be small. The first two terms may be of similar magnitude or either of them may be dominant, when transport in the wave direction is considered.

Transport perpendicular to the wave direction will, of course, be dominated by $\overline{c}\,\overline{v}$ which is the equivalent of the first term in Equation (6.2.3).

By far the majority of the existing sediment transport models are based on this "*cu*-integral approach". However, the alternative models, using the "particle trajectory approach" will, in some cases, lead to simpler and more accurate models.

The basic idea of the particle trajectory approach is illustrated in Figure 6.2.1. It can be seen that the time-averaged transport rate $\overline{Q}$ can be expressed very simply in terms of the time-averaged pickup rate $\overline{p}$, and the average distance l_x travelled by sediment particles.

In a short time interval δ_t around t^i, an amount $p(t^i)\,\delta_t$ of sediment is picked up per unit area. During the time interval from this pickup to the settling of the last grain, these grains will make a total transport contribution of $l_x\, p(t^i)\,\delta_t$. This corresponds to the time-averaged transport rate

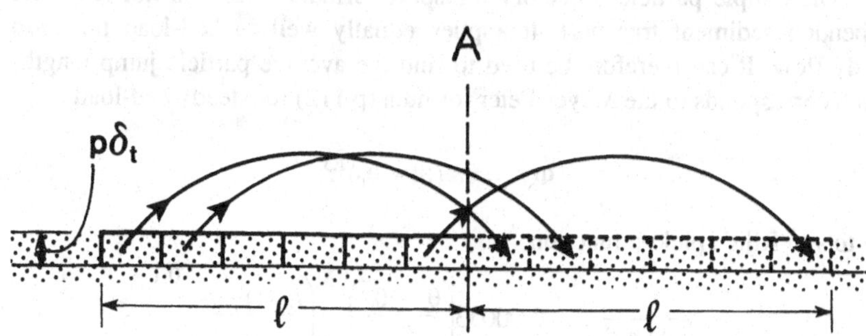

Figure 6.2.1: In a short time interval δ_t the amount of sand picked up per unit area is $p\,\delta_t$. If the average distance travelled by the moving sediment is l_x, the corresponding sediment transport through one unit width of the plane A is seen to be $l_x\,p\,\delta_t$.

$$\overline{Q} = l_x \overline{p} \tag{6.2.4}$$

a remarkably simple result.

The required ingredients to this type of model are obviously the pickup function, which was discussed in Section 5.3, page 222 ff, and an estimate of the average jump length l_x. While the individual particle trajectories may be very complex and highly variable it is generally not too difficult to make a simple description and extract an adequately accurate value of the average jump length l_x.

This was done by Nielsen (1988a) for the case of non-breaking waves over rippled beds. Two particle trajectory models including different amounts of detail were developed for this phenomenon and compared with experimental data. In addition, a classical *cu*-integral model based on the diffusion equation (5.4.27) was considered.

In the simplest particle trajectory model, the "grab and dump model", it was assumed that l_x was simply plus or minus the near-bed water particle semi-excursion ($\pm A$).

It turned out that this simple model was the most consistently accurate of the three. All three transport models will be discussed briefly in the following section.

Example 6.2.1: **The particle trajectory approach to steady bed-load**

The simple particle trajectory transport formula (6.2.4) is not restricted to suspended sediment transport. It applies equally well to bed-load transport in steady flow. It can therefore be used to find the average particle jump length l_x which corresponds to the Meyer-Peter formula (p 112) for steady bed-load

$$\Phi_B = 8\,(\theta' - \theta_c)^{1.5} \qquad (2.3.11)$$

and to van Rijn's pickup function

$$\frac{p}{\sqrt{(s-1)\,g\,d}} = 0.00033\left(\frac{\theta' - \theta_c}{\theta_c}\right)^{1.5}\left(\frac{(s-1)\,g\,d^3}{\nu^2}\right)^{0.1} \qquad (5.3.2)$$

Rewriting these two formulae in the forms

$$Q_B = 8\,(\theta' - \theta_c)^{1.5}\sqrt{(s-1)\,g\,d}\;d$$

where the definition (2.3.10), page 112, for Φ_B has been used, and

$$p = 0.00033\left(\frac{\theta' - \theta_c}{\theta_c}\right)^{1.5}\left(\frac{(s-1)\,g\,d^3}{\nu^2}\right)^{0.1}\sqrt{(s-1)\,g\,d}$$

one finds, for the typical values of $(\theta_c\,, d) = (0.05, 0.2mm)$,

$$l_x = \frac{Q}{p} \approx 441d$$

Thus, for sand which is being picked up at the rate given by van Rijn's pickup formula to yield the bed-load transport rate given by the Meyer-Peter formula, the grains must, on the average, jump a distance equivalent to *441* grain diameters in each jump. Interestingly, this result is independent of the Shields parameter θ'.

6.3 SHORE NORMAL TRANSPORT OVER VORTEX RIPPLES

6.3.1 What the experimental data show

Vortex ripples are common in coastal areas where the flow is wave-dominated and not too violent, see Section 3.4, page 135. They are also the predominant type of bedforms in laboratory flumes. The ripples themselves and the sediment transport above them have therefore been studied in considerable detail.

Sediment transport over ripples is interesting because it is at first sight counter intuitive. That is, the net sediment transport tends to be in the opposite direction to the strongest flow velocities, see Figure 6.3.1. Thus, the net sediment transport over vortex ripples under Stokes waves without boundary layer drift ($u_\infty(t) = U_1 \cos \omega t + U_2 \cos 2\omega t$) is always in the offshore direction, see the tunnel data of Sato (1986). Correspondingly, Inman & Bowen (1963) found that when a current was superimposed on fairly sinusoidal waves over rippled beds the

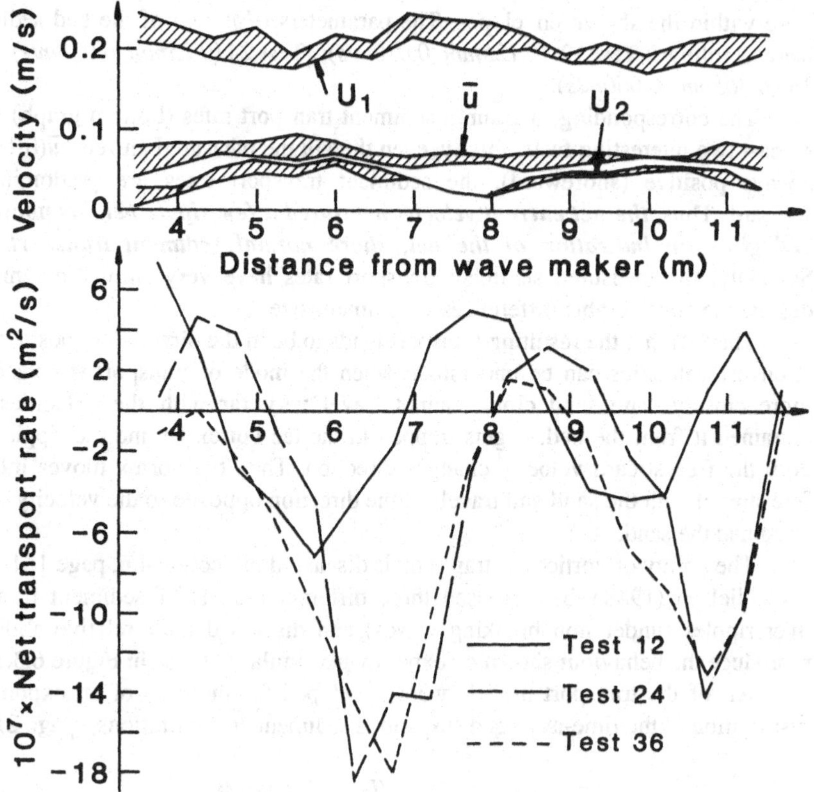

Figure 6.3.1: Measured net sediment transport rates over rippled beds for three different bed sediments under essentially the same flow conditions. Data from Schepers (1978).

net sediment transport tended to be in the opposite direction of the superimposed current.

These general features are shown by the data in Figure 6.3.1 which also show another interesting fact. Namely, that for almost identical flow conditions, the magnitude of the net sediment transport rate over ripples seems to be practically independent of the median size (d_{50}) of the bed sediment. This again is counter intuitive and interesting, particularly in connection with the "CERC Formula Paradox" discussed in Section 6.5.3, page 291.

For the data in Figure 6.3.1 the flow conditions (top graph) were very similar in the three experiments, $(T, D) = (1.5s, 0.3m)$, and the velocity amplitudes U_1 and U_2 and the time-averaged velocity $\overline{u}$, all measured $0.1m$ above the bed, were within the shown envelopes. The parameters (d_{50}, w_o) of the bed sediment were as follows: Test 12 $(0.125mm, 0.010m/s)$; Test 24 $(0.25mm, 0.028m/s)$; Test 36 $(0.465mm, 0.060m/s)$.

The corresponding, measured sediment transport rates (bottom graph) show at least two interesting facts. Firstly, even though the measured current $\overline{u}(0.1m)$ is always positive (shoreward), the sediment transport rates are predominantly seaward. Thus, ***the net current velocity measured a few ripple heights above the bed gives no indication of the net, shore normal sediment transport rate.*** Secondly, the measured sediment transport rates have very similar magnitudes despite the considerable differences in sediment size.

The fact that the resulting transport tends to be in the direction opposite to the strongest velocities can be understood when the mode of transport is considered more closely. The sand cloud cannot travel very far with the velocities that entrained it from the bed. It gets trapped in the lee vortex behind the ripple crest until the free stream velocity changes direction. Then the vortex moves into the free stream with the sand and travels in the direction opposite to the velocity which entrained the sand.

The ability of vortices to trap sand is discussed in Section 4.6, page 181 ff.

Nielsen (1988a+b) discussed three different models of sediment transport over ripples (under non breaking waves) and discussed their relative ability to reproduce the behaviour shown by experiments similar to those in Figure 6.3.1.

All of the transport models were developed for the case of an exponential distribution of the time-averaged suspended sediment concentrations, (page 215).

$$\overline{c}(z) = C_o e^{-z/L_s} = \frac{\overline{p}}{w_o} e^{-z/L_s} \qquad (5.2.4)$$

which would be a reasonable approximation under those experimental conditions, and using the pickup function (p 230)

$$p(t) = V^d\,\delta_T(t-t^d) + V^u\,\delta_T(t-t^u) \tag{5.3.14}$$

For details of the sediment amounts V^d and V^u see page 230.

The delta functions which have zero width are preferred to functions with wider peaks because of mathematical convenience. This has little effect on the final results however, because only the first one or two harmonics make significant contributions to $\overline{\tilde{c}\,\tilde{u}}$ since the higher harmonics of the near-bed velocity are very small.

6.3.2 The gradient diffusion transport model

The first model developed by Nielsen (1988a) was based on the assumption that pure gradient diffusion was the sole sediment distribution mechanism as discussed in Section 5.4.4, p 241.

The sediment diffusivity which corresponds to the assumed $\overline{c}$-distribution, Equation (5.2.4), is the constant $\varepsilon_s = L_s\,w_o$. In this model, the resulting sediment transport rate is calculated from the usual cu-integral expression

$$\overline{Q} = \int_0^D \overline{u}(z)\,\overline{c}(z)\,dz + \frac{1}{T}\int_0^D\int_t^{T+t} \tilde{u}(z,t)\,\tilde{c}(z,t)\,dt\,dz$$

which, with the velocity field given by

$$u(z,t) = \overline{u}(z) + \tilde{u}(t) = \overline{u}(z) + \sum_{n=1}^{\infty} U_n \cos n\omega(t-t_n)$$

and, with the expressions (5.2.4), page 215, and (5.4.38), page 244 for the sediment concentrations inserted, leads to the diffusion based transport formula

$$\overline{Q_D} = \frac{\overline{p}}{w_o}\int_0^D \overline{u}(z)\,e^{-z/L_s}\,dz$$
$$+ \frac{V^d}{w_oT} L_s \sum_{n=1}^{\infty} \frac{U_n}{|\alpha_n|^2}\cos(n\omega t^d + 2arg\alpha_n - n\omega t_n)$$
$$+ \frac{V^u}{w_oT} L_s \sum_{n=1}^{\infty} \frac{U_n}{|\alpha_n|^2}\cos(n\omega t^u + 2arg\alpha_n - n\omega t_n) \tag{6.3.1}$$

where $V^d + V^u = \bar{p}\,T = C_o\, w_o\, T$ and the ratio V^d/V^u was simplistically taken as $(U^+/U^-)^6$, inspired by the form of the pickup function (5.3.11), page 228.

6.3.3 A particle trajectory model for shore normal transport over ripples

The second model, the *convective transport model,* discussed by Nielsen (1988a+b) was of the "particle trajectory type" and based on the assumption that the sediment was distributed vertically by a simple convection process similar to the one described in Section 5.4.3, page 238. For simplicity however, the vertical convection velocity w_c was assumed very large, so that all sand was assumed to reach its entrainment level z_e instantaneously when the lee vortex was released.

In this description the sand which is entrained at time t^i and which totals V^i per unit area, contributes to the total sediment distribution by

$$c^i(z,t^i) \;=\; V^i F(z) \;=\; V^i\, e^{-z/L_s}$$

immediately after the time of entrainment. This contribution subsequently decreases with time as the distribution settles uniformly with speed w_o. Hence, at a later time t, when the distribution has settled the length $w_o(t\text{-}t^i)$ the concentration contribution is reduced to

$$c^i(z,t) \;=\; V^i\, F(z{+}w_o[t{-}t^i]) \;=\; V^i\, e^{-(z+w_o[t-t^i])/L_s} \qquad \text{for } t > t^i \tag{6.3.2}$$

The basic philosophy of particle trajectory models, as outlined in Section 6.2, page 263, is to express transport contributions as the amount of sand entrained times the average distance travelled and then divide by the corresponding time interval to get the average transport rate.

In the case of vortex ripples under periodic waves, where the amounts V^d and V^u respectively, are entrained at the two $u_\infty(t)$-reversals, see Figure 5.3.4, page 229, this corresponds to

$$\overline{Q} \;=\; \frac{1}{T}\,(V^d\, l_x^d \;+\; V^u\, l_x^u) \tag{6.3.3}$$

where l_x^d and l_x^u are respectively the average distance travelled by sand entrained at t^d and t^u. These lengths are calculated as follows.

A sand particle which is entrained to the level z_e will, while it settles at speed w_o be carried the horizontal distance

$$l_x(t^i, z_e) = \int_{t=t^i}^{t^i+\frac{z_e}{w_o}} u(z_e - w_o[t - t^i], t)\, dt \tag{6.3.4}$$

This sand particle hits the bed at the time $t = t^i + z_e/w_o$.

The average distance travelled by particles entrained at time t^i is therefore

$$l_x^i = \int_{z_e=0}^{D} -F'(z_e)\, l_x(t^i, z_e)\, dz_e = \int_{z_e=0}^{D} -F'(z_e) \int_{t=t^i}^{t^i+\frac{z_e}{w_o}} u(z_e - w_o[t - t^i], t)\, dt\, dz_e \tag{6.3.5}$$

where $-F'(z_e)$ represents the probability density of entrainment to the level z_e, see Equation (5.4.8), page 238.

With a velocity field of the form $u(z,t) = \bar{u}(z) + \tilde{u}(t)$, i e, with the boundary layer for the oscillatory velocity component $\tilde{u}$ assumed very thin compared to the sediment distribution, Equation (6.3.5) can be reduced to

$$l_x^i = \frac{1}{w_o} \int_{z=0}^{D} F(z)\, \bar{u}(z)\, dz - \int_{z_e=0}^{D} F'(z_e) \int_{t=t^i}^{t^i+\frac{z_e}{w_o}} \tilde{u}(t)\, dt\, dz_e \tag{6.3.6}$$

where the first term has been transformed using integration by parts.

While this may be a fairly complicated expression, its evaluation is straight forward compared to solving the time-dependent diffusion equation with even a moderately complicated ε_s-distribution.

To illustrate the basic principles of transport modelling with models of the particle trajectory type, we shall evaluate Equation (6.3.6) on the basis of a particularly convenient expression for the velocity,

$$u(z,t) = \bar{u}(z) \pm U_1 \sin \omega(t - t^i) \tag{6.3.7}$$

where the sign "-" is used in connection with the zero down-crossing event (at $t = t^d$) and "+" is used in connection with the zero up-crossing event (at $t = t^u$), see Figure 5.3.4, page 229. With this expression for the velocity inserted into Equation (6.3.6), we find

$$l_x^i = \frac{1}{w_o}\int_{z=0}^{D} F(z)\,\overline{u}(z)\,dz \pm \int_{z_e=0}^{D} \frac{1}{L_s} e^{-z_e/L_s} \int_{0}^{z_e/w_o} U_1 \sin \omega t'\,dt'\,dz_e$$

where $t' = t - t^i$

$$l_x^i = \frac{1}{w_o}\int_{z=0}^{D} F(z)\,\overline{u}(z)\,dz \pm \frac{U_1}{\omega L_s}\int_{z_e=0}^{D} e^{-z_e/L_s}\left(1 - \cos\frac{\omega}{w_o} z_e\right) dz_e$$

which for $D/L_s >> 1$ can be reduced to

$$l_x^i = \frac{1}{w_o}\int_{z=0}^{D} F(z)\,\overline{u}(z)\,dz \pm \frac{U_1}{w_o} L_s \frac{\frac{\omega L_s}{w_o}}{1 + \left(\frac{\omega L_s}{w_o}\right)^2} \tag{6.3.8}$$

Inserting this into the general transport formula (6.3.3) gives

$$\overline{Q_C} = \frac{V^d + V^u}{w_o T}\int_{z=0}^{D} F(z)\,\overline{u}(z)\,dz + \frac{V^u - V^d}{w_o T} U_1 L_s \frac{\frac{\omega L_s}{w_o}}{1 + \left(\frac{\omega L_s}{w_o}\right)^2} \tag{6.3.9}$$

which, since $V^d + V^u = C_o w_o T = \overline{p}\,T$ can be simplified to obtain the convection based transport formula

$$\overline{Q_C} = C_o\int_{z=0}^{D} F(z)\,\overline{u}(z)\,dz + \frac{(V^u - V^d)}{w_o T} U_1 L_s \frac{\frac{\omega L_s}{w_o}}{1 + \left(\frac{\omega L_s}{w_o}\right)^2} \tag{6.3.10}$$

This result is very similar to the result (6.3.1) for the diffusion based model, although the model philosophies are very different. The actual transport rates are practically identical, see Figures 6.3.2 and further examples given by Nielsen (1988a). It would therefore seem logical to use the convection model rather than the diffusion model since Equation (6.3.10) is much simpler than Equation (6.3.1).

The particle trajectory approach, which is used in the convective transport model, has a further advantage however, namely that the expression (6.3.6) can be

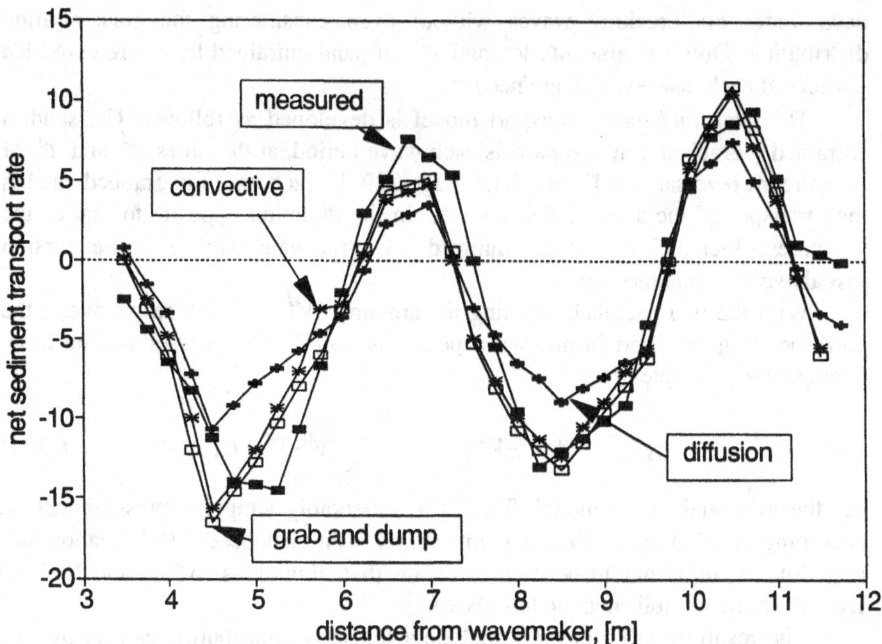

Figure 6.3.2: Estimated sediment transport rates corresponding to the *diffusion* based transport model (6.3.1), the *convective* transport model (Equation 6.3.10), and the simple *grab and dump* model, Equation (6.3.11). For this sand size (d_{50} = *0.125mm*), all of the models give very similar results, and the agreement with the measured transport rates is good. Data from Schepers (1978) Test 12, For the hydrodynamic details see Figure 6.3.1, page 267.

evaluated for almost any convective distribution function $F(z)$ while an analytical solution for the diffusion model is impossible for all but a few ε_s-distributions.

Furthermore, the convective transport model can, with minor modifications, be used on a wave by wave basis for irregular waves, while the analytical solution for the diffusion based transport model relies on the assumption of periodicity.

6.3.4 The grab and dump model for shore normal transport over ripples

With the present level of accuracy of coastal bottom boundary layer calculations, it is difficult to justify very complicated sediment transport formulae. Hence, it is worthwhile checking if further simplification is possible, and indeed it is. It is, in fact, possible to model the shore normal sediment transport over rippled beds under non-breaking waves without even considering the concentration distribution. Only the amounts V^d and V^u of sand entrained by the released lee vortices at each u_∞-reversal are needed.

The *grab and dump* transport model is developed as follows. The sand is entrained ("grabbed") in two parcels each wave period, at the times t^d and t^u of free stream reversal, see Figure 5.3.4, page 229. In each case the grabbed sand is then transported the average distance A, in the direction opposite to that of the velocities which picked it up, and dumped. A is the water particle semi-excursion just above the boundary layer.

With the two parcels containing the amounts V^d and V^u respectively, the corresponding transport in one wave period is $A(V^u\text{-}V^d)$ corresponding to an average transport rate of

$$\overline{Q_G} = A(V^u - V^d)/T = \frac{1}{2\pi}(V^u - V^d)\, U_1 \tag{6.3.11}$$

for the grab and dump model. This is a remarkably simple expression which, according to Nielsen's (1988a) comparisons with Schepers' (1978) laboratory experiments, turns out to be more accurate than Equations (6.3.1) and (6.3.10) from the more complicated models above.

The result (6.3.11) does not include a transport contribution carried by the steady flow component $\bar{u}$, but such a contribution may be included by adding the usual $\int \bar{c}(z)\, \bar{u}(z)\, dz$ provided that $\bar{u}(z)$ and $\bar{c}(z)$ are known.

6.3.5 Comparison of the three models

In Nielsen's (1988a) comparison of the three models of shore normal sediment transport over rippled beds the transport contribution carried by $\bar{u}$ was, in all cases, ignored. Mainly because the distribution of $\bar{u}(z)$ including the boundary layer drift near a rippled bed is virtually unknown, see Section 1.4.5, page 60. The diffusion based solution (6.3.1) was calculated with the use of the first two harmonics of $\tilde{c}(z,t)$ and $\tilde{u}(z)$.

It was found that the three models performed equally well for the finer sand sizes, see Figure 6.3.2. However, only the grab and dump model predicted the right magnitude of the sediment transport rate for the coarsest sand. Thus, for Test 36 shown in Figure 6.3.1, p 267, the diffusion based model and the convective transport model both predicted a range of transport rates which was less than one tenth of the measured range while the grab and dump model predicted the right range, see Figure 6.3.3.

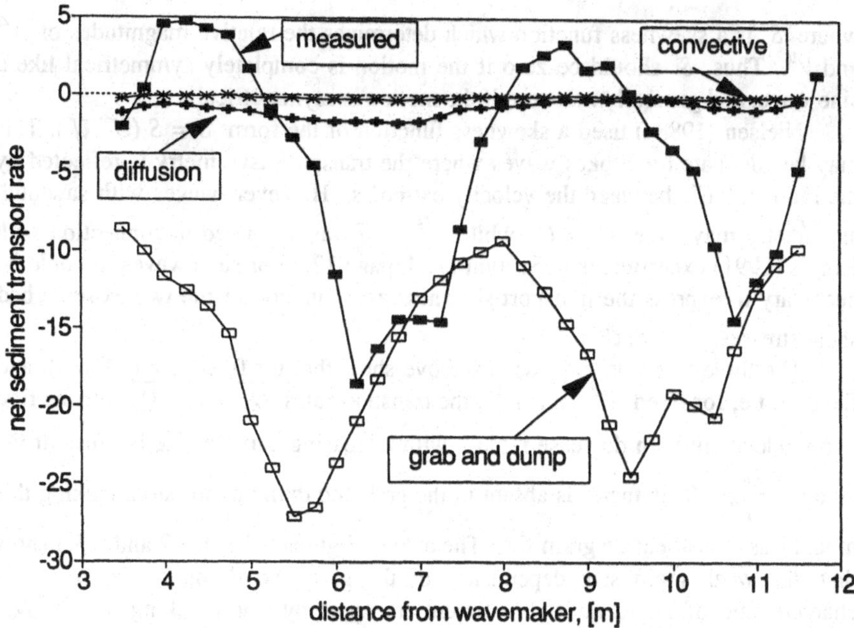

Figure 6.3.3: Estimated and measured sediment transport rates as function of distance from the wavemaker in a wave flume. For this coarse sand (d_{50} = *0.465mm*) only the simple grab and dump model predicts the correct magnitude of the transport rates. Data from Schepers (1978) Test 36. For further details see Figure 6.3.1, page 267.

To understand this, consider the three formulae (6.3.1), (6.3.10) (both with the $\bar{c}\,\bar{u}$-contribution ignored) and (6.3.11) in slightly rewritten and simplified forms. If only the contribution from the first harmonic of $u(z,t)$ is included in the diffusion based transport formula (6.3.1), the three formulae can be written as follows

$$\overline{Q_D} = \frac{\bar{p}}{w_o} L_s U_1 S F_D\left(\frac{\omega L_s}{w_o}\right) \tag{6.3.12}$$

$$\overline{Q_C} = \frac{\bar{p}}{w_o} L_s U_1 S F_C\left(\frac{\omega L_s}{w_o}\right) \tag{6.3.13}$$

$$\overline{Q_G} = \bar{p} \frac{T}{2\pi} U_1 S \tag{6.3.14}$$

where S is a skewness function which determines the relative magnitudes of V^d and V^u. Thus, S should be zero if the motion is completely symmetrical like a sine wave, and generally non-zero if the motion is asymmetrical.

Nielsen (1988a) used a skewness function of the form $S = S(U^+, U^-)$. This may be adequate for Stokes waves where the transport asymmetry is reflected by the ratio U^+/U^- between the velocity extremes. However, waves with sawtooth asymmetry may have $U^+ = U^-$ while $\tau^+ \neq \tau^-$ as discussed in connection with King's (1991) experiments in Section 2.4.4, page 121. For such waves it would be necessary to express the transport skewness as a function of the two extreme bed shear stresses, $S = S(\tau^+, \tau^-)$.

The three transport expressions above show that for fixed $\overline{p}$ $(=C_o w_o)$ and fixed S, i e, for fixed V^d and V^u, the transport rates $\overline{Q_D}$ and $\overline{Q_C}$ for the first two models will both decrease rapidly with increasing particle size because of the factor $\frac{L_S}{w_o T}$. This factor is absent in the grab and dump expression, making this model less dependent on grain size. The data in Figures 6.3.1, 6.3.2 and 6.3.3 show that this weak grain size dependence of the grab and dump model is a true characteristic of shore normal sediment transport by non-breaking waves over rippled beds.

Note that weak grain size dependence is not a unique feature of shore normal transport over ripples in small wave flumes. Also field measurements of the total longshore transport rate or littoral drift show remarkably weak grain size dependence. See Section 6.5.3, page 291.

6.4 SHORE NORMAL SEDIMENT TRANSPORT OVER FLAT BEDS

6.4.1 Introduction

Shore normal sediment transport over flat beds may occur on the beach face, i e, in the swash zone, throughout or in parts of the surf zone and outside the surf zone if the waves are large and the sand is fine.

6.4.2 Shore normal sediment transport in the swash zone

While it is generally recognised that the swash zone contributes a major part of the total littoral drift, and while it is obviously necessary to model the shore normal sediment transport through the swash zone in order to model beach change, it is beyond the present state of the art to model swash zone sediment transport.

Too little is known, at present, about the boundary layer flow and the corresponding shear stresses in the swash zone to even attempt a description of the basic sediment transport mechanisms in this area. See also Figure 6.1.1, page 262.

Furthermore, swash zone sediment transport includes the extra complication of significant flows of water perpendicular to the sand surface. Experience has shown, that increasing the inflow by pumping from the ground water in the beach, tends to increase the rate of beach accretion.

Thus, experiments show that the flow perpendicular to the sand surface in the swash zone affects the rate of beach accretion. The details of the mechanisms are, however, not understood.

6.4.3 Shore normal sediment transport in the surf zone

In the surf zone the bed is often flat or with respect to sediment transport practically flat. That is, the megaripples which are sometimes present have no sharp crests that might cause rhythmic vortex formation, and the boundary layer structure over the megaripples is therefore practically the same as over flat beds.

However, while the bed topography in the surf zone may be simple, the flow structure is certainly not simple. The immediate vicinity of bars offers special problems, see e g Hedegaard et al (1991) and the entrainment under plunging breakers is at present beyond detailed modelling. Even the simplest of hypothetical surf zones, namely a two-dimensional one with constant bed slope and spilling breakers with constant wave height to water depth ratio, causes problems. The main difficulties lie in the modelling of the important seaward bottom current, the undertow, see Section 1.4.5, page 60.

While detailed, quantitative modelling of shore normal sediment transport is very difficult, some aspects, like the position of bars, have been predicted successfully by Boczar-Karakiewicz & Davidson-Arnott (1987) and by Boczar-Karakiewicz & Jackson (1991). They use a fairly simple model of the waves and the shore normal sediment transport which bypasses the intricacies of wave breaking.

The broad question of whether a beach is going to accrete or erode when exposed to a certain set of wave conditions may be answered with a certain amount of confidence according to Kraus et al (1991). Based on a comprehensive review of field and laboratory data they suggested the criterion

$$\text{Beaches erode if} \quad \frac{\overline{H_O}}{L_O} < 0.00070 \left(\frac{\overline{H_O}}{w_O T}\right)^3$$

(6.4.1)

$$\text{Beaches accrete} \quad \frac{\overline{H_O}}{L_O} > 0.00070 \left(\frac{\overline{H_O}}{w_O T}\right)^3$$

where H_O and L_O are the deep water wave height and wave length respectively.

They found the best correlation by using the mean offshore wave height ($\overline{H_O}$). For a Rayleigh distribution, the mean wave height is related to the root mean square and significant heights by $\overline{H} = 0.886\, H_{rms} = 0.626\, H_S$.

6.4.4 Shore normal sediment transport outside the surf zone.

If the waves are big enough ($\theta_{2.5} \gtrsim 1$, see Section 3.4) the sea bed outside the surf zone will be free from vortex ripples and thus, with respect to sediment transport, practically flat. In such cases, it should be possible to model the sediment transport rate with the sheet-flow formulae which were developed in Section 2.4, page 116.

A simpler formula was presented by Ribberink & Al Salem (1990)

$$\overline{Q} = 0.00018\, \overline{u_\infty^3} \tag{6.4.2}$$

where the total net transport $\overline{Q}$ is measured in m^2/s and $u_\infty(t)$ in m/s.

This formula was, however, only designed for $0.2mm$ quartz sand, and it may be under predicting $\overline{Q}$ systematically for shorter wave periods than the ones used by the authors ($6.5s$ and $9.1s$). This was mentioned by the authors in relation to the data of Horikawa et al (1982), and is not unexpected since shorter period means larger d_{50}/A and hence a larger friction factor for fixed u_{rms}, see Section 1.2.5, page 23. It should also be remembered that formulae of this type fail to predict the net sediment transport due to "sawtooth wave asymmetry" as discussed in Section 2.4.4, page 121.

Not all of these limitations apply to the alternative sheet-flow sediment transport formula (2.4.16), which was developed in Section 2.4.4, page 125. It should be equally valid for all (non-cohesive) sediments and for waves of all shapes and periods. However, neither of the two formulae applies, without major modifications, to flows with large steady flow components such as rip currents or in surf zones with a strong undertow.

In such flows, the contribution to the bed shear stress from $\overline{u}(z)$ must be given special consideration. As a first, rough approximation, this problem might be approached as if there was a linear transfer function between near-bed velocities and the bed shear stress. The procedure would then be to subtract the time-average from the near-bed velocities and apply Equation (2.4.15), page 125, to the residual $\widetilde{u}(z_r,t) = u(z_r,t) - \overline{u}(z_r)$ i e

$$\theta'[\widetilde{u}(t_n)] = \frac{\frac{1}{2} f_{2.5} A_{rms}}{(s-1)\, g\, d} \left(\cos\varphi_\tau\, \omega_p\, \widetilde{u}(z_r,t_n) + \sin\varphi_\tau \frac{\widetilde{u}(z_r,t_{n+1}) - \widetilde{u}(z_r,t_{n-1})}{2\, \delta_t}\right) \tag{6.4.3}$$

To this expression, a contribution due to the time-averaged bed shear stress $\overline{\tau}(o)$ must then be added in order to get the total, instantaneous effective Shields parameter

$$\theta'(t_n) = \theta'[\widetilde{u}(t_n)] + \frac{\overline{\tau}(o)}{\rho\, (s-1)\, g\, d} \tag{6.4.4}$$

This total value is then inserted into the sheet-flow sediment transport formula

$$\Phi(t) = \begin{cases} 8\, [\theta'(t) - 0.05]^{1.5} \dfrac{\theta'(t)}{|\theta'(t)|} & \text{for } |\theta'(t)| > 0.05 \\ 0 & \text{for } |\theta'(t)| < 0.05 \end{cases} \tag{2.4.16}$$

The factor *8*, which corresponds to the Meyer-Peter bed-load formula, may be replaced by a factor of the order *12* for fine sand at high flow intensities in accordance with the trends of the data in Figure 2.3.2, page 113, and Figure 2.4.2, page 119.

With respect to obtaining the mean bed shear stress $\overline{\tau}(o)$ from velocity data there is basically two options. If $\overline{u}$ is known only at a single level z_r a rough estimate may be obtained with a simple wave-current interaction model as done in Example 1.5.2, page 90. If a profile of $\overline{u}(z)$ is available, an estimate of the friction velocity $\overline{u_*}$ and hence $\overline{\tau}(o)$ may be obtained from a log-curve fit. In both cases however, new methods should be developed to account for the influence of the momentum transfer term $\rho\,(\overline{\tilde{u}\tilde{w}})_\infty$ which was discussed in Section 1.5.9, page 91. For an example of shore normal sediment transport calculations over a flat bed see Example 2.4.1, page 126.

6.5 SHORE PARALLEL SEDIMENT TRANSPORT

6.5.1 Introduction

Sediment transport in the shore parallel direction or more precisely, the direction perpendicular to the wave motion, is simpler to model than the transport in the wave direction because $\tilde{v} \equiv 0$ and the equivalent formula to Equation (6.2.3), page 264, for the time-averaged sediment transport rate therefore becomes

$$\overline{Q}_y = \int_{z=0}^{D} (\overline{c}\,\overline{v} + \overline{c'v'})\,dz \qquad (6.5.1)$$

where it seems reasonable to assume that the last term $\overline{c'v'}$ is insignificant so that the time-averaged sediment transport rate is given simply by

$$\overline{Q}_y = \int_{z=0}^{D} \overline{c}(z)\,\overline{v}(z)\,dz \qquad (6.5.2)$$

While this formula is valid in principle, both inside the surf zone and outside, the details of the calculations are quite different for the two cases. This is because of the strong influence on both $\overline{c}$ and $\overline{v}$ from the extra turbulence and from additional convective mechanisms which are associated with wave breaking.

The presence of steady flow components $(\overline{u}(z), \overline{v}(z))$ superimposed on the wave motion $\tilde{u}_\infty(t)$ will increase the effective bed shear stresses and hence the magnitude of $\overline{c}(z)$ compared with a pure wave motion. However, direct experimental data which quantify this effect are not yet available.

As an educated guess however, it might be adequate to base the estimation of the reference concentration C_o on a pure-wave formula like Equation (5.3.10), page 228, with $\theta_{2.5}$ replaced by an augmented effective Shields parameter analogous to Equation (6.4.4). For example

$$\theta' = \left\{\left(\theta_{2.5}(\tilde{u}) + \frac{\overline{\tau}_x(o)}{\rho\,(s-1)\,g\,d}\right)^2 + \left(\frac{\overline{\tau}_y(o)}{\rho\,(s-1)\,g\,d}\right)^2\right\}^{0.5} \qquad (6.5.3)$$

A similar concern arises with respect to the sediment concentration distribution. That is, the additional turbulence due to the presence of $\overline{u}(z)$ and/or $\overline{v}(z)$ will stretch the sediment concentration profiles upwards but again, direct experimental quantification of the effect is lacking.

This problem may, in theory, be addressed in fairly simple terms. The natural approach is to replace the pure-wave sediment diffusivity by an eddy viscosity for the combined wave-current flow when calculating the $\overline{c}(z)$-distribution from the combined convection-diffusion Equation (5.4.49), page 248.

However, care must be taken when deriving sediment diffusivities from eddy viscosities in combined wave-current flows. There are various choices of eddy viscosities, see Section 1.5.9, page 91, and their respective relations with the sediment diffusivity are not well understood.

For rippled beds, the increased near-bed sediment concentrations and the increased vertical distribution scale due to current turbulence will be partly balanced by the expected decrease in bedform size and steepness due to the presence of the current, see Section 5.2.5, page 218.

6.5.2 Shore parallel sediment transport outside the surf zone

Outside the surf zone, time-averaged sediment concentrations may be estimated either with the simple exponential model described by Equations (5.2.4)-(5.2.5), pp 215-217, for rippled beds, or in general, with the more detailed combined convection-diffusion model developed in Sections 5.4.5-5.4.9, p 245 ff.

The time-averaged velocities perpendicular to the wave direction are considerably simpler to model than those in the wave direction because $\overline{\tilde{v}\,\tilde{w}} \equiv 0$. This simplifies the equation of motion considerably compared to that for the wave direction, (Equation 1.1.21, p 12), where the corresponding term $\overline{\tilde{u}\,\tilde{w}}$ plays a significant role, see Section 1.5.9, page 91ff.

Thus, the shore parallel current distribution $\overline{v}(z)$ and the shore parallel bed shear stress $\overline{\tau_y}(o) = \rho\,|\overline{v_*}|\,\overline{v_*}$ should be adequately estimated by the method outlined in Example 1.5.2, page 90, with the apparent roughness increase

$$\frac{z_a}{z_o} = F\left(\frac{A\omega}{\overline{v_*}}, \frac{r}{A}, \pm\frac{\pi}{2}\right)$$

determined in agreement with data from perpendicular waves and currents, and from tunnel data where $\overline{\tilde{u}\,\tilde{w}} \approx 0$, see Figures 1.5.13 and 1.5.16, pp 67 and 69.

Example 6.5.1: **Shore parallel sediment transport over a rippled bed**

Consider the problem of estimating the shore parallel sediment transport rate at a position outside the surf zone (no breaking waves). Velocity data from a single current meter at elevation $z_r = 1m$ above the bed is available together with a sample of the bed sediment.

Consider the example of $(\overline{v}(z_r), \tilde{u}_{rms}, T_p, d_{50}, w_o, s) = (0.3m/s, 0.355m/s, 8s, 0.8mm, 0.10m/s, 2.65)$, which correspond to an equivalent velocity amplitude $A\omega = \sqrt{2}\,\tilde{u}_{rms} = 0.502m/s$ and $A_{rms} = \tilde{u}_{rms} T_p/2\pi = 0.639m$.

Using the relevant formulae from Section 2.2, page 102 ff, we find the corresponding parameters: $(\psi, f_{2.5}, \theta_{2.5}) = (19.5, 0.014, 0.136)$, from which we can estimate the ripple geometry with formulae from Section 3.4, page 135 ff.

The relative ripple height is found to be

$$\frac{\eta}{A} = 21\,\psi^{-1.85} = 0.087 \tag{3.4.8}$$

corresponding to an estimated ripple height of $0.056m$, and the ripple steepness is

$$\frac{\eta}{\lambda} = 0.342 - 0.34\sqrt[4]{\theta_{2.5}} = 0.136 \tag{3.4.6}$$

The sediment distribution parameters (C_o and L_s), corresponding to the simple exponential distribution model, Equations (5.2.4) and (5.2.5), page 217, can then be estimated. The reference concentration is

$$C_o = 0.005\,\theta_r^3 = 0.005\frac{\theta_{2.5}^3}{(1-\pi\eta/\lambda)^6} = 0.00035 \tag{5.3.10}$$

and the distribution length scale is

$$L_s = 0.075\frac{A\omega}{w_o}\eta = 0.021m \tag{5.2.5}$$

Hence, the estimated distribution of the time-averaged suspended sediment concentrations is given by

$$\overline{c}(z) = C_o\,e^{-z/L_s} = 0.00035\,e^{-z/0.021} \tag{6.5.4}$$

where z is measured in metres above the ripple crest level. Note that this simple exponential model for $\overline{c}(z)$ is only valid for a certain range of conditions which may be inferred from the Figure 5.2.7, page 216.

The current velocity profile is determined in analogy with Example 1.5.2, page 90. Based on a hydraulic roughness of

$$r \;=\; 8\,\eta^2/\lambda + 5\,\theta_{2.5}\,d \;=\; 0.061m \tag{3.6.14}$$

and $z_o = r/30 = 0.0020m$. The steady friction velocity $\overline{v_*}$ is determined from

$$\overline{v_*} \;=\; \frac{\kappa\,\overline{v}(z_r)}{\ln\dfrac{z_r}{z_o} - \ln F\!\left(\dfrac{A\omega}{\overline{v_*}},\dfrac{r}{A},\pm\dfrac{\pi}{2}\right)} \tag{1.5.50}$$

With the apparent roughness increase given by the simplistic formula (page 89)

$$F\!\left(\frac{A\omega}{\overline{v_*}},\frac{r}{A},\pm\frac{\pi}{2}\right) \;=\; 0.44\,\frac{A\omega}{\overline{v_*}} \tag{1.5.48}$$

and $\kappa = 0.4$ and with other values inserted Equation (1.5.50) becomes

$$\overline{v_*} \;=\; \frac{0.4\cdot 0.3}{\ln\dfrac{1.0}{0.0020} - \ln 0.44\,\dfrac{0.502}{\overline{v_*}}} \;=\; \frac{0.12}{7.73 + \ln\overline{v_*}}$$

This formula converges rapidly to the friction velocity value $\overline{v_*} = 0.029m/s$, corresponding to $F = 0.44\,\dfrac{A\omega}{\overline{v_*}} = 7.62$, and hence (Equation (1.4.5) page 87) $z_a = F\,z_o = 0.015m$, and $l = e^1 z_a = 0.041m$.

With these values inserted, the current velocity distribution given by Equations (1.5.42) and (1.5.43), page 87, becomes

$$\overline{v}(z) \;=\; \begin{cases} \dfrac{\overline{v_*}}{\kappa}\dfrac{z}{l} \;=\; 1.77z\;[m/s] & \text{for } z < 0.041m \\[2ex] \dfrac{\overline{v_*}}{\kappa}\ln\dfrac{z}{z_a} \;=\; 0.073\ln\dfrac{z}{0.015}\;[m/s] & \text{for } z > 0.041m \end{cases} \tag{6.5.5}$$

This current distribution is shown in Figure 6.5.1 together with the sediment concentration profile (Equation 6.5.4), and the distribution of the product $\overline{c}\,\overline{v}$.

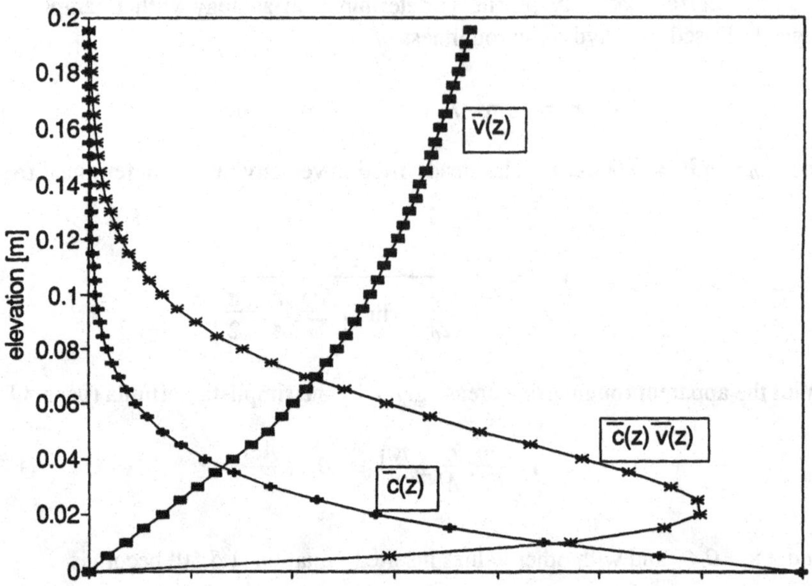

Figure 6.5.1: The sediment concentration distribution (6.5.4) and the current distribution (6.5.5) together with the distribution of the product $\bar{c}\,\bar{v}$ (not to scale). The estimated ripple height is *0.056m.*

The time-averaged sediment transport rate is then calculated from

$$\overline{Q_y} = \int_0^D \bar{c}(z)\,\bar{v}(z)\,dz = \frac{C_o \overline{v_*}}{\kappa\, l}\int_0^l z\, e^{-z/L_s}\,dz + \frac{C_o \overline{v_*}}{\kappa}\int_l^D \ln\frac{z}{z_a}\, e^{-z/L_s}\,dz$$

which can be approximated (within *3%* for $l/L_s > 0.01$) by

$$\overline{Q_y} = \frac{\overline{v_*}}{\kappa} C_o \frac{L_s^2}{l}\left\{1 - e^{-1.9\,(l/L_s)^{0.79}}\right\} \qquad (6.5.6)$$

For $l/L_s \gtrsim 2$, as in the present example, the second term can be neglected completely so, we find the simple formula

$$\overline{Q_y} \approx \frac{\overline{v_*}}{\kappa} C_o \frac{L_s^2}{l} \qquad \text{for } l/L_s \gtrsim 2 \qquad (6.5.7)$$

With the values from above inserted, this gives a net sediment transport rate of $\overline{Q_y} = 2.7 \cdot 10^{-7} m^2/s$.

This transport rate seems to be too small since the Meyer-Peter formula (Equation 2.3.11, page 112), with the effective Shields parameter θ' based on the current friction velocity alone, gives the higher sediment transport rate

$$\overline{Q_y} = 8\left(\frac{\overline{v_*}^2}{(s-1)\,g\,d} - 0.05\right)^{1.5} \sqrt{(s-1)\,g\,d}\;d = 1.3 \cdot 10^{-6}\, m^2/s \qquad (6.5.8)$$

This ought to be a lower estimate, since the additional bed shear stress contribution of the waves is neglected.

Augmentation of the effective bed shear stress in accordance with Equation (6.5.3), page 280, and keeping the bedform shape unchanged leads to an increase of about 36% for the expression (6.5.7). But that is not enough to make it realistic.

Nevertheless, Rasmussen & Fredsoe (1981) used a very similar method to that which lead to Equation (6.5.7) above, and they found good agreement with laboratory experiments. The difference is, that their sand was fine ($d_{50} = 0.18mm$) compared to the sand in the example above ($d_{50} = 0.80mm$).

Thus, we get a similar picture to that for the shore normal sediment transport over rippled beds discussed in Section 6.3, p 266 ff. That is: traditional *cu*-integral formulae like Equation (6.5.7) seem to work for fairly fine sand. For coarse sand, however, the process is a very organised "grab and dump" process and is probably better described by a "grab and dump" model.

A "grab and dump" model for longshore transport over ripples under non-breaking waves may be constructed as follows.

Sand is picked up twice every wave period and the total amount picked up is $V = C_o w_o T$. The average distance l_y travelled in the direction along the ripple crests by that sand would be about $T/2$ times the velocity at a certain near-bed level, say one ripple height. Making use of the velocity distribution (6.5.5), that leads to

$$l_y = \bar{v}(\eta)\frac{T}{2} \approx \frac{\overline{v_*}}{\kappa}\frac{\eta}{l}\frac{T}{2} = \frac{\overline{v_*}}{\kappa}\frac{\eta}{2l}T$$

and hence, (see Figure 6.2.1, page 265) to the transport contribution

$$V\,l_y = C_o w_o T \frac{\overline{v_*}}{\kappa}\frac{\eta}{2l}T$$

from sand entrained in one wave period, or an average grab and dump sediment transport rate of

$$\overline{Q_{yG}} = C_o w_o T \frac{\overline{v_*}}{\kappa} \frac{\eta}{2l} \qquad (6.5.9)$$

With values from above inserted, this gives $\overline{Q_{yG}} = 1.4 \cdot 10^{-5} m^2/s$.

A fine tuning of Equation (6.5.9) and a serious discussion of the limits of applicability for the formulae (6.5.7)-(6.5.9), is hardly possible with the presently available data. However, the discussion in Section 6.3, pp 266 ff, where similar models were compared with laboratory data of shore normal transport over ripples, and the experience of Rasmussen & Fredsoe (1981), indicate that models of the type (6.5.7) are applicable for grain sizes up to $0.25mm$, while the grab and dump models perform reasonably over a wider experimental range, at least $0.085mm < d_{50} < 0.5mm$.

Example 6.5.2: **Shore parallel sediment transport over a flat bed**

Consider a situation outside the surf zone with the same sand and wave parameters as in Example 5.4.3, page 259, but with a shore parallel current of $0.3m/s$ measured at $z_r = 1.0m$, i e, $(\overline{v}(z_r), A\omega, T, d_{50}, s) = (0.3m/s, 1.50m/s, 8s, 0.21mm, 1.65)$.

For these conditions it was found in Example 5.4.3 that the time-averaged sediment concentrations $\overline{c}(z)$ were reasonaly described by the pure convection solution

$$\overline{c}(z) = C_o F(z) = C_o \frac{1}{(1 + z/z_1)^2} \qquad (5.4.15)$$

with $C_o = 0.072$ and $z_1 = 0.0063m$. The hydraulic roughness was found from Equation (3.6.14) to be $0.0026m$ corresponding to $z_o = r/30 = 0.000085m$.

As in the previous example and Example 1.5.2, page 90, the velocity distribution is derived from the information above by first finding the current friction velocity from

$$\overline{v_*} = \frac{\kappa \overline{v}(z_r)}{\ln \frac{z_r}{z_o} - \ln F\left(\frac{A\omega}{\overline{v_*}}, \frac{r}{A}, \pm\frac{\pi}{2}\right)} \qquad (1.5.50)$$

which, with the simplistic formula for the apparent roughness increase (page 89)

$$\frac{z_a}{z_o} = F\left(\frac{A\omega}{\overline{v_*}}, \frac{r}{A}, \pm\frac{\pi}{2}\right) = 0.44\,\frac{A\omega}{\overline{v_*}} \tag{1.5.48}$$

and with the values above inserted, yields

$$\overline{v_*} = \frac{0.4\cdot 0.30}{\ln\frac{1.0}{0.000085} - \ln\left\{0.44\frac{1.50}{\overline{v_*}}\right\}} = \frac{0.12}{9.79 + \ln\overline{v_*}}$$

This converges rapidly to the friction velocity value $\overline{v_*} = 0.020 m/s$. Based on this value, we find

$$z_a = 0.44\frac{A\omega}{\overline{v_*}} z_o = 0.0028m$$

and

$$l = e^1 z_a = 0.0076m$$

From these parameters, the velocity distribution is found by inserting into Equations (1.5.42) and (1.5.43), page 87

$$\overline{v}(z) = \begin{cases} \frac{\overline{v_*}}{\kappa}\frac{z}{l} = 6.58z\,[m/s] & \text{for } z < 0.0076m \\ \frac{\overline{v_*}}{\kappa}\ln\frac{z}{z_a} = 0.050\ln\frac{z}{0.0028}\,[m/s] & \text{for } z > 0.0076m \end{cases} \tag{6.5.10}$$

The net shore parallel sediment transport rate can now be found from

$$\overline{Q_y} = \int_o^D \overline{c}(z)\,\overline{v}(z)\,dz$$

which becomes

$$\overline{Q_y} = C_o\frac{\overline{v_*}}{\kappa}\left\{\int_o^l \frac{z/l}{(1+z/z_1)^2}dz + \int_l^D \frac{\ln\frac{z}{z_a}}{(1+z/z_1)^2}dz\right\}$$

$$\overline{Q_y} \;=\; C_o \frac{\overline{v_*}}{\kappa} z_1 \left\{ \frac{z_1}{l} [\ln \frac{1+z_1}{z_1} - \frac{l/z_1}{1+l/z_1}] + \int_{l/z_1}^{D/z_1} \frac{\ln x}{(1+x)^2} dx + \frac{\ln \frac{z_1}{z_a}}{1+\frac{l}{z_1}} \right\}$$

(6.5.11)

which for $D/z_1 \rightarrow \infty$ and within the range $0.1 < l/z_1 < 80$ may be approximated by

$$\overline{Q_y} \;=\; C_o \frac{\overline{v_*}}{\kappa} z_1 \left\{ 0.52 + 0.38 \cos (0.8 \ln \frac{l}{1.2\, z_1}) + \frac{\ln \frac{z_1}{z_a}}{1 + \frac{l}{z_1}} \right\}$$

(6.5.12)

see Figure 6.5.2. The contents of the brackets in Equation (6.5.12) generally have magnitude about 1 so, for order of magnitude estimates it is reasonable to use

$$\overline{Q_y} \approx C_o \frac{\overline{v_*}}{\kappa} z_1 \tag{6.5.13}$$

for the shore parallel sediment transport rate under non-breaking waves over a flat bed.

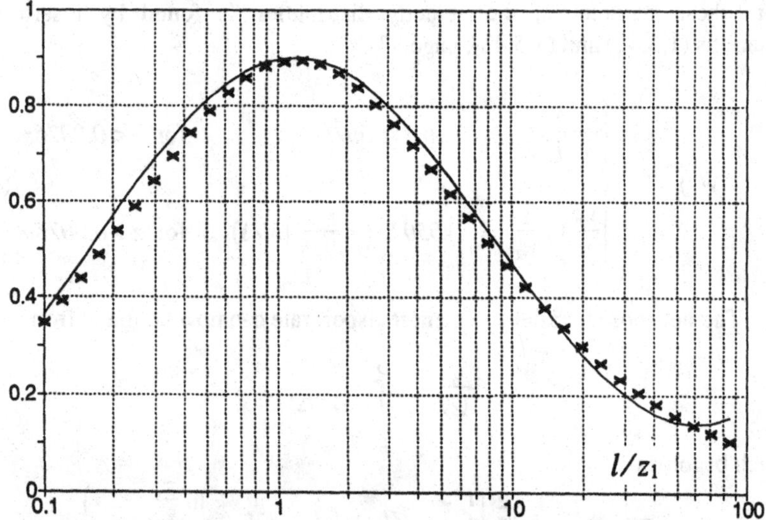

Figure 6.5.2: The first two terms (x) inside the brackets of Equation (6.5.11) can be approximated by the corresponding terms in Equation (6.5.12), (the curve).

With the estimated parameter values inserted, Equation (6.5.12) yields

$$\overline{Q_y} = 2.9 \cdot 10^{-5} m^2/s$$

If the bed shear stress contribution from the current is included, using Equation (6.5.3),page 280, in the calculation of C_o from Equation (5.3.10), one finds $C_o = 0.074$ instead of 0.072 or an increase of 2.8% which is insignificant.

On the other hand, the effect of the current on the concentration distribution through increased diffusivity may be significant. Thus, if the diffusivity $\varepsilon_s(z)$ in the combined convection-diffusion solution (5.4.49), p 248, is assumed equal to the eddy viscosity on which the current velocity distribution (6.5.10) is based, i e

$$\varepsilon_s = \begin{cases} \kappa \overline{v_*}\, l & \text{for } z < l \\ \kappa \overline{v_*}\, z & \text{for } z > l \end{cases} \tag{6.5.14}$$

then, the concentration profile is stretched vertically, compared to the pure wave solution, as shown in Figure 6.5.3. The net shore parallel sediment transport rate,

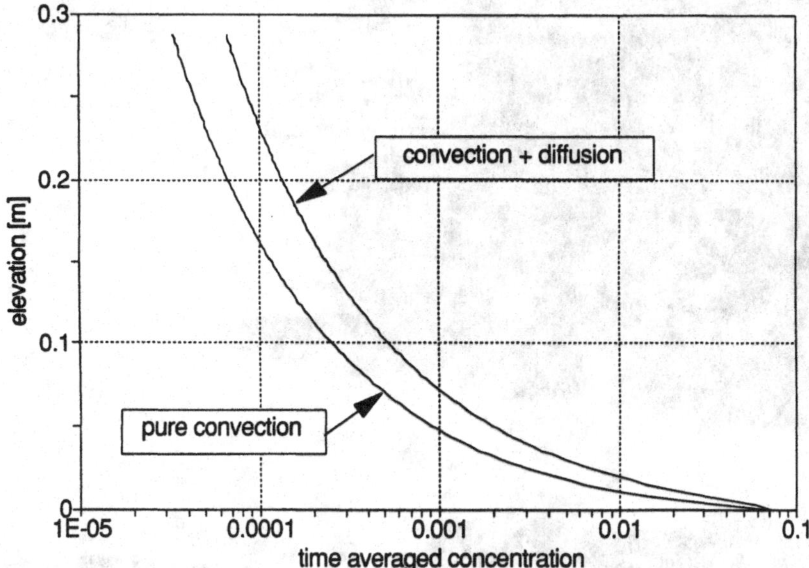

Figure 6.5.3: Estimated time-averaged sediment concentrations based on the pure convection solution (5.4.15) for pure waves and the complete solution (5.4.49) with the diffusivity $\varepsilon_s(z)$ given by Equation (6.5.14). See also Figure 5.4.10, p 260.

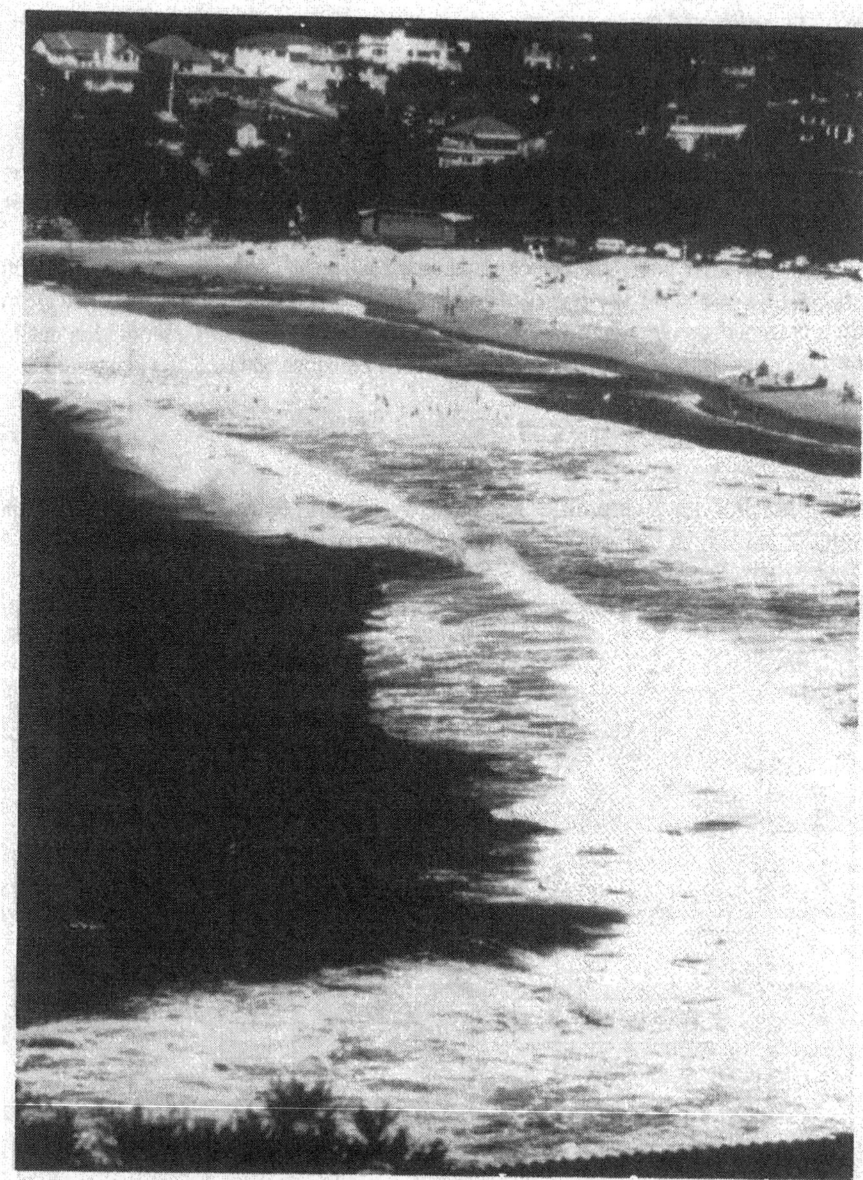

Figure 6.5.4: Sediment suspension in large plunging waves at Whale Beach, Sydney, Australia. Large amounts of sand are brought to the surface near "the plunge point" and spread horizontally along the surface.

which now has to be evaluated numerically, is increased roughly by a factor 2 to about $6 \cdot 10^{-5} m^2/s$.

The fact that the complete solution "convection+diffusion" in Figure 6.5.3, which is based on combined wave-current flow, agrees almost perfectly with the experimental data shown in Figure 5.4.10, page 260, is coincidental. There was no current present in those experiments.

6.5.3 Longshore sediment transport in the surf zone

The modelling of shore parallel sediment transport in the surf zone contains numerous new challenges for a few decades into the future.

The daunting variability in sediment concentrations from point to point in space and from wave to wave in time was illustrated by the field studies of Kana (1979) and Nielsen (1984).

Some significant progress has been made with respect to the modelling of turbulence due to spilling breakers (Peregrine & Svendsen 1978, Svendsen & Madsen 1984 and Stive 1988), and this has been incorporated in gradient diffusion models for suspended sediment by, for example, Deigaard et al (1986).

However, convective (or large scale) distribution mechanisms, such as the strong upward flows which are generated by entrained air behind plunging breakers, see Figure 6.5.4, are obviously important in many surf zones. Another large scale (convective) mixing mechanism in the outer surf zone is provided by the obliquely ascending vortices described by Nadaoka et al (1988). Where these large scale mixing processes are important, a pure gradient diffusion approach will not be adequate.

Comprehensive models of surf zone sediment transport must also include the large contribution from the swash zone, see Kamphuis (1991).

While detailed models are thus still being developed, simpler approaches to longshore sediment transport in the surf zone have been in use for some time.

The most famous of these is the so-called CERC formula which expresses the total, longshore sediment transport rate as a constant times the longshore wave energy flux at the break point. With the use linear shallow water wave theory this can be written

$$\rho\,(s-1)\,g \int_{x_b}^{rl} \overline{Q_y}\,dx \;=\; I \;=\; \frac{K}{16\sqrt{\gamma}}\,\rho\, g^{3/2}\, H_b^{5/2} \sin(2\alpha_b) \qquad (6.5.15)$$

where the subscript "b" refers to the break point, γ is the wave height to depth ratio at the break point, α_b the breaker angle, and "rl" stands for the runup limit. The quantity I is called the immersed weight longshore sediment transport rate.

The value of the constant K for use in connection with the root mean square breaker height for field data is about *0.77* (Shore Protection Manual 1984). Recommended values of K and its scatter have recently been discussed by Bodge & Kraus (1991) in relation to the presently available data.

One of the interesting aspects of the CERC formula is that it seems to work quite well without considering the size of the sand - "The CERC Formula Paradox" (Nielsen 1988b).

The weak grain size dependence was confirmed by Kamphuis (1990) who, based on dimensional analysis and consideration of both field and laboratory data, recommended

$$\frac{\int_{x_b}^{rl} \overline{Q_y}\, dx}{H_{b,rms}^3/T_p} = 2.6\cdot 10^{-3} \left(\frac{H_{b,rms}}{L_{op}}\right)^{-1.25} m_b^{0.75} \left(\frac{H_{b,rms}}{d_{50}}\right)^{0.25} \sin^{0.6}(2\alpha_b) \tag{6.5.16}$$

where m_b is the beach slope at the break point, L_{op} is the deep water wave length corresponding to the peak wave period T_p, and $H_{b,rms}$ is the root mean square breaker height.

This weak grain size dependence of the total longshore sediment transport rate is interesting in relation to the observations of shore normal transport over rippled beds discussed in Section 6.3.1, page 266, and to the corresponding shore parallel transport rates discussed in Example 6.5.1, page 282. In both cases it was found that sediment transport rates over rippled beds seems to be less sensitive to grain size than classical transport models predict.

Classical transport models based on gradient diffusion and the assumption of a flat bed, e g Deigaard et al (1986), predict a very strong grain size dependence, for a given topography.

The most likely reason for the observed weak grain size dependence for the total longshore sediment transport rate seems to lie in the topographical difference between beaches of coarser and finer sand. Beaches of fairly coarse sand ($d_{50} \gtrsim 0.4mm$) tend to develop topographies which are more conducive to sediment transport. An example is shown in Figure 6.5.4. If the sand on that beach had been finer, the beach profile would have been flatter and the waves would have broken as spilling breakers with much less sediment entrainment capability.

CHAPTER 7

LOOSE ENDS AND FUTURE DIRECTIONS

7.1 Introduction

It has been the aim of the previous chapters to point out some of the remaining, unanswered questions in coastal sediment transport modelling as well as to summarise our present knowledge. Identifying loose ends or gaps in our knowledge is the first step to any serious research project. Therefore, the following sections summarise some of the presently unresolved problems and indicate areas in which new insights are urgently needed.

7.2 New models of wave-current boundary layer interaction

The discussion of the concept of eddy viscosity for combined wave-current flows in Section 1.5.3 (pp 68-71) leads to the conclusion that it is time to introduce a new generation of eddy viscosity based models for these flows. The reason is that, with the presently used definitions, the eddy viscosity which applies to the current component of combined wave-current flows, has a number of undesirable characteristics and is unlikely to be related to the sediment diffusivity.

Most of the presently available models which use the eddy viscosity concept are based on the equation of motion

$$\rho \, \nu_C \frac{d\bar{u}}{dz} = \bar{\tau}(z) \qquad (7.1)$$

for the current component. However, the eddy viscosity ν_C which is defined by this equation has some rather unfortunate characteristics. This can be seen by considering the expression

$$\bar{\tau} = \rho\nu\frac{\partial\bar{u}}{\partial z} - \rho\bar{u}\,\bar{w} - \rho\overline{\tilde{u}\tilde{w}} - \rho\overline{u'w'} \tag{1.2.22}$$

for the time-averaged shear stress in a two-dimensional wave-current flow of the form $(u,w) = (\bar{u}+\tilde{u}+u',\ \bar{w}+\tilde{w}+w')$, see page 12.

If this expression for $\bar{\tau}$ is inserted into Equation (7.1) it can be seen that the eddy viscosity ν_c, which applies to the current, contains the following terms

$$\nu_c = \frac{-\bar{u}\,\bar{w} - \overline{\tilde{u}\,\tilde{w}} - \overline{u'w'}}{\frac{d\bar{u}}{dz}} + \nu \tag{7.2}$$

The first deterministic term, $(-\bar{u}\,\bar{w})$, in this expression may often be ignored since $\bar{w}$ must, by continuity, be zero for horizontally uniform flows. The second term, $(-\overline{\tilde{u}\,\tilde{w}})$, however, will often be dominant and this gives the eddy viscosity ν_c a problematic nature.

Hence, ν_c is not a "turbulent eddy viscosity", because the leading contribution $(-\overline{\tilde{u}\,\tilde{w}})$ is deterministic.

It is not isotropic, but depends strongly on the direction of the current relative to the direction of wave propagation.

The eddy viscosity given by Equation (7.2) may in fact become negative if $-\overline{\tilde{u}\,\tilde{w}}$ is dominant and of the form, derived by Longuet-Higgins (1956), (see pages 55 and 56), while $\frac{d\bar{u}}{dz}$ is negative. This corresponds to the commonly observed situation which is shown in Figure 7.1.

The bed shear stress is positive but smaller than the asymptotic value of $-\overline{\tilde{u}\,\tilde{w}}$. See also Figure 1.4.2, page 55.

The steady flow velocity is positive close to the bed, but starts to decrease at the elevation where

$$-\rho\,\overline{\tilde{u}\,\tilde{w}} = \bar{\tau}(z) \tag{7.3}$$

Above this point, the eddy viscosity ν_c, is negative because $\bar{\tau}$ and $d\bar{u}/dz$ have opposite signs.

A negative eddy viscosity is unsatisfactory in itself and even more so in relation to sediment diffusivity. It cannot be avoided however, if the time-averaged equation of motion is written in the commonly used form (7.1).

These conceptual problems with the eddy viscosity are, however, avoided if

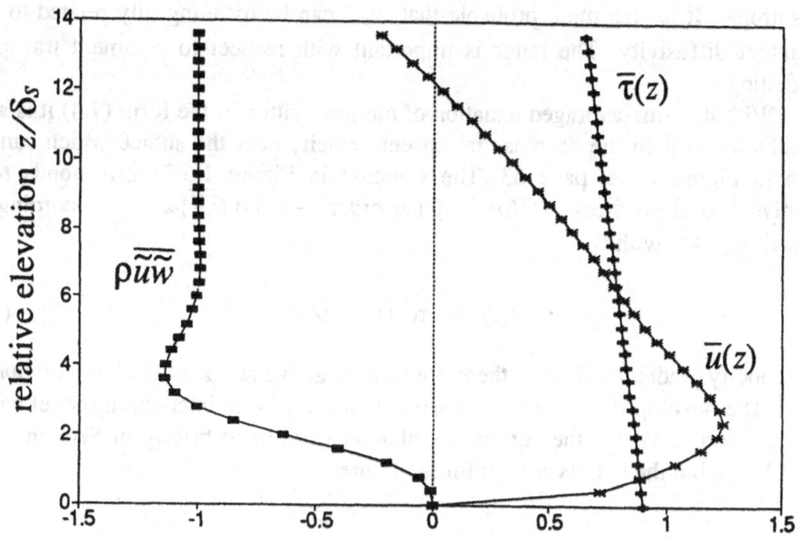

Figure 7.1: In situations where the time-averaged bed shear stress is positive but smaller than the asymptotic value of $-\overline{\tilde{u}\,\tilde{w}}$, the shear stress $\overline{\tau}(z)$ and the current gradient $\frac{d\overline{u}}{dz}$ have opposite signs in the upper part of the flow. This corresponds to negative values of the eddy viscosity defined by Equation (7.1).

the momentum transfer term $-\overline{\tilde{u}\,\tilde{w}}$ is considered explicitly rather than anonymously as part of $\nu_C \frac{d\overline{u}}{dz}$. When that is done, the equation of motion for the current becomes

$$\nu_{C1} \frac{d\overline{u}}{dz} = \frac{\overline{\tau}(z)}{\rho} + \overline{\tilde{u}\tilde{w}} \tag{7.4}$$

With this equation it is easier to explain the situation shown in Figure 7.1. The new eddy viscosity, ν_{C1}, which is given by (for $\overline{\tilde{u}\,\tilde{w}} \approx 0$)

$$\nu_{C1} = \frac{-\overline{u'w'}}{\frac{d\overline{u}}{dz}} + \nu \tag{7.5}$$

is, unlike ν_C, basically a turbulent eddy viscosity and it is unlikely to be strongly anisotropic. It is also more probable that ν_{c1} can be meaningfully related to the sediment diffusivity. The latter is important with respect to sediment transport modelling.

With the time-averaged equation of motion written in the form (7.4) it is also possible to explain the decrease in current velocity near the suface which can be seen in Figure 1.5.2, page 63. The situation in Figure 1.5.2 corresponds to a positive bed shear stress $\overline{\tau}(o)$ of the order $-1.5\,\rho\,(\overline{\tilde{u}\tilde{w}})_\infty$. According to Equation (7.4), with

$$\overline{\tau}(z) = \overline{\tau}(o)\,(1 - z/D) \tag{7.6}$$

the velocity gradient $d\overline{u}/dz$ therefore turns negative at $z \approx D/3 = 100mm$.

The development of a wave-current boundary layer interaction model based on Equation (7.4) for the current distribution is outlined briefly in Section 1.5.9 (pp 91-94), but the details are left for the future.

7.3 Hydraulic roughness of flat sand beds under waves

We have at present no direct information, in the form of detailed velocity measurements, about the hydraulic roughness of flat sand beds under waves, and two types of indirect experimental evidence are apparently conflicting. That is, measured bed-load quantities and bed-load transport rates over flat sand beds under waves indicate much smaller effective bed shear stresses than do energy dissipation measurements. See Section 2.4.2 and 2.4.3 (pp 117-120).

Correspondingly, it was found in Section 3.6.3 and 3.6.4 (pp 149-154), that the roughness values, found by Wilson (1989), for flat sand beds in steady flows were much smaller than the roughness derived from energy dissipation over flat sand beds under waves. Wilson found typical roughness values of the order ten grain diameters while the energy dissipation data of Carstens et al (1969) correspond to roughness values of the order one hundred grain diameters.

Thus, energy dissipation measurements under waves indicate much greater hydraulic roughness for flat, movable beds than other types of experiments.

No direct measurements are available of the wave boundary layer flow over flat, movable sand beds. It is therefore an open question as to whether the large roughness inferred from energy dissipation experiments or the smaller roughness corresponding to sheet-flow sediment transport rates should be applied in modelling the structure of these flows.

It is possible, that the larger roughness indicated by the energy dissipation experiments can be explained in terms of the energy dissipation due to percolation under waves, which is not directly related to the effective bed shear stress. It would, however, be very interesting to see some direct measurements of the flow structure over flat beds of loose sand under oscillatory flows.

7.4 Instantaneous, effective shear stresses on sand beds under waves

As mentioned in the previous section, we have at present no direct measurements of oscillatory boundary layer flow structure over movable sand beds and the various types of indirect evidence are apparently conflicting. It is therefore impossible to estimate instantaneous, effective bed shear stresses for these flows with much confidence.

The instantaneous, effective bed shear stresses are, however, essential for detailed sediment transport modelling. Hence, there is an urgent need for experimental data in this area.

It might be possible to obtain the necessary boundary layer structure data with acoustic velocity probes. Alternatively, the method applied by Lofquist (1986) to measure instantaneous bed shear stresses over rippled beds might be used for flat beds as well.

7.5 The bottom boundary condition for suspended sediment distributions

The bottom boundary condition for suspended sediment distributions presents another urgent problem in sediment transport modelling.

It is by now fairly widely agreed that the description of suspended sediment distributions in non-uniform or unsteady flows requires a different bottom boundary condition than just a quasi-steady version of

$$\overline{c}(o) = \overline{c}(\theta') \tag{7.7}$$

This equation, which is the usual type of boundary condition for steady, uniform flow, expresses equilibrium between suspended sediment concentrations at the bed and the effective Shields parameter θ'. Such an equilibrium does, however, not exist in non-uniform or unsteady flows, see Section 5.3.1 (pp 222-223). A different boundary condition has therefore been suggested by Nielsen et al (1978) and van Rijn (1984).

Their approach is to treat deposition and entrainment separately, and then try to relate the entrainment rate, or pickup rate $p(t)$, directly to the effective, instantaneous Shields parameter $\theta'(t)$.

The pickup function model is, however, still a simplified, formal description rather than an explicit, physical description of the sediment entrainment process. It is, therefore, not obvious how it should be incorporated into the combined convection-diffusion model of sediment suspensions as discussed on page 247.

To resolve this type of problem, a clear physical description is needed for the processes in the layer between the immobile bed and elevations where all sediment moves in suspension according to Bagnold's definition. For the situation shown in Figure 5.2.1, page 208, that is for the elevation range $-5mm < z < 7mm$.

7.6 Distribution models for suspended sediment

A framework for the modelling of sediment suspensions in coastal flows as a combined convection-diffusion process was developed qualitatively in Sections 5.4.5-5.4.9 (pp 245-261).

The new model was found capable of modelling observed differences between sediment distributions in different flows and between the distributions of different grain sizes in the same flow. However, many quantitative details need to be filled in.

One of the advantages of the new combined convection-diffusion approach is that the sediment diffusivity ε_s, which is only responsible for the truly diffusive part of the distribution process in this description, may be closely related to "the eddy viscosity" of the flow. That is often not the case when a natural sediment distribution is modelled in terms of pure gradient diffusion.

Inverted commas are used in relation to "the eddy viscosity" because, the eddy viscosity may be defined in various ways in a combined wave-current flow. Of the two eddy viscosities discussed in Section 7.2, the second, ν_{c1}, seems a far better model for the sediment diffusivity ε_s.

However, even if equality, $\varepsilon_s = \nu_{c1}$, may be assumed, the diffusivity is known no better than the eddy viscosity, and for flat sand beds that is not very well. The problem of properly describing the sediment diffusivity is therefore linked to the uncertainty about the hydraulic roughness of flat sand beds under waves mentioned in Section 7.3 above.

The most direct way to obtain some conclusive information about these matters, is to measure the concentration profiles of different sand sizes together with the flow structure. This is most urgently needed for oscillatory flows over flat sand beds and for combined wave-current boundary layer flows.

After the present manuscript was finished, the paper by Dick & Sleath (1991) appeared, which contains considerable new insights into the problems mentioned in Sections 7.3, 7.4 and 7.5.

REFERENCES

Allen, J R L (1982): *Sedimentary structures: Their character and physical basis*. Elsevier, Amsterdam.

Amos, C L & M B Collins (1978): The combined effect of wave motion and tidal currents on the morphology of intertidal ripple marks: the Wash. *U K J Sedimentary Petrology Vol 48, No 3 , pp 849-856.*

Amos, C L, A J Bowen, D A Huntley & C F M Lewis (1988): Ripple generation under the combined influences of waves and currents on the Canadian continental shelf. *Continental Shelf Res, Vol 8, No 10,* pp 1129-1153.

Arnott, R W & J B Southard (1990): Exploratory flow duct experiments on combined flow bed configurations and some implications for interpreting storm event stratification. *J Sedimentary Petrology, Vol 60, No 2,* pp 211-219.

Asano, T, M Nagakawa & Y Iwagaki (1986): Changes in current profiles due to wave superimposition. *Proc 20th Int Conf Coastal Engineering, Taipei,* pp 925-940.

Bagnold, R A, (1946): Motion of waves in shallow water: Interaction of waves and sand bottoms, *Proc Roy Soc Lond, A 187,* pp 1-15.

Bagnold, R A (1954): Experiments on a gravity-free dispersion of large solid spheres in a newtonian fluid under shear. *Proc Roy Soc Lond A 225,* pp 49-63.

Bagnold, R A, (1956): The flow of cohesionless grains in fluids. *Phil Trans Roy Soc Lond,* No 964, Vol 249, pp 235-297.

Bailard, J A (1981): An energetics total load sediment transport model for a plane sloping beach. *J Geophys Res, Vol 86, No C11,* pp 10938-10954.

Bakker, W T, & T van Doorn, (1978): Near bottom velocities in waves with a current. *Proc 16th International Conference on Coastal Engineering, ASCE, Hamburg,* pp 1394-1413.

Basset, A B, (1888): On the motion of a sphere in a viscous fluid. *Phil Trans Roy Soc Lond, Ser A, Vol 179,* pp 43-63.

Bhattacharya, P K, (1971): *Sediment suspension in shoaling waves,* Ph D thesis, University of Iowa, Iowa City, Iowa.

Bijker, E W (1967): Some considerations about scales for coastal models with movable beds. *Delft Hydraulics Lab Pub No 50.*

Blondeaux, P & G Vittori (1990): Oscillatory flow and sediment motion over a rippled bed. *Proc 22nd Int Conf Coastal Engineering, Delft,* pp 2186-2199.

Boczar-Karakiewicz, B & R D L Davidson-Arnott (1987): Nearshore bar formation by non-linear wave processes - A comparison of model results and field data. *Marine Geology, Vol 77,* pp 287-304.

Boczar-Karakiewicz, B & L A Jackson (1991): Beach dynamics and protection measures in the Gold Coast area, Australia. *Proc 10th Australasian Conf on Coastal and Ocean Engineering, Auckland,* pp 411-415.

Bodge, K R & N C Kraus (1991): Critical examination of longshore sediment transport rate magnitude. *Proc " Coastal Sediments '91"*, A S C E, pp 139 - 155.

Bosman, J J (1982): *Concentration measurements under oscillatory water motion. Delft Hydraulics report M1695 part 2.*

Bosman, J J, E T J M van der Velden & C H Hulsbergen (1987): Sediment concentration measurements by transverse suction , *Coastal Engineering, Vol 11*, pp 353-370.

Bretschneider , C L (1954): Field investigation of wave energy loss of shallow water ocean waves, *Beach Erosion Board, Tech Memo 46.*

Brevik, I, (1981): Oscillatory rough turbulent boundary layers, *J Waterway Port Coastal Ocean Div. ASCE, 107* , pp 175-188.

Bruun , P (1962):Sea level rise as a cause of shore erosion. *Proc A S C E , Vol 88, WW1*, pp 117-130.

Bruun, P (1983): Review of conditions for uses of the Bruun Rule of erosion. *Coastal Engineering* , Vol 7, pp 77-89.

Bruun, P (1990): *Port Engineering, IV edition, Vol 2*, Gulf Publishing Company.

Cacchione, D A & D E Drake (1982): Measurements of storm generated bottom stresses on the continental shelf. *J Geophys Res, Vol 87, No C3*, pp 1952-1960.

Carstens, M R, F M Neilson & H D Altinbilek (1969): Bedforms generated in the laboratory under an oscillatory flow, *C E R C Tech Memo 28.*

Carter T G, P L-F Liu, and C C Mei (1973): Mass transport by waves and offshore bedforms. *J Waterways Harbours & Coastal Division, A S C E, Vol 99, WW2*, pp 165-184.

Christoffersen, J B & I G Jonsson (1985): Bed friction and dissipation in combined current and wave motion, *Ocean Eng, Vol 12, No 5*, 387-423.

Clift, R, R J Grace, & M E Weber, (1978): *Bubbles, drops and particles*. Academic Press, New York

Clifton, H E (1976): Wave-formed sedimentary structures - A conceptual model. In Davis & Ethington ed: *Beach and nearshore sedimentation.* S E P M special publication 24, pp 126-148.

Coffey, F C (1987): *Current profiles in the presence of waves and the hydraulic roughness of natural sand beds*. Ph D Thesis, Univ of Sydney, 252 pp.

Coffey, F C & P Nielsen (1984): Aspects of wave-current boundary layer flows. *Proc 19th Int Conf Coastal Engineering, Houston Texas*, pp 2232-2245.

Coffey, F C & P Nielsen (1986): The influence of waves on current profiles. *Proc 20th Int Conf Coastal Eng, Taipei*, pp 82-96.

Coleman, N L, (1970): Flume studies of the sediment transfer coefficient, *Water Resource Res, 6(3)*, pp 801-809.

Dean, R G & R A Dalrymple (1991): *Water wave mechanics for engineers and scientists.* World Scientific, Singapore.

Deigaard, R, J Fredsoe & I B Hedegaard (1986a): Suspended sediment in the surf zone. *J Waterway, Port, Coastal and Ocean Eng, ASCE, Vol 112, No 1*, pp 115-128

Deigaard, R, J Fredsoe & I B Hedegaard (1986b): A mathematical model for littoral drift. *J Waterway, Port, Coastal and Ocean Eng, ASCE, Vol 112, No 3*, pp 351-369.

Deigaard, R, P Justesen & J Fredsoe (1991): Modelling undertow by a one-equation turbulence model. *Coastal Engineering, Vol 15*, pp 431-458.

Delft Hydraulics (1989): Bedforms, near-bed sediment concentrations and sediment transport in simulated regular wave conditions. *Report No H840.*

Dick, J E & J F A Sleath (1991): Velocities and concentrations in oscillatory flow over beds of sediment. *J Fluid Mech, Vol 233*, pp 165-196.

Dingler, J R (1974): *Wave formed ripples in nearshore sands.* Ph D thesis Univ of California, San Diego, 136 pp.

Downing, J P, R W Sternberg & C R B Lister (1981): New instrumentation for the investigation of sediment suspension processes in shallow marine environments. *Marine Geology, Vol 42*, pp 14-34.

Du Toit, C G, & J F A Sleath, (1981): Velocity measurements close to rippled beds in oscillatory flow, *J Fluid Mechanics 112*, pp 71-96

Engelund,F (1970): Instability of erodible beds. *J Fluid Mech, Vol 42*, pp225-244.

Engelund, F (1981): Transport of bed load at high shear stress. *Progr Rep 53*, Inst Hydrodynamic and Hydraulic Eng, Tech Univ Denmark, 31-35.

Engelund, F & E Hansen (1972): *A monograph on sediment transport in alluvial streams.* Teknisk Forlag, Copenhagen.

Fernandez-Luque, R (1974): *Erosion and transport of bed-load sediment.* Dissertation, KRIPS Repro BV, Meppel, The Netherlands.

Fredsoe , J (1984): Turbulent boundary layer in wave-current motion, *J Hydr Eng, A S C E, Vol 110, HY8*, 1103-1120.

Gibbs, R J, M D Mathews & D A Link (1971): The relationship between sphere size and settling velocity. *J Sed Petrology, Vol 41*, pp 7-18.

Gilbert, G K (1914): The transportation of debris in running water. *U S Geol Survey, Prof Paper 86.*

Grant, W D, & O S Madsen, (1979): Combined wave and current interaction with a rough bottom, *J Geophysical Res, 84*, 1808

Grant, W D, & O S Madsen, (1982): Movable bed roughness in unsteady oscillatory flow. *J Geophys Res, Vol 87*, pp 469-481.

Grant, W D, J A Williams, S M Glen, D A Cacchione & D E Drake (1983): High frequency bottom stress variability and its prediction in the Code Region. *Woods Hole Oceanogr Inst, Tech Rep* 83-19.

Guy, H P, D B Simons, & E V Richardson (1966): Summary of alluvial channel data from flume experiments, 1956-1961. *U S Geological Survey, Prof paper 462-I,* Washington D C.

Hallermeier R J (1980): Sand motion initiation by water waves: Two asymptotes. *J Waterway, port,coastal and ocean Div, A S C E, Vol 106*, 299-318.

Hallermeier R J (1981): A profile zonation for seasonal sand beaches from wave climate. *Coastal Engineering, Vol 4*, pp 253-277.

Hallermeier, R J (1981b): Seaward limit of signficant sand transport by waves. *CETA 81-2, CERC*, Ft Belvoir, Va.

Hammond, T M & M B Collins (1979): On the threshold of transport of sand-sized sediment under the combined influence of unidirectional and oscillatory flows. *Sedimentology, Vol 26*, pp 795-812.

Hanes, D M (1990): The structure of events of intermittent suspension of sand due to shoaling waves, in *The Sea, Vol 9 Part B,* John Wiley & Sons, pp 941-954.

Hanes, D M & A J Bowen (1985): A granular fluid model for steady intense bed-load transport. *J Geophys Res, Vol 90, No C5*, pp 9149-9158.

Hanes, D M & D A Huntley (1986): Continuous measurements of suspended sand concentration in wave dominated nearshore environment. *Continental Shelf Res, Vol 6, No 4*, pp 585-596.

Hanes, D M & D L Inman (1985a): Observations of rapidly flowing granular-fluid materials. *J Fluid Mech, Vol 150*, 357-380.

Hanes, D M & D L Inman (1985b): A dynamic yield criterion for granular-fluid flows. *J Geophys Res, Vol 90, No B5*, 3670-3674.

Hansen, J B & I A Svendsen (1986): Experimental investigation of the wave and current motion over a longshore bar. *Proc 20th Int Conf Coastal Eng, Taipei*, pp 1166-1179.

Hardistry, J & J P Lowe (1991): Experiments on particle acceleration in stationary and oscillating flows. *J Hydraulic Engineering, A S C E, Vol 117.*

Hattori, M, (1969): The mechanics of suspended sediment due to standing waves, *Coastal Engineering Japan, 12*, 69-81

Hedegaard, I B, R Deigaard & J Fredsoe (1991): Onshore/offshore sediment transport and morpho- logical modelling of coastal profiles. *Proc Coastal Sediments '91, Seattle*, pp 643-657.

Ho, H W, (1964): *Fall velocity of a sphere in an oscillating fluid.* PhD thesis, University Iowa, Iowa City, Iowa

Homma, M, K Horikawa & R Kajima (1965): A study of suspended sediment due to wave action. *Coastal engineering in Japan, Vol 8*, pp 85-103.

Horikawa, K, & A Watanabe, (1968): Laboratory study on oscillatory boundary layer flow, *Coastal Engineering Japan 11, 13-28*

Horikawa, K, A Watanabe & S Katori (1982): Sediment transport under sheet flow condition. *Proc 18th Int Conf on Coastal Eng, Capetown*, pp 1335-1352.

Inman, D L (1957): *Wave generated ripples in nearshore sands*. Beach Erosion Board, U S Army Corps of engineers, Tech Memo 100.

Inman, D L & A J Bowen (1963): Flume experiments on sand transport by waves and currents. *Proc 8th Int Conf Coastal Engineering*, A S C E, pp137-150.

Iwagaki Y & T Kakinuma, (1967): On the bottom friction factors off five Japanese coasts, *Coastal Eng Japan*, Vol 10, 13-22.

Jackson, P S (1981): On the displacement height in the logarithmic velocity profile. *J Fluid Mech Vol 111*, pp 15-25.

Jackson, R G (1976): Sedimentological and fluid-dynamic implications of the turbulent bursting phenomenon in geophysical flows. *J Fluid Mech, Vol 77*, pp 531-560.

Jansen, R H J (1978): The in situ measurement of sediment transport by means of ultrasound scattering. *Delft Hydraulics Lab Publication No 203.*

Jensen, B J (1989): Experimental investigation of turbulent oscillatory boundary layers. *Series Paper 45, Institute of Hydrodynamics and Hydraulic Engineering (ISVA), Technical University of Denmark.*

Jonsson, I G (1966): Wave boundary layers and friction factors, *Proc 10th Int Conf Coastal Eng, Tokyo*, 127-148.

Jonsson, I G (1980): A new approach to oscillatory rough turbulent boundary layers, *Ocean Engineering 7*, 109-152

Jonsson, I G (1990): Wave current interactions, in *The Sea, Vol 9 Part A*, John Wiley & Sons, pp 65-120.

Jonsson, I G & N A Carlsen, (1976): Experimental and theoretical investigations in an oscillatory turbulent boundary layer, *J Hydraulic Res, 14*, 45-60

Justesen, P (1988): Turbulent wave boundary layers, *Series Paper 43*, Inst Hydrodynamic & Hydraulic Eng , Tech Univ Denmark,

Kajiura, K, (1968): A model of the bottom boundary layer in water waves, *Bulletin Earthquake Res, Institute, 46*, 75-123.

Kalkanis, G (1957): Turbulent flow near an oscillating wall. *Beach Erosion Board, Tech Memo 97.*

Kalkanis, G (1964): Transportation of bed material due to wave action. *U S Army C E R C, Tech Memo 2.*

Kamphuis, J W, (1975): Friction factors under oscillatory waves, *J Waterway Harbours Coastal Engineering Division, ASCE, 101*, 135-144.

Kamphuis, J W (1990): Littoral transport rate. *Proc 22nd Int Conf Coastal Eng, Delft*, ASCE, pp 2402- 2415.

Kamphuis, J W (1991): Alongshore sediment transport rate distribution. *Coastal Sediments '91*, ASCE , pp 170 - 183.

Kana , T W (1979): *Suspended sediment in breaking waves. Tech rep 18-CRD*, Department of Geology, University of South Carolina.

Kemp, P H, & R R Simons, (1982): The interaction between waves and a turbulent current: waves propagating with the current, *J Fluid Mechanics 116*, 227-250.

Kemp, P H, & R R Simons, (1983): The interaction between waves and a turbulent current: waves propagating with the current, *J Fluid Mechanics 130*, 73-89.

Kennedy J F (1963): The mechanics of dunes and antidunes in erodible-bed channels. *J Fluid Mech*, Vol 16, pp 521-544.

Kennedy, J F & M Falcon (1965): Wave-generated sediment ripples. *M I T, Hydrodynamics Laboratory Report No 86*, 55pp.

King, D B Jr (1991): *Studies in oscillatory flow bedload sediment transport.* Ph D Thesis, University of California, San Diego (Scripps), 183pp.

References

Kline, S J, W C Reynolds, F A Schraub & P W Rundstadler (1967): The structure of turbulent boundary layers. *J Fluid Mech, Vol 30*, 741-777.

Kraus, N C, M Larson & D L Kriebel (1991): Evaluation of beach erosion and accretion predictors. *Proc "Coastal Sediments '91"*, A S C E, pp 572-587.

Lamb, H (1936): Hydrodynamics, 6th ed, Cambridge Univ Press

Lambrakos, K F, D Myrhaug & O H Slaattelid (1988): Seabed current boundary layers in wave-plus-current flow conditions. *ASCE, J Waterway Port Coastal and Ocean Eng, Vol 114, No2*, pp 161-174.

Le Mehaute, B (1976): An introduction to hydrodynamics and water waves. Springer.

Lian, Qi Xiang (1990): A visual study of the coherent structure of the turbulent boundary layer in flow with adverse pressure gradient. *J Fluid Mech, Vol 215*, 101-124.

Lofquist K E B (1978): Sand ripple growth in an oscillatory flow water tunnel. *C E R C tech paper 78-5*.

Lofquist, K E B (1980): Measurements of oscillatory drag on sand ripples, *Proc 17th Int Conf Coastal Eng*, Sydney, 3087-3106.

Lofquist, K E B (1986): Drag on naturally rippled beds under oscillatory flows, *Misc Paper CERC-86-13*.

Longuet-Higgins, M S (1953): Mass transport in water waves. *Phil Trans Roy Soc Lond, Vol 245 A*, pp 535-581.

Longuet-Higgins, M S (1956): The mechanics of the boundary-layer near the bottom in a progressive wave, Proc 6th Int Conf Coastal Eng, Miami, 184-193.

Longuet-Higgins, M S (1981): Oscillating flow over steep ripples. *J Fluid Mech, Vol 107*, pp 1-35.

Longuet-Higgins, M S (1983): Wave setup, percolation and undertow in the surf zone. *Proc Roy Soc Lond, Vol A 390*, pp 283-291.

Lundgren, H, (1972): Turbulent currents in the presence of waves, *In Proc. 13th Conference on Coastal Engineering,Vancouver, ASCE*, 623-634.

Lundgren, H & T Soerensen (1956): A pulsating water tunnel. *Proc 6th Int Conf Coastal Engineering*, Miami.

McFetridge, W F & P Nielsen (1985): Sediment suspension by non-breaking waves over rippled beds. *Tech Rep UFL/COEL-85/005*, Coastal & Oceanographical Engineering Dept, Univ of Florida, Gainesville.

Madsen, O S & W D Grant (1976): Sediment transport in the coastal environment. *Report No 209, Palph M Parsons Lab, M I T*.

Magnus, G, (1853): *Poggendorfs Annalen der Physik und Chemie, Vol 88, No 1*.

Manohar, M (1955): Mechanics of bottom sediment movement due to wave action. *Tech Memo 75*, Beach Erosion Board, U S Army corps of engineers, Washington D C.

Meyer-Peter, E & R Muller (1948): Formulas for bed-load transport. *Proc Int Ass Hydr Struct Res, Stockholm.*

Miller, M C & P D Komar (1980): A field investigation of the relationship between ripple spacing and near-bottom water motions. *J Sed Petrology, Vol 50, pp 183-191*.

Murray, S P (1970): Settling velocities and vertical diffusion of particles in turbulent water, *J Geophysics Res. Vol 75, No 9*, 1647-1654.

Myrhaug, D (1982): On a theoretical model of rough turbulent wave boundary layers. *Ocean Eng, Vol 9, No 6*, pp 547-565.

Myrhaug, D & O H Slaattelid (1989): Combined wave and current boundary layer model for fixed, rough seabeds. *Ocean Engineering Vol 16, No 2*, pp 119-142.

Nadaoka, K, S Ueno & T Igarashi (1988): Sediment suspension due to large eddies in the surf zone. *Proc 22nd Int Coastal Eng Conf*, Torremolinos, pp 1646-1660.

Nakato, T, F A Locher, J R Glover, and J F Kennedy (1977): Wave entrainment of sediment from rippled beds, *Proc. ASCE, 103 (WW1)*, 83-100.

Natarajan, P (1969): *Sand movement by combined action of waves and currents*. Ph D thesis, University of London.

Nielsen, P, (1979): Some basic concepts of wave sediment transport, *Ser. Paper 20*, Inst Hydrodyn Hydraul Eng, Tech Univ Denmark, 160 pp.

Nielsen, P (1981): Dynamics and geometry of wave generated ripples. *J Geophys Res, Vol 86, No C7*, pp 6467-6472.

Nielsen, P (1983): Entrainment and distribution of different sand sizes under water waves. *J Sedimentary Petrology, Vol 53, No 2*, pp 423-428.

Nielsen, P (1983): Analytical determination of nearshore wave height variation due to refraction, shoaling and friction. *Coastal Engineering, Vol 7*, pp 233-251.

Nielsen, P (1984a): On the motion of suspended sand particles. *J Geophys Res, Vol 89, No C1*, pp 616-626.

Nielsen, P (1984b): Field measurements of time-averaged suspended sediment concentrations under waves. *Coastal Engineering, Vol 8*, pp 51-72.

Nielsen, P, (1985): On the structure of oscillatory boundary layers, *Coastal Engineering 9*, 261-276.

Nielsen, P (1986): Suspended sediment concentrations under waves. *Coastal Engineering, Vol 10*, pp 23-31.

Nielsen , P (1988a): Three simple models of wave sediment transport. *Coastal Engineering, Vol 12*, pp 43-62.

Nielsen, P (1988b): Towards modelling coastal sediment transport. *Proc 21st Int Conf Coastal Eng, Torremolinos*, pp 1952-1958.

Nielsen, P (1990): Coastal bottom boundary layers and sediment transport. In P Bruun ed *Port Engineering (4th edition), Vol 2*, pp 550-585.

Nielsen, P, I A Svendsen & C Staub (1978): Onshore-offshore sediment transport on a beach. *Proc 16th Int Conf Coastal Eng*, Hamburg, pp 1475-1492.

Nielsen, P, N R Sena & Z J You (1990): The roughness height under waves. *J Hydraulic Res, Vol 28, No 5*, pp645-647.

Nikuradse, J (1933): Stromungsgesetze in glatten und rauhen rohren. *V D I Forschungsheft 361*, Berlin.

References

Owen, P R (1964): Saltation of uniform grains in air. *J Fluid Mech, Vol 20, No 2*, pp 225-242.

Peregrine, J H & I A Svendsen (1978): Spilling breakers, bores and hydraulic jumps. *Proc 16th Int Conf Coastal Eng, Hamburg*, pp 540-551.

Rasmussen, P & J Fredsoe (1981): Measurements of sediment transport in combined waves and current. Inst Hydrdyn & Hydraul Eng (ISVA), Tech Univ Denmark, *Progress report 53*, pp 27-30.

Raudkivi, A J (1988): The roughness height under waves. *J Hydraulic Res, Vol 26, No 5*, pp569-584.

Reizes, J A, (1977) A numerical study of the suspension of particles in a horizontally flowing fluid, *paper presented at 6th Australasian Hydraulics and Fluid Mechanics Conference, Adelaide* .

Ribberink, J S and A Al-Salem (1989): Bed forms, near-bed sediment concentrations and sediment transport in simulated regular wave conditions. *Delft Hydraulics Report H840, part 3*.

Ribberink, J S and A Al-Salem (1990): Bed forms, sediment concentrations and sediment transport in simulated wave conditions. *Proc 22nd Int Conf Coastal Engineering*, Delft, pp 2318-2331.

Riedel, H P (1972): *Direct measurement of bed shear stress under waves*. Ph D Thesis, Dept Civ Eng, Queen's University, Kingston, Ontario.

Roelvink, J A & M J F Stive (1989): Bar generating cross shore flow mechanisms on a beach. *J Geophys Res , Vol 94, No C4*, pp 4785 - 4800.

Sato, S (1986): Oscillatory boundary layer flow and sand movement over ripples. *Ph D thesis, Dept of Civ Eng, Univ of Tokyo*.

Sato, Y & K Yamamoto (1987): Lagrangian measurement of fluid particle motion in an isotropic turbulent field, *J Fluid Mech*, Vol 175, 183-199.

Sawamoto, M & T Yamashita (1986): Sediment transport rate due to wave action. *J of Hydroscience and Hydraulic Eng, Vol 4, No 1*, pp1-15.

Schepers , J D (1978): *Zandtransport onder invloed van golven en een eenparige stroom bij varierende korreldiameter*. M Eng Thesis, Delft Univ of Technology.

Schlichting, H (1979): *Boundary layer theory, 7th ed*, McGraw-Hill, New York.

Shields, A (1936): Anwendung der Aehnlichkeitsmechanik und Turbulenzforchung auf die Geschiebebewegung. *Mitt Preuss Versuchsanstalt fur Wasserbau und Schiffbau, No 26, Berlin*.

Shore Protection Manual (1984), U S Army Coastal Engineering Research Centre, Vicksburg Mississippi.

Simons, R R, A J Grass & A Kyriacou (1988): The influence of currents on wave height attenuation. *Proc 21st Int Conf Coastal Engineering*, Malaga.

Slaattelid, O H, D Myrhaug & K F Lambrakos (1990): North Sea bottom steady boundary layer measurements. *J Waterway, Port, Coastal and Ocean Engineering, Vol116, No 5*, pp 614-633.

Sleath, J F A (1970): Measurements close to the bed in a wave tank. *J Fluid Mech, Vol 42*, pp 111-123.

Sleath, J F A (1978): Measurements of bed-load in oscillatory flow. *Proc A S C E, Vol 104, No WW4*, 291-307.

Sleath, J F A (1982): The suspension of sand by waves. *J Hydraulic Res, Vol 20, No 5*, pp 439-452.

Sleath, J F A (1984): *Sea Bed Mechanics*, Wiley Interscience.

Sleath, J F A (1985): Energy dissipation in oscillatory flow over rippled beds, *Coastal Eng*, Vol 9, 159-170.

Sleath, J F A (1987): Turbulent oscillatory flow over rough beds, *J Fluid Mech*, Vol 182, 369-409.

Sleath, J F A (1990): Bed friction and velocity distributions in combined steady and oscillatory flow. *Proc 22nd Int Conf Coastal Eng, Delft*, pp 450-463.

Sleath, J F A (1991): Velocities and shear stresses in wave-current flows. *J Geophys Res, Vol 96, No C8*, pp 15237-14244.

Snyder, W H & J L Lumley (1971): Some measurements of particle velocity autocorrelation functions in a turbulent flow, *J Fluid Mech*, Vol 48, 41-71.

Southard, J B, J M Lambie, D C Federico, H T Pile, & C R Weidman (1990): Experiments on bed configurations in fine sands under bidirectional purely oscillatory flow, and the origin of hummocky cross-stratification. *J Sedimentary Petrology, Vol 60, No 1*, pp 1-17.

Spalart, P R & B S Baldwin (1987): Direct simulation of a turbulent oscillating boundary layer, *NASA Tech Memo 89460*, Ames Res Centre, Moffett Field, Ca.

Staub, C, I G Jonsson & I A Svendsen (1984): Variation of sediment suspension in oscillatory flow. *Proc 19th Int Conf Coastal Eng*, Houston, pp 2310-2321.

Stive, M J F (1988): Cross-shore flow in waves breaking on a beach. *Dissertation, Delft University of Technology.*

Svendsen, I A & P A Madsen (1984): A turbulent bore on a beach. *J Fluid Mech, Vol 148*, pp73-96.

Svendsen, I A, H A Schaffer & J Buhr Hansen (1987): The interaction between the undertow and the boundary layer flow on a beach. *J Geophys Res, Vol 92*, pp 11845-11856.

Swart , D H (1974): Offshore sediment transport and equilibrium beach profiles. *Delft Hydr Lab Publ No 131.*

Taylor, G I (1921) Diffusion by continuous movement, *Proc Lond Math Soc, Vol 20*, 196-211.

Tooby, P F, G L Wick, & J D Isacs, (1977): The motion of a small sphere in a rotating velocity field: A possible mechanism for suspending particles in turbulence, *J Geophysical Res, 82 (15)*, 2096-2100.

Trowbridge, J & O S Madsen (1984): Turbulent wave boundary layers I. Model formulation and first order solution. *J Geophys Res, Vol 89, No C5*, pp 7989-7997.

van Doorn, T, (1981): Experimental investigation of near-bottom velocities in water waves without and with a current. *TOW Report M 1423 part 1, Delft Hydraulics Laboratory*.

van Doorn, T, (1982): Experimenteel onderzoek naar het snelheidsveld in de turbulente bodemgrenslaag in een oscillerende stroming in een golftunnel, *TOW-Report M 1562-1a*, Delft Hydraulics Laboratories.

van Doorn, T, (1983): Computations and comparisons with experiments of the bottom boundary layer in an oscillatory flow. *TOW-Report M 1562-2*, Delft Hydraulics Laboratories.

van Rijn, L C (1984): Sediment pickup functions. *J Hydraulic Eng, Vol 110, No 10*, pp 1494-1502.

van Rijn, L C (1984): Sediment transport, Part III: Bedforms and alluvial roughness. *J Hydraulic Eng, Vol 110, No12*, pp 1733-1754.

van Rijn, L C (1986): Applications of sediment pickup function. *J Hydraulic Eng, A S C E, Vol 112, No 9*, pp 867-874.

Vincent, C E & M O Green (1990): Field measurements of the suspended sand concentration profiles and fluxes and of the resuspension coefficient γ_0 over a rippled bed. *J Geophys Res, Vol 95, No C7*, pp 11591-11601.

Wilson, K C (1966): Bed-load transport at high shear stress. *J Hydraulics Div, A S C E, Vol 92, No HY6*, pp 49-59.

Wilson, K C (1989): Mobile bed friction at high shear stress. *J Hydraulic Eng, A S C E, Vol 115, No 6*, pp 825-830.

Wright, L D, P Nielsen, N C Shi & J H List (1986): Morphodynamics of a bar-trough surf zone. *Marine Geology, Vol 70*, pp 251-285.

Yalin, M S (1977): *Mechanics of sediment transport, 2nd Ed.* Pergamon Press, London, 312 pp.

Yalin, S & R C H Russell (1962): Similarity in sediment transport due to waves. *Proc 8th Int Conf Coastal Eng*, Mexico City, pp 151-167.

AUTHOR INDEX

SUBJECT INDEX